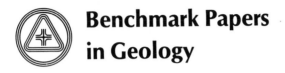

Benchmark Papers
in Geology

Series Editor: Rhodes W. Fairbridge
Columbia University

A selection from the published volumes in this series

Volume

A complete listing of volumes published in this series begins on p. 423.

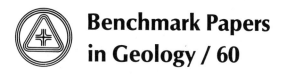

Benchmark Papers
in Geology / 60

A BENCHMARK® Books Series

RIFT VALLEYS
Afro-Arabian

Edited by

A. M. Quennell

Consultant Geologist
Bristol, England

Hutchinson Ross Publishing Company

Stroudsburg, Pennsylvania

Copyright ©1982 by **Hutchinson Ross Publishing Company**
Benchmark Papers in Geology, Volume 60
Library of Congress Catalog Card Number: 81-6251
ISBN: 0-87933-383-9

84 83 82 1 2 3 4 5
Manufactured in the United States of America.

LIBRARY OF CONGRESS CATALOGING IN PUBLICATION DATA

Main entry under title:
Rift Valleys, Afro-Arabian.
  (Benchmark papers in geology ; 60)
  Includes bibliographical references indexes.
  1. Great Rift Valley—Addresses, essays, lectures.
I. Quennell, Albert Mathieson. II. Series.
QE606.5.G74R53   556   81-6251
ISBN 0-87933-383-9   AACR2

Distributed worldwide by Van Nostrand Reinhold Company Inc.,
135 W. 50th St., New York, NY 10020

CONTENTS

Contents

Contents

SERIES EDITOR'S FOREWORD

The philosophy behind the Benchmark Papers in Geology is one of collection, sifting, and rediffusion. Scientific literature today is so vast, so dispersed, and, in the case of old papers, so inaccessible for readers not in the immediate neighborhood of major libraries that much valuable information has been ignored by default. It has become just so difficult, or so time consuming, to search out the key papers in any basic area of research that one can hardly blame a busy person for skimping on some of his or her "homework."

This series of volumes has been devised, therefore, as a practical solution to this critical problem. The geologist, perhaps even more than any other scientist, often suffers from twin difficulties—isolation from central library resources and immensely diffused sources of material. New colleges and industrial libraries simply cannot afford to purchase complete runs of all the world's earth science literature. Specialists simply cannot locate reprints or copies of all their principal reference materials. So it is that we are now making a concerted effort to gather into single volumes the critical materials needed to reconstruct the background of any and every major topic of our discipline.

We are interpreting "geology" in its broadest sense: the fundamental science of the planet Earth, its materials, its history, and its dynamics. Because of training in "earthy" materials, we also take in astrogeology, the corresponding aspect of the planetary sciences. Besides the classical core disciplines such as mineralogy, petrology, structure, geomorphology, paleontology, and stratigraphy, we embrace the newer fields of geophysics and geochemistry, applied also to oceanography, geochronology, and paleoecology. We recognize the work of the mining geologists, the petroleum geologists, the hydrologists, and the engineering and environmental geologists. Each specialist needs a working library. We are endeavoring to make the task of compiling such a library a little easier.

Each volume in the series contains an introduction prepared by a specialist (the volume editor)—a "state of the art" opening or a summary of the object and content of the volume. The articles, usually some twenty to fifty reproduced either in their entirety or in significant extracts, are selected in an attempt to cover the field, from the key papers of the last century to fairly recent work. Where the original works are in foreign languages, we have endeavored to locate or commission translations. Geologists, because of their global subject, are often acutely aware of the oneness of our world. The selections cannot therefore be restricted to any one country, and whenever possible an attempt is made to scan the world literature.

To each article, or group of kindred articles, some sort of "highlight commentary" is usually supplied by the volume editor. This commentary should serve to bring that article into historical perspective and to emphasize its particular role in the growth of the field. References, or citations, wherever possible, will be reproduced in their entirety—for by this means the observant reader can assess the background material available to that particular author, or, if desired, he or she too can double check the earlier sources.

A "benchmark," in surveyor's terminology, is an established point on the ground that is recorded on our maps. It is usually anything that is a vantage point, from a modest hill to a mountain peak. From the historical viewpoint, these benchmarks are the bricks of our scientific edifice.

RHODES W. FAIRBRIDGE

PREFACE

This volume brings together articles with editor's comments in order to provide a background for study of the geology and tectonics of rift valleys or rift systems within the Afro-Arabian series of rift systems. In general, the grouping of this material is on a geographic basis, commencing in the south, but this is not necessarily arbitrary. In any geographic region, crustal structure and geodynamic processes tend to belong to one class, therefore, a geographic grouping will generally conform to a classification that is based on geotectonic setting.

The development of techniques of geologic and geophysical observation in recent years has given an impetus to radical development in geodynamic theory. Progress in technology and theory has overtaken progress in geologic mapping; both field observation and basic geology not only must continue as the launching platform of sound geodynamics research but also must be returned to for the proper testing of hypotheses.

Ideally, this volume should contain articles and descriptions extracted only from objective accounts of geologic exploration and mapping. While reports and maps by state geological surveys conform to this description, the reports of university expeditions and projects generally concentrate attention on some specific subject of research.

Articles of a generalized and speculative nature, reflecting the particular viewpoint of an author and in some cases written to promote its acceptance, are inclined to be selective in their presentation of geologic background and should be treated appropriately. Nevertheless, it is on such descriptions that further research is often based and projects planned. In this volume, where articles of this nature contain recorded observations that are known to be reliable, the observations have been excerpted and reproduced, but without the outdated hypotheses and conclusions. This latter material, usually pertaining to geodynamics rather than descriptive geology, is discussed, however, along with alternative hypotheses.

It so happens that in recent years political and economic factors have impeded and even halted progress in field mapping and investigations and the related laboratory research. This is the case for much of the length of the Afro-Arabian series rift systems. Many records of field and laboratory work nevertheless remain to be tapped. Their existence and easy accessibility is not always appreciated especially by those who have taken up the subject in recent years. It is hoped that my contribution will help ease this situation.

The Introduction systematizes and relates different aspects of rift valley study. An outline of a proposed classification is included. The gulf that separates irreconcilable views on earth models extends also to those on the nature of rift valleys and their origin. A classification must therefore allow for this.

Donald Holloway, until recently on the staff of the library of the Institute of Geological Sciences, London and now with Charter Consolidated Limited, gave valuable assistance in the selection and assembly of material for inclusion in this volume.

A. M. QUENNELL

CONTENTS BY AUTHOR

RIFT VALLEYS

INTRODUCTION

Rift valleys are tectonic land (or oceanic) forms—elongated depressions whose origins are primarily tectonic, not erosional. Tectonic forms—fault scarps and, in some cases, monoclinal scarps—are introduced into the landscape and, until eroded, are easily recognizable.

In its simplest and classical form a rift valley is a graben, a depression bordered by opposed normal fault scarps. The resulting graben structure is two symmetrically opposed fault planes enclosing a down-faulted block. By its nature the length can hardly be other than a multiple of the width.

A rift valley is a graben whose width has the order of magnitude of that of the local thickness of the brittle crust. Therefore, beneath a rift valley the crust is affected for its full thickness. A graben of lower order of magnitude, by decrease in angle of dip of the fault planes or failure of the faults to penetrate the zone of plasticity, must be confined to the upper layers of the crust.

The original definition of a continental rift valley was given by Gregory in 1894. The "centre of the arch has fallen in, forming one of those valleys of subsidence with long steep, parallel walls which Professor Suess has called 'Graben'. . . 'rift valleys' as they may be called" (Gregory, 1894, p. 295). However, with time his term has come to have a wider connotation. This was inevitable with the growth of knowledge and the evolution of tectonic hypotheses that accompanied changes in the favored earth model. Examples of forms, other than graben, now included are valleys of asymmetry or fault-angle depressions and features resulting from major wrench (transcurrent) faults, sometimes described as rifts, which have developed simulated graben.

The terminations of individual rift valleys are generally tectonic or structural forms. In the case of continental rift valleys, they can be: (1) cross faults; (2) pitching elongated blocks; (3) the dying out of one fault by decrease in throw and the transition into an asymmetric rift valley or a region of block faulting; (4) the lateral transition into another rift valley placed en echelon; and (5) the disappearance beneath a younger belt of folded rocks.

1

Oceanic rift valleys belong to an entirely different category. They are midoceanic ridge crest features. They are still regarded, by those who deny seafloor spreading, as being of the same nature as graben, the result of limited horizontal extension. However, the generally held current view is that the oceanic rift valleys—as distinct from oceanic trenches—that, together with the transforms (see below), make up the oceanic rift systems, are the topographic expression of the belt of active separation lying along the crests of the midoceanic ridges (Heezen and Ewing, 1961; Heezen, 1969).

These oceanic rift valleys, although in some cases bearing superficial resemblance, are tectonically and structurally different from continental rift valleys. Whereas the type continental rift valley has a graben structure—a downfaulted crustal block previously in continuity with the flanks—the oceanic rift valley has a floor of material exposed when the flanks are drawn apart under the influence of seafloor spreading. The floor consists of diapirically emplaced igneous rock, whereas the uplifted flanks, made up of the same but older material, are not faced by the scarps of opposed normal faults but by step-faulted blocks. This is one extreme of the oceanic rift valleys, the general case; but in some localities the graben form is replaced by a horst form with the diapirically emplaced material rising to a level above that of the crests of the flanks. In the case of oceanic rift valleys their terminations are against transforms or transform faults.

RIFT ZONES AND RIFT SYSTEMS

A continental rift zone is made up of rift valleys that are structurally connected by recognizable faults, giving rise to an en-echelon pattern. A continental rift system, on the other hand, consists of rift valleys or rift zones, tectonically, not structurally, connected, arranged in a belt, along the length of which are operative the stresses that resulted in the rift valley formation.

If the stress is tensile and transverse to the zone, the rift valleys will be of the single graben type with normal faults, but if the tensile stress is effective over a wide belt, then block faulting will result. If shear stress acts along a zone, a wrench fault system with accompanying rift-valley topographic forms may result.

An oceanic rift system consists of rift valley segments, offset but connected by transforms. It is now well established that an unbroken rift valley extending along a single continuous midoceanic ridge does not in fact exist. Along some lengths of the midoceanic ridges, spreading by movement on transforms dominates.

The term "rift" in this context is used to qualify valley, zone, or system and its use in this manner is acceptable.

CRUSTAL SETTING

Rift valleys are found in differing crustal settings. The model, derived from laboratory experiments, of a slab of elastic or brittle crust, homogeneous and isotropic, resting on a plastic or ductile layer, subjected to uniform horizontal tensile stress over its whole thickness so as to reproduce a rift valley structure, is no longer credible. Not only is the crust anisotropic, heterogeneous, and of varying thickness, but the applied stress is not uniform and deformations result in changes in axis orientation. From these factors arises the wide variation in rift valley structural and geomorphic forms.

The basic distinction is between continental and oceanic settings. Continental settings include: (1) platform, that is, areas of generally crystalline rocks that have undergone cratonization in varying degree. Orogenic belts may have retained anisotropism sufficiently to influence the trend of later rift faulting and flexuring, whereas some areas have lost any significant trend, and the strike of rift faults will be determined by other factors; (2) continental shelf, extending to the continental slope, whether forming coastal plains or lying beneath shelf seas. The "basement," which may be craton margin or younger geosynclinal belt, is mantled with young, virtually unfolded sediments. Rift-valley topographic forms, if they ever existed, will have been infilled and masked by the younger sediments. Prerift sediments may have been down-faulted. The rift or graben structure will be present; (3) cordilleran fold belts, intercratonic or marginal, of generally younger, unaltered sediments and volcanics. Within these belts conditions for graben faulting may occur. These can be a change from compressional (folding) to tensional (normal faulting) stress. In some cordillera, a median zone may be present where rift valleys can develop.

The oceanic crustal setting is along the crests of the midoceanic ridges, in zones of vulcanism and seismicity. Unlike continental crustal areas, where in general there is little lateral change with time, the ocean floor is changing constantly, symmetrically outward from the oceanic ridges as seafloor spreading operates. The essential differences between continental and oceanic crustal structure relevant to this study are that, whereas continental brittle crust is completely different in nature and properties from the underlying plastic layer, the oceanic crust of the ridges is actually derived from that which underlies, by diapiric or intrusive processes.

The intercontinental crustal setting is found where continental crust has fractured and separated so as to expose the deeper layers of the lithosphere. The intercontinental zones are the prolongation of active oceanic ridge zones. The subcrustal conditions are of the same nature, and, with increasing separation, the pattern of the oceanic rift system—rift valleys and connecting transforms—appears. However, not all crustal separations have active spreading as their cause. Some may operate passively.

TECTONIC REGIMES

Of the four tectonic regimes—orogenic, epeirogenic, taphrogenic, and lineagenic—conditions for the origin of rift valleys are most favorable in the taphrogenic where, in the continental setting, horizontal (tangential) tensile stress operates. Brittle continental crust will fracture, resulting in either the graben type of rift valley, or block faulting; or, where crustal separation follows fracture, in intercontinental and eventually in oceanic rift valleys, that is, rift seas succeeded by rift oceans. In plate tectonic terms, these are constructive plate margins.

Major wrench faulting can result in rift valleys and these belong to the lineagenic regime, a term originated by Hills (1963, p. 315). In plate tectonics these correspond to conservative margins.

Mention is made above of conditions of tension succeeding the compressional phase during which young geosynclinal sediments have been folded into orogenic belts (cordillera). Thus rift valleys also can belong to the orogenic regime. Destructive plate margins (subduction zones) are supposed to underlie or to have underlain orogenic belts.

Some rift systems lie around the flanks of major continental swells but their fault scarps must not be confused with the erosion scarps of a new cycle of planation. Rift valleys can originate under epeirogenic conditions wherever tensile stress exists.

Some authors (for example, Dixey, 1959) who reject plate tectonics go further. Under the title vertical tectonics, which they equate with epeirogeny, they regard all rift valleys, even oceanic, as either graben subsidences or as "lag" areas flanked by uplifts.

Taphrogeny, a term used by Krenkel (1925, p. 240) to mean "tearing apart," can conceivably result in the freeing of adjacent blocks that have been constrained from isostatic adjustment. Normal block faulting of rift valleys could therefore be a result of operation of taphrogeny, that is, of primarily horizontal and only subsequently vertical forces.

VOLCANISM

Contrary to what is sometimes stated, volcanic activity is not an invariable accompaniment of rift valley formation. It is or has been the essential accompaniment of oceanic rifting by seafloor spreading but, along the continental rift zones, it is found generally where doming or arching has been the cause of the rift faulting.

Chemically there is strong contrast between the potash-poor tholeiites of the oceanic systems and the strongly alkaline volcanics and intrusives of some continental systems. This has been cited as evidence against the hypotheses that continental rift systems are embryonic rift seas or rift oceans.

CLASSIFICATION

There is increasing divergence between the earth models of two schools of thought, and any classification system should provide for this. The division is between those tectonic hypotheses in which vertical (radial) crustal movement predominates and those in which horizontal (tangential) movement is of greatest importance.

A practical classification system that would be of greatest value must be based not on genetic, geodynamic, or tectonic considerations, but on geomorphology, geology, crustal setting, and structure (as distinct from tectonics). Milanovsky (1972) has developed a classification system compatible with the views of the first school. The classification system in Table 1 provides for both schools, but gives greater credence to the second. A classification should not endeavor to provide for all possible forms (and thus leave pigeon holes empty) but should contain provision only for the rift valleys and rift systems for which examples can be cited.

Table 1 Classification of Rift Valleys, Rifts, and Rift Systems

Crustal setting	Structural style	Selected examples
Oceanic		
	Normal	North and South Atlantic, Indian, and South Pacific oceans
	Oblique	Mid-Atlantic and Southwest Indian Ocean
Intercontinental		
	Normal to oblique	Red Sea, Gulf of Aden
	Oblique to strike-slip	Gulf of California, Gulf of Aqaba
Continental		
Cordilleran (orogenic)	Graben	Rio Grande, Andean Chain
	Strike-slip	Rocky Mountain trench (?)
Shelf (continental margin)	Graben	North Sea
	Block-faulted	Southern California shelf (?)
	Strike-slip (wrench)	San Andreas, Alpine Fault

(continued)

(continued)

Crustal setting	Structural style	Selected examples
Platform (anorogenic)	Isolated	
	Graben-extension	Rhine
	Graben-crestal	Kenya
	System	
	Graben-extension	Gulf of Suez
	Fault-angle	Manyara, North Malawi
	Block-faulted	Basin-Range, Central Tanzania
	Strike-slip (wrench)	Dead Sea system and Gulf of Aqaba, Great Glen
	Volcano-tectonic	
	Crestal	Kenya, Kivu
	Triple-junction	Ethiopia

Source: Adapted from Rift Valleys, in *The Encyclopedia of Structural Geology and Plate Tectonics,* C. Seyfert, ed., Hutchinson Ross, Stroudsburg, Pa., in prep.

REFERENCES

Dixey, F., 1959, Vertical tectonics in the East African rift zone, *20th Intern. Geol. Congr. Proc. Assoc. Afr. Geol. Surv.,* pp. 359–375.

Gregory, J. W., 1894, Contribution to the physical geology of British East Africa, *Geog. Jour.* **4**:290–315, 408–424, 505–514.

Heezen, B. C., 1969, The world rift system: An introduction to the symposium, *Tectonophysics* **8**:269–279.

Heezen, B. C., and M. Ewing, 1961, The mid-oceanic ridge and its extension through the Arctic basin, in *Geology of the Arctic,* University of Toronto Press, Toronto, pp. 622–642.

Hills, E. S., 1963, *Elements of Structural Geology,* Methuen, London, 483p.

Krenkel, E., 1925, *Geologie Afrikas,* vol. 1, Borntraeger, Berlin, 1918p.

Milanovsky, E. E., 1972, Continental rift zones: Their arrangement and development, *Tectonophysics* **15**:65–70.

Part I

HISTORICAL RETROSPECT

Editor's Comments
on Paper 1

1   WILLIS
East African Plateaus and Rift Valleys: Historical Retrospect

Willis's historical review (Paper 1), the first chapter in the classic account of the rift valleys of East Africa by this renowned structural geologist, needs little comment. It traces the history of the theory of rift-valley formation from the earliest times until the 1930s and links investigations and speculations regarding the Rhiengraben and the African rift valleys. The last section "Analysis of Existing Theories," has been omitted because, although of historical interest, it deals in part with hypotheses now discarded and is not important in discussion. Bailey Willis was a protagonist of the "compression" hypothesis of Wayland (1929).

REFERENCE

Wayland, E. J., 1929, Rift valleys and Lake Victoria, *15th Intern. Geol. Congr. Proc.* **2:**323-353.

1

EAST AFRICAN PLATEAUS AND RIFT VALLEYS

HISTORICAL RETROSPECT

Bailey Willis

INTRODUCTION

The Carnegie Institution geological expedition to East Africa had as its objective the study of the rift valleys, those peculiar depressions which lie like inverted mountain ranges sunk in the plateaus. At an early stage of the study it became apparent that the plateaus themselves are the question. They are of exceptional extent and notable elevation. They characterize Africa, which thus differs from other continents where plains and mountain chains prevail. Thus the problem shifted to the query: Why plateaus? And to the related question: Why rift valleys in plateaus?

It is a fact that the surface of East Africa is flat and high. Escarpments there are, and their bold fronts, cut by deep ravines, resemble serrate mountain crests as seen from below; hence such names as Livingstone Mountains; but when one has climbed to the summit he finds a broad highland, a plateau. Hills rise from its flat surface and some are of imposing height, but they are but remnants of an older plateau, which has over wide areas been eroded down to the level of the existing younger one. There are no mountain chains or ranges in Africa, except in the Atlas of Morroco and the Cape ridges of the extreme south.

Africa thus differs from Europe, Asia, the Americas, and Australia in some fundamental way which is expressed in the elevation of large portions of its great mass, as if by a vertical force; whereas in other continents narrow belts are folded up, as if by a horizontal force.

Horizontal force of one type or another has been the active cause of rift valleys, according to previous interpretations. The simplest and earliest thought was that Africa had been pulled apart. A rift valley represented a crack or split. Or if arched up and then stretched, it corresponded with a dropped keystone. On the other hand it was argued that horizontal pressure, causing shortening, had forced the margins of a rift valley up and the valley strip down. Thus the earlier explanations appealed to tension and the later to compression, and the alternative has long divided and still divides two schools of theory in geology. Each school can cite evidence of facts. There are rift valleys where there has been tension. There are also rift valleys where there has been compression. But in each such case the effects of the horizontal pulling or pushing are subordinate to those of the vertical lift that has raised one section or another more than an adjoining one.

9

The observation that rift valleys differ among themselves demands elasticity of thinking in seeking to interpret them. The different aspects require different explanations through diversity of mechanical conditions or assumption of diverse stresses or both. There is but one postulate which can not be changed: the action of the forces must follow the immutable laws of mechanics and physics, as applied to the deformation of the earth's crust; that is, to changes of form and position of rockmasses in the outer shell of the earth.

To give these generalizations more definite form we may briefly review the principal steps in the history of theories relating to rift valleys in Europe, Arabia and Africa.

Theories are long lived. They outlive the basic assumptions from which they logically sprang and like some old patriarch continue to dominate the generation of younger thought. Thus it is with certain elements of current theories in geology, including those relating to rift valleys. We turn back a century.

THE COLLAPSED ARCH

A hundred years ago the question of the fact or possibility of absolute, not merely relative, elevation of the earth's crust was being discussed pro and con by the most eminent geologists. There was in progress a prolonged controversy, for a vivid account of which we are indebted to de Lapparent.[1] It goes back to Leopold von Buch, who had called attention to the positions assumed by upturned strata on either side of a central core of older rocks in mountain chains in general, in such a manner that common sense appeared to require that they must have been raised by a force acting from below upward. His disciple, Elie de Beaumont, while supporting this view, modified it by suggesting that the lifting force originated in the horizontal pressure resulting from the shrinking of the cooling globe and the adjustments of the crust to the diminishing diameter. He was opposed by Constant Prévost, who maintained that the salient relief of the folds would be less than the subsidence of the crust in general and that consequently there could be no action of absolute uplift. The discussion was continued for many decades, though with modifications demanded by advancing knowledge of the stability of the earth. Edouard Suess adopted the view that uplift against gravity could not occur. His epoch-making work, *Die Entstehung der Alpen*, which contains the seeds of most of the speculations subsequently elaborated in *Das Antlitz der Erde*, presented the hypothesis of horizontal displacements and folding, which was followed five years later by the positive negation of absolute uplift:[2]

---

[1] A. de Lapparent, *Soulèvements et Affaissements*, Revue des Questions Scientifiques, vol. XIV, 5-33, 1898.

[2] E. Suess, *Ueber die vermeintlichen säcularen Schwankungen einzelner Theile der Erdoberfläche*, Verh. der K. K. Geol. Reichsanstalt, 1880.

"There are, however, no vertical movements of any kind of the firm crust, with the exception of those which result directly from the development of folds."

In 1825 Elie de Beaumont was on the threshold of that brilliant career which placed him among the leaders of French scientists. Twenty-seven years old, he had conceived ideas native to his philosophical and mathematical mind regarding the distribution of mountain chains and their relative ages, embracing the entire globe. During two decades he fitted facts to his theory and planted firmly the concept that the features of the earth's surface are arranged according to a geometrical plan.[1]

A century ago it was already recognized that mountain chains have been upfolded, and it was not difficult to conceive that the thin crust covering the supposed molten, fluid interior had yielded from time to time as it adjusted itself to its shrinking support. De Beaumont's concept of the conditions is clearly expressed in a mathematical discussion of the rate of cooling of the entire globe as compared with that of the crust.[2] Upon certain assumptions which he regarded as reasonable he concluded:

"The annual cooling of the surface will have been greater than that of the entire mass of the globe during the lapse of 38,359 years, dating from the beginning of cooling, and from that epoch on the mean annual cooling of the mass will surpass that of the surface, and will surpass it more and more."

He regarded it as probable that more than 38,359 years had elapsed since the moment of the beginning of solidification of the crust. Hence it followed that the more rapidly cooling mass would from then on shrink away from the crust and the latter would from time to time collapse in wrinkles upon its foundations. He remarked logically that the older mountain chains thus produced would be lower, as we observe them to be, because at the moment of their formation the crust was thinner. He conceived the yielding of the crust to have been a sudden event, a catastrophe involving the entire surface of the globe and resulting in mountain ridges, which for any one collapse were all parallel among themselves. It followed that, by observing the trend of a mountain range, one could assign it to its proper system and could determine the relative age by fixing that of the system on geological evidence of unconformities, etc.

One reads with wonder the coordination of facts accomplished by de Beaumont. Like his successor Suess, he marshalled the observations of mountain trends for the entire world. Having himself contributed important data for the construction of the geologic map of France and having extended his

[1] Elie de Beaumont, Mémoire; *Annales des Sciences naturelles,* vol. XVIII-XIX, 1829-1830.
Ibid, Note sur les systèmes de montagnes les plus anciens de l'Europe, Bull. Soc. Géol. de France, 2d series, vol. IV, 864-991, 1846-47.
[2] *Ibid, Note sur le rapport qui existe entre le refroidissement progressif de la masse du globe terrestre et celui de sa surface,* Comptes-rendus hebdomadaires des Sciences de l'academie des Sciences, vol. XIX, 1327, 1844.

personal studies over much of western Europe, he reached out to Russia, to Massachusetts and Virginia, and even to Australia, China and Japan, through the observations of others. To satisfy his standards of mathematical accuracy he applied the principles of spherical trigonometry to the determination of parallelism among mountain chains, and he calculated the average trend of a range or system to minutes of arc. But the deductions were wholly artificial, the work of his imagination. Not only was the mathematical analysis far too refined for the gross observations, but it overlooked the fact that the rocks of the crust are much too weak to accumulate and transmit over any large part of the earth the uniform stresses which the hypothesis assumed. De Beaumont fell into the error common to many similar studies in that he assumed an unnatural homogeneity of the globe, postulated an impossible uniformity of reactions to heterogeneous changes. His brilliant researches are almost forgotten and may be regarded as among the curiosities of science, except that the concept of world-wide geometrical patterns still holds a place among current theories, as may be seen in connection with the orientation of supposed rifts in Africa.

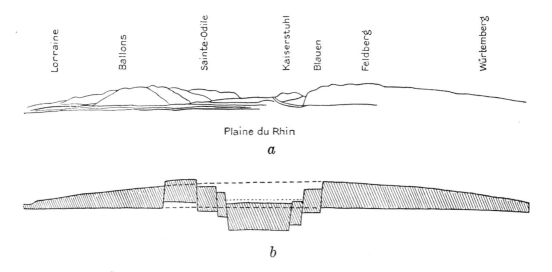

Fig. 1, a and b—Diagrams of the Rheingraben. After Elie de Beaumont.

The rift valley known to de Beaumont and his contemporaries is that stretch of the Rhine Valley from Basel to Bingen which lies between the plateaus of the Vosges in France and the Schwartzwald in Germany. De Beaumont observed a certain unity in the structure of the two, now separate plateaus, of which he wrote:[1]

[1] Elie de Beaumont, *Sur les systèmes de montagnes les plus anciens de l'Europe*, Bull. Soc. Géol. de France, 2d ser., vol. IV, 1846-47, page 906.

"The rocks of the Vosges and the Black Forest have been compressed in a very general folding (ridement) which affected all the ancient terranes of a large part of Europe and impressed upon them the common strike of East 20° to 40° North, which I have observed in the gneisses, schists, and other old rocks, whose contiguous bands constitute the mass of the Vosges."

The general folding or *ridemont* referred to by de Beaumont in the preceding quotation was the intimate structure of the gneiss and schist. He also conceived, however, that the earth's crust had been broadly arched, that such an arch had spanned the valley of the Rhine above Bingen, and that the trough or graben, as it is now called, in which the river flows was formed by the collapse of the arch. The concept is clearly expressed in his drawings (figs. 1, a and b, above) published in volume one of the Explication de la Carte géologique de la France, 1841, from which they are copied. The observations on which his views were based are detailed in his account of the Vosges plateau and the adjacent Rhine valley.[1]

Verweke introduces a detailed account of the Rheingraben [2] with the following references to the views of Elie de Beaumont:

"The Middle Rhine Valley is in a geologic sense, according to its structure, a graben, and its walls are the Middle Rhine Mountains.

"E. de Beaumont evidently sketched the structure of the southern portion of the graben and its walls in the drawing which is here reproduced" (fig. 1a).

"In an outlook from the Röthifluh above Solothurn, on July 28, 1836, just at sunset and under a cloudless sky, he saw, beyond the crest of the Jura and the plain of Belfort, the steep, southeastern slope of the Vosges and also their continuation northward along the Rhine Valley as far as Odilienberg. The profile of the Schwartzwald stood out clearly, presenting the gradual slope of the broad plateau toward Schwaben, which is slightly dissected only near Belchen, almost on the margin of the Rhine plain. The Feldberg scarcely detached itself from this line, but the striking feature was the descent of the blue profile to the Rhine Valley. The view swept over this continuous plain, in which the Kaiserstuhl rises like a molehill from the broad graben."

It is evident that the enthusiastic German student of de Beaumont's work had visited the Röthifluh with the master's drawing in hand and had thus identified the features of the opposed escarpments of the Vosges and Schwarzwald, which produce upon the observer the impression of a depressed block between the masses from which it has sunk away. Verweke continues:

"It was with this view that Elie de Beaumont connected his well-known explanation, which he first put forward in 1827 and illustrated by the graphic diagram [fig. 1b]. In his mental picture he saw the gap of the Rhine Valley filled by a moun-

[1] Elie de Beaumont, *Observations géologiques*, etc., Annales des Mines, 2d series, pages 5-82, 1827.
[2] L. van Verweke, *Die Entstehung des Mittelrheintales und der mittelrheinischen Gebirge*, Mitt. der Gesellsch. für Erdkunde und Kolonialwesen zu Strassburg, 1913, viertes Heft.

tain mass of the height of the Schwartzwald and the Vosges, saw the two united to a broad, continuous arch, which sloped very gently toward Lorraine on the one side and toward Würtemberg on the other. It appeared to him that there was lacking only the keystone of this arch, which in collapsing had produced the Rhine Valley, leaving the buttresses standing on either side.

"The drawing was simple and as easily understood as the underlying idea and aided materially toward recognition of the latter.

"De Beaumont imagined the arch to have been extended in the course of the Rhine Valley; higher in the south than in the north. The elevating force had been centrifugal [1] and had accumulated directly in the elevated mass. The outer layers of this section must have been extended to cover the larger area, they must have been rent apart, and portions of the crust were thus given opportunity to sink into the depths."

Verweke, himself, adhered to the view of de Beaumont, that the Rheingraben represents a dropped keystone, and supported his thesis by a detailed account of the geology of the region.

Although ideas have changed with the advances in knowledge of the earth's crust and of the interior, those of de Beaumont have had many adherents and may be said still to hold a place among current theories of the origin of rift valleys. De Lapparent, whose youth overlapped on the later years of de Beaumont and who was a contemporary of Suess, the author of the concept of collapse without prior folding, strongly supported the earlier explanation. In a long discussion in which he opposed Suess he referred to the Rheingraben in the following terms: [2]

"That this valley is the result of a subsidence of the central part of the ancient massif of the Vosges and Black Forest; that this subsidence assumed the form of parallel, sunken strips, en echelon, bounded by equivalent fractures, that is something which no one would dispute and which Elie de Beaumont long ago clearly explained. But that there was simply a subsidence of the central section, while the external lips of the great fracture remained standing at their original level, that is something we believe one may be permitted gravely to doubt."

De Lapparent [3] thus contrasted the idea of a broken arch, which had been raised above the original level of the crust, with that of a crust standing at its original level in the Vosges and Black Forest, while the intervening section subsided. After citing many examples of uplift of ancient, marine sediments and other illustrations to support the idea of elevation, to which Suess was opposed, he concludes:

[1] This statement appears to have been an inference on the part of Verweke and in contradiction of de Beaumont's idea of a shrinking and wrinkling crust. The stresses in the crust in that case would develop as horizontal compression and the arching would result from lateral, not from vertical, centrifugal forces. De Beaumont himself uses the phrase "*le sol des Vosges et de la Forêt-Noire avait été compris dans un ridement tres general*" in the passage quoted above.

[2] A. de Lapparent, *Conference sur le Sens des Mouvements de l'Ecorce terrestre*, Bull. Soc. Géol. de France, 3d Ser., vol. xv, 1886-87, page 220.

[3] *Op. cit.* page 230.

"Thus, having passed in review all the districts within French territory, I believe myself in the right in affirming the conclusion that: not one of them justifies the theory of the Horst;[1] that they all bear the deeply impressed effects of lateral compressions; that in all of them, in the development of the relief, the preponderance of the uplifting components has been demonstrated in such manner that gravity has not come into play until the prior elevation and fracture of the massif gave it the opportunity to manifest itself."

In the course of his discussion de Lapparent refers to the Jura and the Alps and uses the term "pli" or fold, clearly expressing the idea of folding in the modern sense as an antecedent of rift valley faulting. He thus places himself among those who hold the concept of a fallen arch or anticline. The latter term is somewhat unfortunate. An anticline, strictly speaking, is a fold in which strata dip away from a common crest; it is an arch of stratified rocks. As it happens, the rocks adjacent to rift valleys are rarely stratified and true anticlinal structure is exceptional in connection with such troughs. The term anticline is, however, given a broader meaning by writers who disregard mechanical conditions and has been applied to any height of land from which there are divergent slopes. The marginal lips of rift valleys frequently exhibit this relation. They are highest next to the valley and slope down, away from it. Thus the intervening trough occupies the position of a sunken keystone and suggests the former existence of an arch, erroneously denominated an anticline.

De Martonne, writing entirely upon hearsay evidence of early explorers in Africa, was perhaps the first to suggest the relation between uparched zones and the rift valleys of that continent; he remarked: [2]

"It is a striking fact that each of the rift valleys lies on the major axis of an elevated zone, while between the two uplifts extends a central sunken area, so that the position of Lake Victoria appears to be definitely fixed."

He does not elaborate the idea, but de Lapparent, publishing 11 years later, refers as follows to the same conditions: [3]

"East Africa is a great, super-elevated arch (voussoir), whose axis, directed from north to south, has been appreciably bent and which is flanked on the sides by two ridges, more abrupt and more localized, or at least by two violent folds, whose axes, by their subsidence, have each one given rise to a graben or trough. In this manner, there has well occurred the collapse of a zone, but only there where the earlier elevation had been exceptional."

[1] According to Suess, a mass, block, or column of the earth's crust which has stood firmly at its original level, while adjacent segments have subsided, leaving it outstanding.

[2] E. de Martonne, *Die Hydrographie des oberen Nil-beckens*, Zeitsch. d. Gesellsch. fur Erdkunde, Berlin, vol. XXXII, 1897, page 315.

[3] A. de Lapparent, *Soulèvements et Affaissements*, Revue des Questions scientific, vol. XIV, 1898.

From the French theorists we turn next to the German, whose observations and deductions dominate the field.

The exploration of the German colony of Tanganyika Territory and adjacent areas contributed greatly to our knowledge of the regions in which the African rift valleys occur and gave rise to expressions of opinion regarding the origin of those peculiar features. The fieldwork was carried out by several geologists during the last decade of the last century and the first of this, but the publication of results was cut off in 1914, except in so far as manuscripts then in hard were subsequently sent to press. Most of these writers entertained the idea of a collapsed arch or anticline in one form or another.

Uhlig, who explored the southern end of the Gregory Rift Valley in detail and traversed the middle section, inclined at first to the opinion that the great depression was due to overthrusts, as noted on a subsequent page, but later adopted the idea of uplift and collapse. Regarding the latter interpretation he wrote, speculating:[1]

"Is it to be assumed that the landmass of old crystalline rocks on either side of the graben owes its actual elevation to a renewed folding of the ancient, almost leveled continent, which seems very improbable; or may one accept the formation of many horsts, which have been welded together by young volcanic material to form a superficially united mass; or finally, should one assume a single broad updoming; in any case the present form of the entire region is that of a broad, low back, with north-south axis, in the summit of which the Great Rift Valley lies depressed. I, myself, for various reasons which if here detailed would carry me too far, assume that we have to do with one great dome-like arch (Aufwölbung)."

Uhligh does not commit himself further as to the conditions which determined the subsidence of the graben along the axis of the dome, but discusses at some length the manner of eruption and the relative ages of the volcanics, with which his immediate paper was concerned.

We have next to consider an application of the anticlinal theory which appears to involve an extension of the term anticline to structures that are not anticlinal and to spring from a misunderstanding of the mechanics of deformation of the crust. De la Beche in 1832 defined an anticlinal line as "that line from which the strata dip on either side." The concepts of "strata" and "dip" are implicit in the definition, to which general usage has conformed for a century. Neither strata nor dip occur in the crystalline complex of which the earth's crust is composed, except locally. A complex can not fold. When compressed it shears. Abendanon,[2] through observations on the folded strata in the Szechuen basin of China, became satisfied that folding is the dominant

---

[1] C. Uhlig, *Beitrage zur Kenntniss der Geologie und Petrographie Ostafrikas*, Centralblatt fur Mineralogie, etc., 1912, page 565.
[2] E. C. Abendanon, *Die Grossfalten der Erdrinde*, Leiden, 1914.

structural development in the earth's crust, and he extended the idea to comprise the entire thickness. He thus imagined that a given segment might be compressed between two others, which he assumed to be subsiding in response to gravity, and might be bowed upward. The changes of form in the compressed segment would consist of shortening in the deeper portion and of extension in the outer part. The extension would result in the subsidence of a wedge, *i.e.* in the formation of a rift valley. His idea of a great fold of the earth's crust, *eine Grossfalte der Erdrinde*, is expressed in the following paragraph, which sums up a discussion of many diverse views regarding the rift valleys in Africa:[1]

"To what conclusion do all these citations lead us? Once again to the necessity of assuming a *Grossfalte*, although one of peculiar form, which extends from about 36° north latitude to 16° and perhaps even to 28° south latitude in East Africa, in a sinuous line along a meridian. This it seems to me clearly follows from the account by de Lapparent. Furthermore we have the depression of the Congo Basin on the west and also the one west of the Nile and that of the Mediterranean; on the east lies the deeper depression of the Indian Ocean and Arabia." (Arabian Sea?)

After further discussion of de Lapparent's views, Abendanon continues:

"The increase of dimension in consequence of the anticlinal stretching is made evident in the most convincing manner by the numerous graben or depressions, which constitute an eastern and a western fosse, . . . Throughout the length of the *Grossfalte* we observe as direct and indirect effects of the extension (Distraktion) in the anticlinal zone the occurrence of seismotectonic and volcanic phenomena, the latter more than 1000 km. from the coast."

This is the anticlinal theory of rift valleys on a grand scale.

THRUST STRUCTURES

In so far as the views which attribute rift valleys to some kind of doming and collapse of an arch involve the assumption of horizontal compression, they are to that extent related to those which invoke thrust faulting. Solomon of Heidelberg reasoned thus regarding the development of the Rheingraben.[2] Through borings drilled in the margins of the graben it appeared that younger strata occur beneath older, in such a relation as to demonstrate the presence of overthrusts.[3] There is no doubt of the correctness of the observations, but the effects may be regarded as subsidiary to conditions of vertical displacements.

---

[1] *Op. cit.* page 126.
[2] W. Solomon, *Ueber die Stellung der Randspalten des Eberbacher und des Rheintalgrabens,* Zeitschr. der deutsch. geol. Gesellsch., vol. 55, page 403, 1903.
Ibid, Die Erborung der Heidelberger Radium Sol-Therme und ihr geologischen Verhältnisse, Abhändl. der Heidelberger Akademie der Wiss., math.-naturwiss. Klasse, vol. 14. Abhandl., 1-105, 1927.
[3] See *Geologic Structures,* Willis and Willis, 2d edition, pages 87-89, for fuller account of Solomon's evidence and interpretation.

Uhlig, to whose studies in Africa we have already referred, held no fixed opinion regarding the origin of the rift valleys, but his early inclination was toward the recognition of overthrusts. He wrote: [1]

"I have shown that the volcanoes are in part older, in part younger than the great tectonic lines of the region. I have intentionally avoided going into the recently much discussed question of the causes. It would certainly be difficult to make it appear in any degree plausible that a number of volcanic outbreaks or other activities of magmatic energy occurred *first*, just in this gigantic meridional line, and that thereafter there followed the development of the great faults. Just as little, however, is it demonstrable on the evidence of our region that the loci of eruption are intimately related to the tectonic lines. . . . The essential postulate, equally for the faulting as for the volcanic eruptions, involves the processes which gradually brought about a change in the conditions affecting this section of the earth's crust to a great depth, for tension or compression, perhaps in connection with thermal changes.

"I have thus far referred only to faults, without going into the question of the manner of movement of the blocks of the crust along these lines, without suggesting that they were normal faults or thrusts. The investigation of such questions is exceedingly difficult even in regions which are a hundred times as well known and in which the work of observation is much easier than it is in graben and fault steps. Nevertheless, I would like to present a few considerations which bear evidence in favor of the hypothesis that the graben and faulting have resulted from overthrusting."

Uhlig proceeds to cite local occurrences, too detailed to be followed in this context, but suggesting the following reasons for preferring the hypothesis of overthrusting to that of faulting, namely: (1) In certain districts of the southern part of the Great Rift Valley, where great vertical displacements have been produced, volcanic eruptions, such as would presumably occur along tension faults, are lacking; (2) the observation that ancient mica schist overlies relatively very young lavas east of the great escarpment on the western side of Lake Natron in the Great Rift Valley indicates an overthrust of 2 and 1/2 km.; (3) the notable elevation of the plateau between the Great Rift Valley and the Indian Ocean appears to be of the nature of uparching due to horizontal pressure; (4) the elevated margins of the rift valleys suggest uparching along their trends in a manner consistent with the hypothesis of overthrusting.

Regarding the specific, though even yet unverified, observation of overthrusts of considerable displacement, he appears to have entertained no doubts, but was willing to admit that they might be secondary structures (see page 262 in the descriptive section) and he later inclined rather to the recognition of a broad uparching followed by collapse of the crest, as already noted.

The idea that compression may have been the principal cause of the African rift valleys appears not to have been seriously considered until it was again brought forward by Wayland to explain the Albert trough. He postulated a gradual compressive movement to which he attributed the development of the

[1] Carl Uhlig, *Die sogenannte Grosse Ostafrikanische Graben*, Geogr. Zeitschr., Leipzig, vol. XIII, 1907, page 500.

"Uganda-Congo dome and the Victoria Nyanza depression" and assumed subsequent movements, as follows:[1]

"After a period of rest the movements reasserted themselves to a smaller degree but with greater rapidity, resulting in fractures which gave rise to the uplift of Ruwenzori, the outpouring of its attendant lavas and the (?fissure) lavas of Karamajo to the north of the Lake Victoria depression. Then, after a second and comparatively short rest, came the great compressive movement—of which that resulting in the formation of the great block mountain, above referred to, was the immediate forerunner—giving rise to a great system of reverse faulting, which, starting in a virtually homogeneous medium, ran to the surface and produced the structural feature of the Albertine (and presumably the major part of the crest of the Great) Rift Valley."

It appears from Wayland's discussion of the mechanics of the postulated faulting that he regarded the faults as shearing planes, which defined wedges of the crust; and that the uplift of Ruwenzori as well as the depression of the Albert trough followed from the action of the compression on the wedge-shaped blocks, whose thinner edges pointed respectively downward and upward.

Willis in 1929 examined the structural evidences of uplift and depression in Mount Ruwenzori and the Albert trough and confirmed Wayland's interpretation of the mechanical conditions. He found that Mount Ruwenzori presents the form of an uparched peneplain, of Miocene age, with monadnocks, and that the uplift may with reason be attributed to the action of compression upon a nucleal mass of greenstone, presumably thinning downward. He found a subordinate overthrust in the escarpment on the south shore of Lake Albert and hypothetically considers it probable that the Albert depression is located by the same body of greenstone, which is thought to pitch under the site of the lake and to have been forced down by horizontal compression against its upper, oval section.[2] He had previously developed the idea that the Dead Sea trough is a ramp valley, a type of depression attributed to compression, though differing in some respects from the Albert trough.[3]

TENSION-GRABENS

Even as Elie de Beaumont influenced French thought in geology during the middle half of the Nineteenth Century, so has Edouard Suess dominated the thinking of German geologists since 1875, the date of publication of *Die Entstehung der Alpen*, the initial summation of his views on mountain uplift. The outstanding idea of Suess's speculation was the denial of the possibility of any

[1] E. J. Wayland, *Some Account of the Geology of the Lake Albert Rift Valley*, Geogr. Jour., vol. LVIII, 1921.
[2] Bailey Willis, *Living Africa*, 1931, page 47 and also in this volume, plate 17.
[3] Bailey Willis, *The Dead Sea Problem; Rift Valley or Ramp Valley*, Bull. Geol. Soc. America, vol. 39, 490-542, 1928.

uplift of the earth's crust against the force of gravity, except by horizontal pressure and folding, as exemplified in the Alps and similar mountain ranges.

The question of uplift of mountains and plateaus is a very ancient one. The French geologist, Ami Boué,[1] in 1832, cited the observations of Stenon, an Italian, who in 1669 maintained that the strata which are now tilted were formed in a horizontal position; that the mountains we now observe have not existed since the beginning; that some of them are the results of igneous eruptions; and that they take diverse directions on the surface of the globe. De Beaumont himself, in 1829, wrote in support of the view that elevations of the surface had been actually raised.[2] It was this view that Suess contravened. His idea was briefly stated in 1880 to the effect that there could not have been any movements of the earth's crust from below upward, with the exception of those which may have been produced indirectly in the development of folds.[3]

As applied to rift valleys or graben the negation of uplift left no alternative except that of subsidence in response to gravity. The concept is most clearly stated by Neumayer, an intimate collaborator with Suess.[4] Having described the Alps and other folded ranges, he devotes a chapter to sunken fields (*Senkungsfelder*) and by contrast states:

"There is a great group of phenomena, among which faulting holds a foremost place, the subsidence of strips of the earth's mass along displacements, which points to the action of a vertical force, and which we can ascribe only to the effect of the weight of the sunken mass itself."

To illustrate this idea he refers among other examples to the plateau of the Black Forest as:

"a column of old rocks, from which the land on both sides has sunk away along great fissures, in such a manner that the margins step down as so many successive fault blocks. . . . We must imagine the whole broad region between the central plateau of France and that of Bohemia to have been a continuous plateau, covered with a thick mantle of sediments, which hid the old crystalline foundations. In this broad tableland there developed a vast system of fissures, along which subsidences occurred, and these were of such an arrangement that there remained standing a row of firm columns, the Bohemian massif, the Black Forest, the Vosges, the central plateau of France, while between these 'Horsts' the other sections sank down, step below step, into the depths. The portions which remained elevated were most vigorously attacked by erosion, the destructive action of running and seeping waters and of frost, and the entire covering of sediments was removed. On the lower lying sections the oldest strata remained, as for example the Buntsandstein of Hornisgrunde, and younger deposits were laid down as an undisturbed cover and became thickest where the particular section had sunk to greatest depth."

[1] Ami Boué, *Observations on the Progress of Geology*, Bull. Soc. géol. de France, vol. 2, 1831-32, page 124.
[2] Elie de Beaumont, Ann. des Sciences nat., vol. XVII-XIX, 1829-30.
[3] E. Suess, Verh. der K. K. Geol. Reichsanstalt zu Wien, 1880, page 180.
[4] Melchior Neumayer, Erdgeschichte, Leipzig, 326-331, 1887.

After citing several broad sunken areas, such as the Ægean archipelago and the Mediterranean, in which the island of Crete is described as an upstanding horst, Neumayer refers specifically to the typical graben as follows:

"A peculiar type of subsidences is presented by the sunken grabens, in which a long, narrow strip of land has sunk down between two parallel faults. The Rhein Valley between the Black Forest and the Vosges, the Leinethal near Göttingen, the Red Sea, the trough of the Jordan and Dead Sea, and presumably that of Lake Tanganyika in central Africa are examples of this phenomenon."

The affirmations of these statements are generally true; the areas described are depressed. But the negation of uplifts is not true, as was shown by Davis [1] and is now generally recognized on physiographic evidence, even by the followers of Suess. Suess's concept of a sunken graben continues, however, to affect speculation regarding rift valleys and it is desirable to quote his own account; it is difficult, however, to do so adequately without too greatly expanding this summary. It was his habit to assemble all the literature on a region under study, to store his prodigious memory with the descriptions and to form a definite mental picture of the land and its geologic characteristics. Diversity of languages was no embarrassment to the master of all European idioms. Geographic remoteness suggested no strangeness to the man who from his Viennese study had explored the whole world through the observations of others. He readily conceived comparisons between African, Asiatic, American and European lands and their features and made a synthesis of their elements. In his writings he skips from continent to continent in search of illustrations and for the average reader he is as difficult to follow as an airplane for a man on foot. We may be content to cite the summary of ideas with which he closes a detailed account of the African Rift Valleys, so far as they were known to him in 1891: [2]

"The so-called parallel horsts, such as the Vosges and Schwartzwald, Lebanon and Anti-Lebanon, are lacking for the graben of the African continent. Extensive blocks form the borders of the depressions, such as the Somali block, the Abyssinian, and the Arabian blocks, which although, no doubt, horsts of a kind are not to be compared with those. So far as the facts are known to me from the foregoing descriptions, it would appear that the opening of fissures of such magnitude can be explained only by the action of a tension, directed perpendicularly to the trend of the split, the tension being relieved in the instant of bursting, that is of opening of the fissure. In the description of the sunken trough of the Gunnison Valley in the Great Basin, which is 20 miles long and nowhere more than 3 miles wide, Dutton says: 'It appears to be a very clear case of a block dropping through the drawing apart of the strata and sinking to fill the gap thus produced.' [3] Thus we may well assume for

[1] W. M. Davis, *The Bearing of Physiography upon Suess's Theories*, Amer. Jour. Sci., vol. XIX, 1905.

[2] E. Suess, *Die Brüche des östlichen Afrikas*, Denksch. der Kaiserl. Akadem. der Wiss. zu Wien; math.-phys. Klasse, vol. LVIII, 1891, pages 580 *et seq.*

[3] C. E. Dutton, *Geology of the High Plateaus of Utah*, 1880, page 34.

the development of meridional fissures of such magnitude the existence of a tension in the outer part of the globe, in the direction of the parallels of latitude. The crack may heal, the tension may in time be renewed, and a new split and new depression may occur.

Under this assumption of a sudden release of accumulated tension one may also include those divergencies which are noted at the northern end of the Lebanon in the north and, on a much larger scale, at the northern end of Nyassa in the south.

Similarly, under this assumption, we may cover the striking fact that the water parting so often lies very close to the edge of the graben, as described with astonishment by Stanley and other travellers. One would be mistaken if one deduced from this relation the inference that the development of the fissure had been preceded by uparching, as is, we may concede, the case with radial fissures on the moon, at least for the middle, though not for the individual rays. A splitting of the earth's crust by elastic masses, protruding from within, would probably not produce troughs with occasional volcanoes, such as we see here or for example in Iceland, but would fill up with lava all of the depressions that might form. Immense floods of lava of this character are indeed observed in eastern India, in Arabia, Abyssinia, and far down in South Africa. But the faulting of the Ghats in eastern India is younger than these lava floods and most of the graben which we here observe are certainly younger than these lavas and sunk in them. It seems to me that the effects, which must certainly accompany the opening of fissures of such magnitude, suffice to explain the actual position of the water partings. The bursting of the earth may well be accompanied by a certain upward movement of the lips, that is the plateau margins, suddenly set free."

There is much more in Suess's comprehensive discussion of the African Rift Valleys and their homologues which throws light on his intimate knowledge of the literature and original interpretation of the mechanical processes as he understood them; but the above quotation suffices to express the central idea of a bursting tension and resultant subsidence. His concluding paragraph, however, contains an admonition which is pertinent to the general discussion. He says:

"In all representations of this kind we have, however, to guard against the assumption of a geometrical arrangement of any kind; indeed, considering the almost incomprehensible variety of the occurrences, a systematic search for regularity is not without danger, because the inquiring mind is so easily led astray from the path of sound synthesis."

Eighteen years later, upon the publication of the third volume of his comprehensive description of the globe, *Das Antlitz der Erde*, Suess reiterated the concepts already cited and supplemented them by suggesting a causal condition. Under the heading "Outlines of the East African Graben," he briefly reviews the observations and continues:[1]

"One should not imagine for oneself any too systematic concept of these graben, as of a strip of the earth which is sunk between two parellel faults. . . A more cor-

[1] E. Suess, *Das Antlitz der Erde*, vol. III, part 2, 1909, page 316.

rect idea will be obtained if one conceives that step faults descend from both sides in succession toward the middle of the valley and that along them many long, downward pointing wedges have sunk unequally. In this manner horsts may remain standing up in the area of subsidence, as in the cases of the Kamasia Range, west of Lake Baringo, and the saddle south of the Albert-Edward lakes. Obvious is the lessening of the breaking down in northern Syria, where the virgation and many other circumstances indicate an advance of the activity from south to north.

"Any attempt at an explanation on local grounds, on the assumption of any particular, downward divergent attitude of the fault planes, etc., vanishes in view of the extraordinary extent. An occurrence which is manifested over more than 52° of latitude must originate in a condition of the planet itself. We reach the conclusion that throughout this great area there must have existed a tension in the outer shell of the globe, which acted in a direction at right angles to that of the fissures, here at right angles to the meridian.

"That is tearing apart by contraction, and in such manner that the cracks opened from above downward."

Having thus defined his concept of the cause and character of the fissures, Suess continues:

"This condition [the opening of the cracks from above downward] is significant in considering the occurrence of volcanoes. There are great examples of graben, as for instance Tanganyika, 700 km. long, in which, although there are hot springs, there are no volcanoes. On the other hand, there are in Syria individual, quite minor displacements which are accompanied by chains of volcanic phenomena. If one considers the volcano Kenia or the group of the Kirunga [Virunga or Mfumbiro] volcanoes on Lake Kivu one searches in vain for rifts (Gängen) or volcanic chains, which might indicate a connection with other volcanoes. Nevertheless, a general view of the whole shows the connection and also the dependence upon the great tearing apart.

"The fractures, in opening from above downward, and the cracks between the subsiding strips have in some places facilitated the eruptions, in others they have not; on many the eruptions are still active. Whether large or small volcanic piles develop may probably be dependent upon quite subordinate movements. Along the axis of the East African graben there is evidently the most vigorous volcanic activity. In the graben itself stand the least number of great cones, and presumably just because here where the greatest number of tectonic displacements occur no channel endures long, but new ones are constantly opening."

After enumerating certain other so-called grabens in West Africa and summing up his understanding of the general knowledge of the rifts in Africa, Suess observes (p. 320):

"All the above described fractures and graben, with the exception of the Rheingraben, lie in a tableland which in Africa and India, perhaps over its entire extent, had been eroded before Lower Gondwana time [late Paleozoic] and since then has not been folded.

"Many of these fractures are remarkable for their length and, with the exception of the slightly curved Tanganyika trough, for their straightness. Many adhere in a striking manner to the course of a meridian. They are in much the larger part accompanied by volcanoes. But the noteworthy fact is that they are limited in occurrence to a definite part of the earth's surface. Nothing of the kind is to be found in Asia, with the exception of Syria and the Peninsula of India, nor in America. Fissuring due to tension (disjunctive lines) is also to be observed in regions of folding, but there their form is always curved, corresponding with the strike of the folds."

In summing up the views of the student who first grasped the magnitude of the African rift valleys and inspired many others to explore them, we may enumerate four sequential concepts, namely: (1) The continuity of the graben, essentially along a meridional line, through the great distance from Syria to southeast Africa, 52° of latitude; (2) the postulate that the rifts must be due to a tension, which developed in the outer crust of the earth as a result of contraction; (3) the subsidence of strips isolated by faults, while adjacent masses, "horsts," remained standing on firm foundations as elevated blocks or columns; (4) the general if not universal occurrence beneath the outer crust of a potential source of volcanic activity and lavas. None of these postulates can now be said to have been sustained by subsequent research, except in a limited manner and with important qualifications.

Obst, a naturalist in the broader sense, though not a specialist in structural geology, adhered to the tension theory and supplemented it with the suggestion that the action might have been due to subsidence of the Indian Ocean Basin. His observations were restricted to, but covered in some detail the southern portion of the Great Eastern (Gregory) Rift Valley, south and west of Lake Manyara, where the faulting presents a number of eastward facing cliffs, without westward facing opposites. Primarily a physiographer and anthropologist, he describes the landscape aspects and the people, but relates them to the geologic foundations and has in mind the general problem of rifting. Referring to the idea that the faulting may have resulted from the collapse of a far-reaching meridional fold, he puts it aside and concludes:[1]

"While in continuous contact with this problem, there has grown up in our mind an hypothesis, which seems not only more probable then the theory of folding, but also presents a more immediate and fundamental cause. We incline to the idea that the great disturbance in East Africa and the young volcanic eruptions may have been related to the subsidence of the Indian Ocean, that is to say, the great graben faults and step faults are to be regarded as tension effects. It will be the work of a future publication to set forth this idea more in detail. [The events following 1913 appear to have prevented the fulfilment of this purpose.] It will suffice here to say that we regard the Great East African marginal fault as the oldest of these tension faults and entertain the view that the process of parting extended farther and farther westward during the Tertiary and Diluvium in consequence of continued subsidence in the region of the Indian Ocean, producing cracks with approximately meridional trends and a breaking down in strips in that portion of the continent

[1] E. Obst, *Der östliche Abschnitt der Grossen Ostafrikanischen Störungszone*, Mitt. der Geogr. Gesellsch. in Hamburg, vol. XXVII, 1913, page 187.

which adjoins the Indian Ocean. The ultimate result of this great process of tension and parting is that structural landscape of plateaus in steps, which is presented to our view in the great peneplain of central and eastern Africa. . . . The elevation of the lips of the fractures, which often occurs, we incline to attribute to the active effort of the earth's crust toward isostatic equilibrium. In consequence of the breaking down of long strips there develops a disturbance of the static balance of the crust; the sunken blocks produce an increase of density, a local compression, which in turn results in a pressure that is directed upward and tends to restore the equilibrium. This upward pressure produces the elevation of the margins of the blocks which have remained standing in place. Thus we may explain without difficulty the swelling of the lips of the sunken graben, and the swelling, which in the case of step faults is to be observed only on the margin of the uppermost block, is also explained by the undeniable relation between the throw of the displacement and the amount of the swelling."

Suess had died the year before this explanation by Obst was published and there is no available record of discussion between the originator of the tension theory and the observer who extended it by suggesting an immediate cause. But it is well known that Suess, the author of the concept of Gondwana land, believed in the relatively recent subsidence of that continental mass and the consequent formation of the Indian Ocean Basin. Although Obst does not refer to Suess, the parallelism of thought is obvious. Obst presumably was acquainted with that passage in *Das Antlitz der Erde* in which Suess, after enumerating the African troughs, says:[1]

"It can not be denied that along the margin of the Indian Ocean there appear lines, which show a relationship (to the African faults). . . . The positions of these lines influenced the outlines of the ocean. To be sure, it is not to be assumed that the subsidence might be attributed to them, but rather that they have resulted from other causes and that they have been used by the subsidence and have fixed its boundaries."

We pass on to one of the most devoted followers of Suess, one whose name is identified with the study of rift valleys, J. W. Gregory. The constructive imagination of the German philosopher early fired the enthusiasm of the young Scot, who retained throughout a long and distinguished career his belief in the master's theories. We have elsewhere in this volume briefly outlined Gregory's initial explorations in the Great Eastern (Gregory) Rift Valley and are here concerned with the evolution of his theoretical concepts. His first impressions were given in his narrative book, *The Great Rift Valley, 1896*, and earlier in an address before the Royal Geographical Society in 1894. We quote from the latter. Adhering to the idea of continuity postulated by Suess, he said:[2]

[1] Vol. III, Part 2, page 320, 1909.
[2] J. W. Gregory, *Contributions to the Physical Geography of British East Africa*, Geogr. Jour., vol. IV, 1894, page 290.

"From Lebanon, then, almost to the Cape there runs a deep and comparatively narrow valley, margined by almost vertical sides, and occupied by the sea, by salt steppes and old lake basins, and by a series of over twenty lakes, of which only one has an outlet to the sea. This is a condition of things absolutely unlike anything else on the surface of the earth. It is, therefore, only natural to inquire whether this great valley consists merely of a series of independent basins formed by depression or erosion, the linear arrangement of which is but an accident, or whether it is all due to a common cause, which formed a great earth crack or depression at some comparatively recent period of the earth's history."

Of these two alternative explanations, Gregory discarded the one which implied activity of local forces under local conditions and adopted the view of a general common cause. All his subsequent reasoning was influenced by that assumption. Seeking for rift valleys of comparable dimensions he found them only in the moon, regarding which he remarks:

"The nearest approach in size can probably be found on the moon, whose clefts or rills no doubt represent long, steeply walled valleys and present to us much the same aspect as this East African valley would do to any inhabitants of our satellite. Not the least interesting of the points raised by the African-Red Sea-Jordan depression is the possibility that it may explain the nature of those lunar clefts which have so long been a puzzle to astronomers."

That section of the Gregory Rift Valley which Gregory explored is peculiarly characterized by volcanic eruptions, both fissure and central eruptions constituting a long and still active succession. They could not fail to take a prominent part in any theory of the origin of the valley in connection with which they are so abundant. He reasoned accordingly.

Gregory was also impressed by the unusual height of the plateaus immediately bordering the rift valley, a condition that is particularly evident in the northern portion which he traversed. The Aberdare range is a half dome, which attains an altitude of more than 12,000 feet (3500 meters) and faces the valley with stepped precipices; opposite to it the Nakuru plateau presents a similar form, which rises to 10,000 feet (3000 meters). The split dome stands out as an evident feature of large dimensions, whether conceived in the constructive imagination or viewed on a general hypsometric map. He became convinced of its reality and causal relation to the rift valley.

The distribution of the Gondwana flora convinced him of the former existence of the continent of that name, as postulated by Suess and Neumayer on the site of the Indian Ocean, and the diversity of Jurassic faunas along the northern and southern portions of the East African coast was to his thinking proof of its existence up to that period.

In searching for a cause he looked for coincidence of rift valley faulting with other great disturbances and consequently considered the age of the faulting of prime significance. While recognizing the youthfulness of many of the fault movements and assigning them to the Pliocene and Pleistocene, he thought that there had been a long period of eruptions and faulting prior to the Pliocene. At first he dated the beginning in Cretaceous time and later accepted evidence of its post-Eocene age.

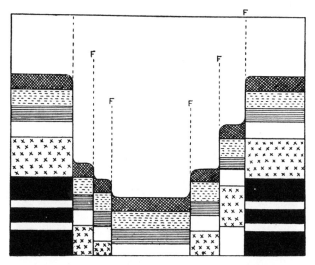

FIG. 2—Idealized rift valley section. After Gregory.

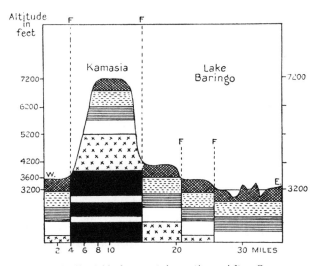

FIG. 3—Idealized block mountain section. After Gregory.

Furthermore he had clearly observed in the escarpments of the Great Rift Valley the repeated and general occurrence of stepfaults, of a type which is intimately, and to his thinking exclusively associated with horizontal tension.

With these elements of fact and theory he built the hypothesis of the origin of the rift valley, which is associated with his name. It can best be summarized in his own words from the article written 27 years after his first studies and after a second visit to the scene of his early exploration.[1]

"The Great Rift Valley therefore extends from Lebanon to the Sabi River, and its branches reach eastward to the Gulf of Aden, and westward include Tanganyika and the upper Nile, and the rift valleys of lakes Moero and Upemba in the Central Congo. This wide-spread valley system is obviously not the result of some local fracture. Its length is about one-sixth of the circumference of the Earth. It must have some world-wide cause, the first promising clue to which is the date of its formation."

Gregory then discusses the physiographic and structural evidences of youth and age and concludes:

"The evidence seems clear that the Red Sea trough was formed by earth-movements at successive dates, beginning not later than the Oligocene and continuing until quite recent times."

The context makes clear that he regarded the foundering of the Red Sea trough as one incident of the process of rifting throughout the whole extent of the Great Rift. He continues:

"The history of the Rift Valley in British East Africa is dependent in the main on the volcanic history of the country; the two are connected, as the subsidence of the earth blocks doubtless forced up the lavas along the fractures.

"The first stage in the formation of the Rift Valley was the uplift of a long, low arch with the axis trending north and south. . . .

"The second stage was the cracking of the sides of the arch as the lateral pressure was reduced, and the top sank as the keystone of an arch sinks if the end supports give way. The sinking of the keystones of the East African arch into the plastic material below forced some of it up the adjacent cracks, through which it was discharged in volcanic eruptions.

"To determine the dates of these events, we must consider the relations of East Africa to the Arabian Sea, and therefore go back to the time when East Africa and India were included in a continent which extended from Brazil to Australia. This continent, Gondwanaland, lasted through the Carboniferous period and through the two succeeding periods, the Permian and Trias, so that no marine beds belonging to them were deposited on the mainland of tropical or southern Africa. . . .

"In the next two periods, the Jurassic and Cretaceous, Gondwanaland began to break up, and the sea invaded the coast-lands of India and of East Africa. The Jurassic deposits were laid down quietly, undisturbed by volcanic eruptions; but the active earth-movements in Upper Cretaceous times, and the foundering of the floor of the Indian Ocean between East Africa and India, led to volcanic outbreaks on a colossal scale. . . .

"The subsidence of the Arabian Sea and the outflow of the vast quantities of lava left the East African arch insufficiently supported and the top of it sank between parallel fractures. This subsidence happened along the first of the Rift Valley faults."

[1] J. W. Gregory, *The African Rift Valleys*, Geogr. Jour., vol. LVI, page 31, et seq., 1920.

Following this definite outline of his hypothesis of the antecedents and conditions of development of the rift valley system, Gregory follows Suess both intellectually and in geologic theory in linking the Great Rift Valley with the "mountain chains due to the last great uplift of fold mountains" in Europe, Asia, and the Americas. He contrasts the African structures with the cordilleras of North and South America and deduces effects of world-wide stresses. His final paragraph runs:

"The essentially different character of the contemporary earth-movements of Africa and of the Pacific borders is explained by their antipodal position. Africa was antipodal to the Pacific, and it is in accordance with the well-known antipodal relation of ocean to continent that while the Pacific was sinking and the crust beneath it undergoing compression, its antipodal land should be rising and subject to tension. Africa as a whole has remained throughout the Kainozoic era as a raised and, judging from its river system, a still rising plateau; but as the crust on both sides sank, it was left unsupported laterally. Instead of its main highland axis being laterally compressed like the coastlands of the Pacific, Africa was in tension and torn by north and south fractures, along which the sinking of a strip of the crust formed the longest meridional land valley on earth. The Great Rift Valley is thus due to a long series of earth-movements which began in the Cretaceous, and to faults formed at intervals between the Oligocene and the Pliocene. It owes its unique character to its position antipodal to the Pacific, and its course to the wrench in the crust of the Eastern Hemisphere between the segment pressing northward against Europe and that pressing southward in Asia toward the deepening basin of the Indian Ocean."

The above was written in 1920, shortly after Gregory's return from his second visit to East Africa, after the lapse of twenty-seven years following his early exploration. In 1929, when I had the pleasure of discussing the problems with him before I visited East Africa, and again in 1930, when he welcomed me on my return, his broad speculations regarding the world-wide causes of African rifting remained essentially the same. They had, indeed, developed to include similar explanations of the origin of the Atlantic and Pacific Ocean basins, which he regarded as due to foundering of continental areas.[1] His geologic philosophy was also hospitable toward the theory of continental drift. His thought had long been stimulated by imagination; his vision grasped the globe; local features had become details in the general concept and were to be explained in terms consistent with the theory that encompassed the whole. If that theory proved to be erroneous, some other explanation would have to be found, but believing that theory to contain the principle of verity, he thought the explanation of African Rift Valleys as phenomena of tension due to foundering of other portions of the crust to be the most probable.

[1] J. W. Gregory. *The Geological History of the Atlantic Ocean.* Quart. Jour., Geol. Soc. London. vol. LXXXV, 1929, and *Geological History of the Pacific Ocean; ibid.* vol. LXXXVI. 1930.

It is logical that the theory of continental drift should find its application to the rift valleys of East Africa. If continental masses have split and separated, the outstanding example of a splitting continent should not be overlooked. Given initially the idea that rift valleys are usually, if not always, bounded by normal, gravity faults, all the conditions of a satisfactory hypothesis may seem to be met. This view is entertained by Krenkel, the leading student of the rift valleys among German geologists and author of the most complete résumé on the subject.[1] His field observations were carried out in 1908 and 1914 and his manuscript was finished in 1921, though he felt at that time that there was need of more work on the problem. His approach to it was enthusiastic. We quote from the introduction to his work:

"Two African days are for me not to be forgotten. The one—it was April 24th, 1908—afforded me an outlook over the Great Rift Valley in the cool of the morning. During many months spent in German East Africa I had looked forward eagerly to this hour when I should see the wonderful landscape made familiar by the descriptions of v. Höhnel and J. W. Gregory. The Uganda railway had slowly advanced over the rising ground without opening any far-reaching view until the steep descent on the eastern margin of the great Graben was reached. There, suddenly, there was unveiled one of the most enchaining of landscape views: real, far-reaching African vastness and sharpness of outline decked with a richness of unusual ornament, one of the most impressive occurrences of the tremendous working of geologic forces. Steeply from our feet the wall of the Graben sinks down many hundred meters, breaking up into sharp-edged steps in the descent. Toward the north and south the floor of the wide rift stretches its length into the mists of the plateau, the clear yellow tones of the steppe contrasting with the green woods that clothe its escarpments. Great volcanoes tower up, unchanged in form; around them spread bright sands with dull green flecks of stunted acacias. And on the far side of the valley floor rises the west wall of the Graben, climbing upward in gigantic steps of equal boldness. In spite of the distance they seem close at hand. . . . Though later wanderings there by lakes and over volcanoes, under burning suns, might weary, though the stepped walls might be veiled during the rainy season by dark clouds and tatters of mist drifting along the Graben, never could the memory of this first view of the sundered strip of the earth's crust be dimmed.

"That other day: a September evening in 1915. It was war time. The task of the explorer, upon which I with my wife had entered in the spring of 1914, had given place to stern service in the defense force. A short train of the Tanganyika railway, our faithful aid in the war, carried me through the monotonous steppe from Tabora to Kigoma. In the late twilight of the tropic evening there glinted as from a great depth, through a vista in the greenwood that borders the Tanganyika Graben and which was so grateful after the heated hours, the gleaming flood of the lake, but only for an instant. An hour later, under the pale light of the moon, the lightly moving waters of the Bay of Kigoma lay spread out before the astonished eye, which, having during a year in the arid interior become a stranger to the sight of living waters, found itself overpowered by this abundance. Slowly the spotlight on the post on the Entenschnabe [a promontory in Kigoma bay] swung over the rounding bay and its surroundings, reminding us that the peaceful scene on the shore of

[1] E. Krenkel, *Die Bruchzonen Ostafrikas*, Berlin, 1922.

the superb lake was set in a frame of grave events. During three-quarters of a year many a voyage on the lake and many a fatiguing march on its shores displayed the ever-changing suggestion and beauty of Tanganyika, whether in the heat of summer days or in heavy thunder storms, which permitted the farther shore, that was often hidden for weeks, to emerge as if within reach.

"What a contrast to the Great Rift Valley! There a steppe, yellow grass and thornbush, volcanoes and lava flows, here an unending reach of water with unusual fauna, green woods, banana groves, and oil palms on the shores. Both, however, similarly framed within protecting walls of steeply rising mountain margins. Two African scenes of the most individual character: fundamentally different, it is true, in landscape features, but in their inner cores owing their existence to the same geological laws of development."

Krenkel refers to the fact that his notes and collections were captured and lost and that the publication of his manuscript in 1922 had been delayed by untoward circumstances.

We leave the enthusiastic young geologist of 1908 and the poet-soldier of 1914 to turn to the deliberate opinions of the mature student of 1921. In the latter part of the work from which the above introduction is translated, he gives an historical outline of the various theories, upon which we have drawn for many references, and then sums up at considerable length his opinion of their respective validities. He puts aside quite unequivocally (page 161) all those hypotheses that assume an *anticlinal* uplift as a precedent of rifting. We need not here pursue his analysis of that relation, as we shall treat it elsewhere and with the same conclusion. He describes the relatively high margins of the graben as a common, though not universal and consequently not essential, phenomenon and explains them as perhaps, in some cases, effects of an effort toward restoration of isostatic equilibrium, but more generally as a result of a tilting of the marginal blocks, which he conceives to have been sucked down in the widening split in the earth's crust (page 166). He concludes:

"The tectonic setting of the East African fault zones, whether considered in detail or as a whole, admits of only one explanation: they are zones of tearing apart of the crust, produced by a directed tension. Only as tension phenomena can the wealth of observed structural facts and changes be brought into orderly relations. Accompanying these deep-seated developments as inevitable accompaniments are disturbances of gravity, earthquakes, and volcanic activity. Forces that produce partings (disjunctive forces) alone produce structures like the rift valleys, with their series of splits, that extending to great depths converge downward and are unequally filled with light-weight superficial rocks. The separation is, however, most evident in those gaping rifts whose depths are filled with water. The action of compressive forces is nowhere recognizable. Overthrusts of significant displacement toward the axes of splitting are everywhere lacking.

"The tearing apart has resulted in a certain areal expansion of the central African landmass, which has been most strongly affected. Any estimate of the amount can only be tentative, presenting only a rough approximate value. . . . Nevertheless the attempt must be made in order to obtain some idea of the magnitude of the expansion. Several estimates for the Tanganyika graben result, for example, in a parting coefficient (Lockerungskoeffizient) for its marginal blocks of at least 5 per cent. The parting is, therefore, quite considerable. To the end that it may occur, we must assume powerful tensile stresses, which must accumulate in order that they may ultimately be released by the sudden, energetic development of the faulting."

Following this exposition of the tension hypothesis, as he developed it from his observations in Africa, Krenkel describes the arcs which characterize the alignments of the western and eastern zones of rifting, and he cites other occurrences of similar arcuate forms. From these forms he draws the inference that:

"For East Africa it appears that arcuate arrangement of many systems of faults indicates a control of the splitting by certain stresses which determined the directions."

Seeking the cause of these determining stresses he first considers, but sets aside as inconsistent with the facts, the idea that the continental mass of Africa might have been squeezed from north to south and so spread from east to west by the folding of two east-west mountain chains, the one in Asia Minor, the other at the Cape of Good Hope. This is an application of the mechanical principle that compression is necessarily accompanied by tension normal to its direction, but, he concludes, the jaws of the vise are too small and too far apart. Nevertheless the exploring mind, surveying the relations of continents, ocean basins, and folded mountain chains, feels that there must be some general condition, some grand movement or effect, by which to account for the Great Rift of Suess's constructive imagination. Thus Krenkel is led to assert:

"In spite of all these objections there must be some relation between the intense fracturing of Africa and the development of the Eurasian chain of folded mountains. [The chain comprising the Himalayas, Caucasus, Alps, and Pyrenees.]

"Two great primeval masses stand, in relation to one another, under the control of movements whose direction is determined by the unavoidable balancing of the altered position of the Pole, which has been changing since Tertiary time. The Eurasian mass is pushing southward, with a pronounced and powerful component toward the southwest in its east wing, eastern Asia. The African primal mass, on the other hand, slides northward, with a partial deflection toward the northwest. Between the two masses lie the sediments of the Mediterranean geosyncline, compressed into the young, folded ranges of the Old World, they being passive strips of the crust, like all folded mountain chains. The movements of the two primeval masses toward the Mediterranean continue in opposition to each other and result in continued deformation in the region of pressure between them.

"The northward drift of the original African mass does not, however, progress equally in all parts in complete harmony of rate and direction, without obstruction or friction. Inequalities of movements in the major mass and its parts lead necessarily to tearing apart of the crust. The eastern wing especially is hindered in its progress in the indicated direction of sliding and peculiarly affected by a series of occurrences, which originate in its geologic history.

"Thus, it is established that eastern Africa extended considerably eastward before the Tertiary period. It broke up and sank in blocks into the depths, losing more and more in area as the waters of the Indian Ocean advanced. The cause of the breaking up and subsidence is to be found in magmatic wanderings beneath the upper crust, of whose sunken blocks the floor of the Indian Ocean consists. The displacement of magma in the realm of the Indian Ocean extended farther, affected its very foundations in some portions, and worked in on the eastern edge of the African continent, drawing and splitting it off. It is furthermore certain that in the interior of East Africa—to a much greater degree than in all the rest of the continent—during the Tertiary and Quaternary immense masses of Atlantic rocks, the magmas of the rift valley region, rose to and in part flooded the surface. It is conceivable that their deep-rooted path of ascent, forming a united intrusive column below and branching widely above—and including older magma channels—may have given East Africa a firm anchorage in the subterrane. Thus hindered from taking part unobstructedly in the movement of the primal mass, East Africa must again have been subjected to tension stresses."

After elaborating the idea of horizontal tension as a result of continental sliding within Africa itself and including its neighboring continents, Krenkel considers the evidence of vertical movements in a manner that would have delighted Elie de Beaumont, who championed the theory of "les soulèvements," but would not have been approved by Suess, whom Krenkel in so many other respects accepts. He says:

"In order to explain the profound splitting of East Africa, one might also consider the following: East Africa may be regarded as a region of uplift. The elevation has not been equal in all parts; it was here greater, there less, and may have been interrupted by short pauses or reversals. It began energetically in the passage from Jura to Cretaceous, as can be proved, and has the character of a very prolonged movement, which is still in progress."

He cites the evidence of the uplift of the Abyssinian highland in support of the general contention and then suggests magmatic intrusion as a cause of elevation:

"One of the causes of this elevation of Africa may be uplift by magma: Atlantic rocks [the alkali-rich volcanics of Atlantic facies] appear in rising movements, uplifting their cover. Their high level, not far below the surface, is indicated by many occurrences—above all by the fissure eruptions—which one might even designate as African or Atlantic vulcanism.

"The uplift of the long East African strip, stretching out along the meridian, between the lagging blocks on either side, might afford the possibility of extension of the domed-up surface. This might result in the development of zones characterized by splits, which might to a certain degree spread open to occupy the space given them. It is, however, very doubtful whether the uplift could suffice to produce such an elongation that the cracks would be drawn in."

33

In succeeding pages Krenkel discusses among other topics the future development of the Great Rift as a continental parting and concludes:

"Hidden in the broken regions and disturbed zones of East Africa there lie, as one realizes in viewing the many-colored scene, a wealth of problems still unsolved and questions of the most diverse character. The attempt to coordinate their development with a simple process of movements of the great primeval masses, during the recent past, may be incomplete and may raise doubts. In comparison with the many-sided structures of the Alps, the inconspicuous features of these regions, with their simplicity of ornament, have not received the attention that they deserve through their bearing on important problems of the constitution of the earth."

In his comprehensive work on Africa, *Die Geolgie Afrikas* (1925), Krenkel confines himself to descriptions of the rift valleys, which are accompanied, however, by maps and diagrams that illustrate the idea of faulting according to systems of coordinate, parallel fractures. But he has not yet said his last word on the subject. Even as this volume goes to press he is extending his studies of Africa in the field and is gathering data for further explanation of those problems which he regarded as still unsolved.

Having thus brought this brief résumé of theories of the origin of rift valleys down to the present we may close it, referring the reader who desires more intimate knowledge of the hypotheses to the original works. This account makes no claim to be more than a summary of the outstanding contributions and does not include references to a great many other writers, who have simply applied the same or similar hypotheses to the African or to other occurrences. The object has been to cite the author of an idea with sufficient fullness to present his thought fairly.

ANALYSIS OF EXISTING THEORIES

There are apparently four ways in which force may be exerted to cause a parting of the earth's crust in such a manner that two or more adjacent strips become displaced and a rift valley is formed. They are:

(a) By tension in a horizontal plane.
(b) By compression in a horizontal plane.
(c) By depression, vertically in the direction of gravitative attraction.
(d) By uplift in a vertical direction opposed to gravity.

It may be said that all of these methods of disrupting the crust have been considered and each one of them embodied in some theory of the origin of rifts. It is also true that a large measure of the controversy which the subject involves has arisen from the exclusive adoption of one process and only one as a cause, to the negation of all others. In the actual variety of effects the causes listed above cooperated in various ways.

Part II

GEOMORPHOLOGY

Editor's Comments
on Paper 2

2   JOHNSON
Geomorphologic Aspects of Rift Valleys

A continental rift valley is a tectonic land form. It is recognized initially from its geomorphic expression, that is, faulting that has interrupted the course of a geomorphic cycle. Tectonic features — fault scarps and related forms — actively impress themselves on an existing landscape and are not merely the passive revelation of internal structure.

Paper 2 by Johnson has been surpassed in some respects, such as its inclusion of the now discarded thrust-fault model. There have also been some modifications in terminology, for example, "fault-line 'erosion' scarp" and "fault-line valleys" as erosion features on shear zones. However, it offers clear models of fault scarps, their development, modification, and disposition to form continental rift valleys in a subaerial environment.

Cotton (1950) supplements Johnson, especially with regard to forms other than simple graben. Johnson uses the term "tilt block" where Cotton uses "fault angle" and "fault-angle depression," self-defining terms. Cotton's paper also contains an extensive list of references. Temperley (1966) adds to the study of rift-fault scarp faces and their development. The application of projected-profile and stream-profile techniques in relating rift-valley landforms to planation surfaces is demonstrated by Quennell (1958).

Oceanic rift-valley morphology requires a different approach because neither the originating tectonic (endogenetic) processes nor the consequent erosion and sedimentation (exogenetic) processes are comparable with those of continental rift valleys. In the continental forms, the flanks are only incidentally drawn apart by the amount of horizontal component of fault movement and have floors of downfaulted blocks or infilling sediments. The oceanic forms, on the other hand, are derived from the uplift and separation of the flanks and floors of underlying material. Detailed geomorphological study of the oceanic forms is carried out by bathymetric survey using echo-sounding, underwater photography, and other techniques. So far only limited lengths of the oceanic ridges have been observed directly, chiefly where research has been carried out in the northern Atlantic and the eastern Pacific. Among the papers that give accounts and illustrations based on recent investigations are those by Needham and Franchetau (1974) and Heirtzler and van Andel (1977).

Continental rift valleys in general are tectonic landforms set in a landscape of erosional relief. Because of this, study of the evolution of landscape is essential if the history of a rift valley is to be understood.

Continental (anorogenic) rift valleys are usually set in a terrain of cratonized metamorphic rock with or without a cover of sediments or volcanics. Because of resistance to erosion, evolution of landscape has taken place slowly, and, in Africa, surfaces, either mature or senile, belonging to cycles as old as Jurassic, are preserved. The recognition of these planation surfaces, which can be either extensive or remnants, and the geomorphic cycles or partial cycles to which they belong, has an essential place in the study of these continental rift valleys.

It is important, but rarely stated, that for the use of the term "cycle" to be justified there must be a return "full circle" to a semblance of a former state. Hence the term "geomorphic cycle" is here understood as including not only an erosional phase but an initiating diastrophic phase, which is comprised of earth movements and rift faulting.

King (1962) established a pattern of continental cyclic landscapes for the African continent and demonstrated its relevance to other continents. His work supplements and clarifies the earlier work of Dixey and others. He introduced pediplanation as the dominant process in African geomorphology (King, 1962, Ch. 5). This derives from the theories of W. Penck. On the other hand, Dixey (1946) places the emphasis on the Davisian cycle operating by peneplanation. The production of a planation surface within the savanna cycle is of restricted importance (Cotton, 1961). However, the choice of planation process has only indirect importance in rift-valley study. Tectonic landforms are studied not as such but as providing the key to the nature of the faulting that produced them.

The planation surfaces that are part erosional, part aggradational, are variously described as: erosion level, erosion bevel, erosion surface, peneplain, pediplane, and pediplain, all of which have genetic implications. Planation surface is a term that is neutral in all respects (Small, 1970). The naming and dating of such land surfaces is important for our present purpose. Cahen (1954, p. 432) suggests three ages: local, initial and terminal. King (1962, p. 222) prefers a geographic name with statement of geologic time range of the formation process, which is, in fact, the interval elapsing between the end of one major diastrophic phase and commencement of the next. Dixey (1956) and Shackleton (1951) name the surface from the geologic age of the oldest sediments lying on it. In practice, imprecise geochronologic terms have come into general usage and are easily understood. For example, "African" (King, 1962, 1963) is equivalent to "sub-Miocene" (Pulfrey, 1960), or "Miocene" or "mid-Tertiary" (Dixey, 1956); whereas "Gondwana" (King, 1962, 1963) is equivalent to "late-Jurassic" (Dixey, 1956).

King (1962, p. 225) describes the African cycle planation as "the fundamental landsurface from which nearly all existing African (and world) scenery has subsequently been carved." He extends this concept to other continents. Older "cycle" upland surfaces usually occur as remnants, while "below" the African surface are the broad valley plains and rolling topography. Earth movements have generally been uniform uplifting (epeirogenic) or arching (cymatogenic). There are, however, some depressed areas of limited extent.

From late Palaeozoic until the present day, the only cycle that can be established as having run its full course, that is, to have reached old age or continent-wide planation before interruption, is the one that produced the African or mid-Tertiary planation surface. It is, in practice, easily recognizable from its wide extent with residuals rising above, and with flat-floored valleys or coastal plains lying below the general level. The earlier Gondwana planation surface of Jurassic age, now seen as forming upland remnants, is unlikely to have been of such perfection over the whole of the Gondwana continent.

Partial or interrupted cycles that produced partial planation surfaces occurred not only between the Gondwana and African cycles, but also subsequent to the latter (Small, 1970, p. 263). The most important of the post-African partial cycles is the end-Tertiary, which is typified by planed valley floors or pediments formed by scarp retreat. Dixey (1939) introduced the term "valley-floor peneplain," a contradictory term.

Two basic concepts—base level and grade of load-carrying streams—have been disregarded by Dixey in assessing the relative importance of erosion and faulting in the derivation of the present forms of Nyasa and Tanganyika. Where the level of the floor of a rift valley is now below the level of the threshold over which an outlet stream obviously formerly flowed—the Lukugu in the case of Tanganyika and the Shire in the case of Nyasa—only differential movement or faulting, rather than scouring of weak sedimentary fill by erosion, could have produced the present lowered rift floor.

Unequal movement on marginal faults will give rise to tilting of the floor and unequal uplift of the flanks. If the outlet stream flows across a flank, changes in lake level will result, with possible drowned topography on one margin and retreated shore lines on the other. The classic cases are south Lake Tanganyika and north Lake Nyasa. Changes can also occur along the length of a rift valley and from side to side.

The mid-Tertiary or Miocene planation surface was used as a datum surface or plane of reference by Willis (Paper 1). By relating this plane to the levels of the lake, he monitored the movements of the flanks. He paid special attention to Lake Tanganyika (pp. 182–226).

Any study of rift valley structure and tectonics is incomplete without appreciation of their relation to geomorphic history.

REFERENCES

Cahen, L., 1954, *Géologie du Congo Belge,* Vaillant-Carmanne, Liège, 577p.

Cotton, C. A., 1950, Tectonic scarps and fault valleys, *Geol. Soc. America Bull.* **61:**717–758.

Cotton, C. A., 1961, The theory of savanna planation, *Geography* **46:**89–101.

Dixey, F., 1939, The early Cretaceous valley-floor peneplain of the Lake Nyasa region, *Geol. Soc. London Quart. Jour.* **95:**75–108.

Dixey, F., 1946, Erosion and tectonics in the East African Rift System, *Geol. Soc. London Quart. Jour.* **102:**339–388.

Dixey, F., 1956, Erosion surfaces of Africa: Some considerations of age and origin, *Geol. Soc. S. Africa Trans.* **59:**1–16.

Heirtzler, J. R., and H. T. van Andel, 1977, Project Famous, mid-Atlantic ridge between 36°30′ and 37°, 1971–1974, *Geol. Soc. America Bull.* **88:**481–608, 609–736.

King, L. C., 1962, *The Morphology of the Earth,* Oliver & Boyd, Edinburgh, 669p.

King, L. C., 1963, *South African Scenery,* 3rd ed., Oliver & Boyd, Edinburgh, 308p.

Needham, H. D., and J. Franchetau, 1974, Some characteristics of the rift valley in the Atlantic Ocean near 30°48′ North, *Earth and Planetary Science Lett.* **22:**29–43.

Pulfrey, W., 1960, Shape of the sub-Miocene erosion bevel in Kenya, *Kenya Geol. Surv. Bull.* **3:**1–18.

Quennell, A. M., 1958, The structure and geomorphic evolution of the Dead Sea Rift, *Geol. Soc. London Quart. Jour.* **114:**1–24.

Shackleton, R. M., 1951, A contribution to the geology of the Kavirondo Rift Valley, *Geol. Soc. London Quart. Jour.* **106:**345–392.

Small, R. J., 1970, *The Study of Landforms: A Textbook of Geomorphology,* Cambridge University Press, Cambridge, 486p.

Temperley, B. N., 1966, The faced scarp structure and the age of the Kenya rift valley, *Overseas Geol. Mineral Resour.* **10:**11–29.

2

Reprinted from *15th Intern. Geol. Congr. Proc.*, **2**:354-373 (1930)

GEOMORPHOLOGIC ASPECTS OF RIFT VALLEYS

BY

DOUGLAS JOHNSON,
Columbia University, New York City.

In his field work the geologist is confronted by topographic forms which are merely the end product of a certain geological history. Every geologist. makes use of these forms, to greater or less extent, in deciphering past events. Unfortunately, different geological histories give end products resembling each other so closely that the most conscientious observer is in danger of misinterpreting the record. For this reason it is important that investigators make conscious effort to visualize all possible combinations of events capable of producing similar topographic forms. Only thus will he seek, and haply find, those minor but critically important differences in form or structure which give invaluable clues to the true history of a region.

The foregoing generalizations, which form the basis of what CHAMBERLIN has well called the method of multiple working hypotheses, apply with peculiar force to the results of fracturing and faulting of the earth's crust. The object of this paper is not to present a complete discussion of fault forms, nor to describe the criteria by which the different histories of similar forms may be discriminated; it is merely to emphasize the variety of events possibly responsible for forms which are, or which may be, classed as rift valleys. Since such valleys are, according to the most restricted definition of the term, the direct result of faulting, and consequently involve, among other things the interpretation of such apparently simple forms as scarps developed along fault planes, it will be profitable to begin our discussion with a review of some well established principles relating to these forms.

A fault scarp, in the generally accepted sense of that term, is a line of cliffs or an escarpment directly due to faulting. (Fig. 1). Typically it shows marked indifference to previously established drainage lines, and may be notched by a special variety of hanging valley, these latter being quite as likely to drain away from the scarp fault into the block, as toward the scarp, since they are merely remnants of the pre-fault valley system. Ponded drainage and deflected streams are normal features where the faulting is of recent date. An excellent illustration of a fault scarp, notched by a hanging valley which drains from the scarp into the upraised block, is found on the eastern side of the Klamath rift basin in Southern Oregon. The drainage formerly passing through the hanging valley was deflected south when relative uplift of the block placed the scarp as a barrier athwart its course. Ponded drainage is found in Parowan Valley, Utah, where a stream apparently flowing directly across a small rift valley and long able to maintain its antecedent course despite continued relative uplift of one of the bounding blocks, has recently been partially obstructed by further uplift.

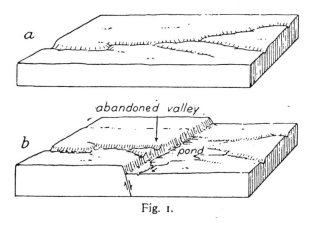

Fig. 1.

A fault scarp may in time be worn far back from the fracture which gave birth to it, and its formerly smooth face may be deeply dissected by stream erosion; yet it is a true fault scarp, young, mature or old according to the stage of its dissection; because it was originated directly by faulting. On the other hand a smooth undissected scarp, coinciding exactly with a fault plane, and having the same stratigraphic horizon at summit and base, clearly marking the precise amount of displacement, frequently is an erosion feature and in no sense a true fault scarp. In the third diagram of Figure 2 we have an escarpment developed along a fault, which might quite naturally be mistaken for a fault scarp due to comparatively recent displacement of the rock; yet the history of this form is more complex. An ancient displacement, possibly dating far back in geological time, produced a true fault cliff shown in the upper diagram. Long continued erosion reduced the country to a peneplane, and the fault scarp was completely obliterated, as shown in the second diagram. In a new cycle of erosion weak rocks were stripped away from one side of the fault plane, giving an escarpment due wholly to recent erosion limited by a fracture of very ancient date. For such a form DAVIS has suggested the name "fault-line scarp." The term has been criticised because it might equally well signify a true fault scarp, which likewise borders a fault line. " Fault-line erosion scarp " is less concise, but also less open to misinterpretation. In whatever manner we designate this form, we must not confuse it with the resurrected fault scarp, which has had a still different history. In the case of the fault-line erosion scarp there has been no burial of a pre-existing scarp, and hence there can be no resurrection. The only pre-existing scarp (a true fault scarp) was annihilated. The present scarp was not a scarp at all until erosion of weak rocks first gave birth to it.

The distinction between fault scarp and fault-line erosion scarp is no mere matter of academic interest; it is the indispensable prerequisite to a correct interpretation of geologic history. In the Colorado plateau province of America are magnificent escarpments scores or even hundreds of miles in length, and hundreds or even thousands of feet high. The same lime-stone capping the uplifted block is normally found at the foot of the scarp covering the down dropped block. The scarps are little dissected, and are located directly on the fault planes. Hence DUTTON in his classic " Tertiary History of the Grand Canyon District," described them as young fault scarps, assigned them a date later that the great peneplanation of the area, and interpreted the Colorado River as an antecedent stream which had persisted in its course while the fault blocks were raised across its path. More recent studies by DAVIS and others have demonstrated that the escarpments are for the most part faultline erosion scarps, that the faults are very ancient instead of recent, that they antedated the period of peneplanation instead of following it, and that the Colorado River was probably superposed from the peneplane instead of being antecedent to the still earlier faulting. Failure to recognise the distinction between fault-line erosion scarps and fault scarps inevitably led so able an inves-tigator as DUTTON to give an erroneous chonology for some of the major events in the geologic history of the region, and caused him to overlook phenomena of critical significance which he otherwise certainly would have looked for, and most probably would have found.

If one examines the third diagram of Figure 2 it will be seen that the scarp faces toward the down dropped block, and therefore in this respect gives a repetition of the earlier topography consequent upon the initial faulting. Such a scarp has been called a reconsequent or *resequent fault-line scarp,* to distinguish it from the scarp which would result if the rocks capping the down thrown block had been the more resistant. In the latter case, removal of relatively weaker material from the raised block would give a scarp facing backward, or opposite to the direction which was con-sequent on the initial faulting. Such an opposite-consequent, or (in briefer form) *obsequent fault-line scarp,* must carefully be discriminated from true fault scarps caused by recent reversal of the direction of movement along an earlier fault plane. The terms " resequent " and " obsequent " easily suggest re-following the initial orientation, and following an opposite orien-

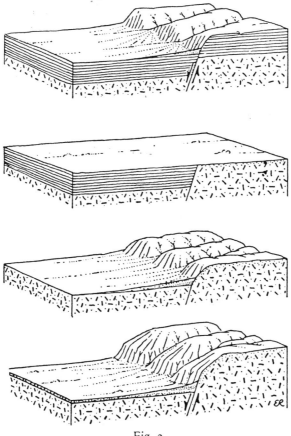

Fig. 2.

tation; and I have found both DAVIS's discrimination of the two types of fault-line erosion scarp, and his terminology, most helpful in regions like the Colorado plateau where both types are present. It is worth noting, also, that when studying the shorelines of the New England and Arcadian regions, where fault scarp shores have been reported in large numbers, I found not a single true fault scarp, although fault-line erosion scarps are fairly abundant.

The fourth diagram of Figure 2 shows that renewed faulting has given a scarp of double slope, the base of the scarp being steeper, the upper part more gently inclined. In this case it is evident that the upper portion is a fault-line scarp due to erosion, while the lower part is a fault scarp due directly to movement along the fracture plane. An excellent example of such a composite scarp is found in that part of the Hurricane Ledge just south of the Virgin River in Arizona.

Figure 3 shows a similar topographic form, but the history responsible for it is just the reverse. Here the upper part of the scarp is a true fault scarp, as can be seen from the first and second diagrams; while the base is a fault-line erosion scarp, caused by later removal of weaker beds, some remnants of which remain in places to tell the story. Similar double scarps also occur where a true fault scarp has experienced two periods of uplift separated by an interval of erosion; or where stripping of weak beds to give a fault-line erosion scarp has occurred in two successive cycles; or where the upper beds in either fault scarp or fault-line erosion scarp weather more readily than those below.

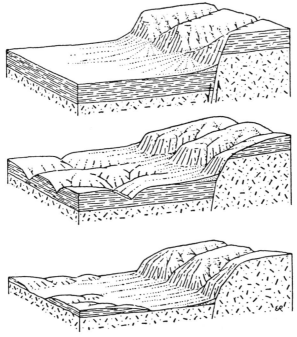

Fig. 3.

Both fault scarps and fault-line erosion scarps may be more or less completely buried by accumulating alluvial or other debris. If alluvium washed from the uplifted block shown in the first diagram of Figure 4 ultimately spreads across the fault plane into the valleys of that same block as shown in the second diagram, the remnants of the scarp still rising above the alluvial accumulation will appear irregular or ragged, with embayments of the sloping piedmont plain between projecting spurs of the upland. Now let renewed erosion strip away the alluvium, and we shall have a resurrected fault scarp, or a resurrected fault-line erosion scarp as the case may be. A key to the true history will sometimes be found in teraces or benches of alluvium high up on the valley walls, as shown in the third diagram. If the alluvium is more completely removed, the true history may only be deciphered with the aid of more remote data bearing on the sequence of events in the region as a whole.

The Central Plateau of France, recently described most admirably by BAULING, furnishes many examples of resurrected fault scarps, and possibly also of resurrected fault-line erosion scarps. The impressive escarpment west of Clermont-Ferrand is a beautiful illustration of the resurrected fault scarp. In this particular case the scarp was first buried by Tertiary alluvium (Fig. 5a), after which one of the Auvergne volcanos poured out a stream of lava which flowed down the alluvial slope and far across the buried fault scarp (b). Later erosion removed much of the alluvium, but the hard lava cap preserved a portion of it (c) in such manner as to form an easy key to the past history. This classic example might therefore be taken as the type of resurrected fault scarps. In the south-western United States a lava flow has similarly preserved such a key to the past history of a part of the Hurricane Ledge which is a fault-line erosion scarp.

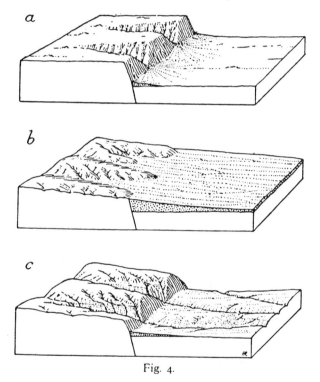

Fig. 4.

Renewal of faulting during the accumulation of alluvium on the down dropped block appears to be a rather common occurrence. This may give a compound scarp in which the upper part is a peculiar type of resurrected fault scarp, the lower part an ordinary fault-line erosion scarp. It will be seen from Figure 6 that a true fault scarp (a) is buried as in (b), after which the buried scarp is again exposed to view, not by erosion as in cases previously considered, but by a renewal of the faulting, as shown in diagram (c.) Further accumulation of alluvium again buries the scarp (d.) Then the renewed uplift indicated by the arrow in (e) resurrects the upper portion of the scarp " a," while stripping away of the alluvium for the first time gives a fault-line erosion scarp at the base, " b." Assuredly the history is complicated, and one is tempted to say that these minor episodes have only theoretical interest for those who have time to waste on geological cross-word puzzles. To this it must be answered that the history outlined is the normal, expectable history of a growing fault; that the succession of events in nature is ordinarily longer and more complicated than those sketched here; that the resurrected portion of the scarp at intervals contributes debris of its rocks to the alluvium during periods when the future fault-line erosion scarp does not; and that there is thus preserved in the alluvium a decipherable record of past events which may have produced important changes in the topography and in the stratigraphy of regions extending far beyond the limits of the fault zone. The scarp west of Clermont-Ferrand, already cited, may belong to this complex type of resurrected fault scarp. For a long period of time the history of the Connecticut Valley in New England has been misinterpreted because critical evidence of successive movements along the boundary fault on its eastern side was not sought for and hence not found. Only recently have special studies been directed to this phase of the problem, and evidence of repeated faulting found in debris washed into the fault basin from cliffs exposed to weathering in Triassic times. Existing difference of opinion as to the history of this great fracture will doubtless be reconciled when the complex record of successive faultings, alluviations and erosions is more fully deciphered.

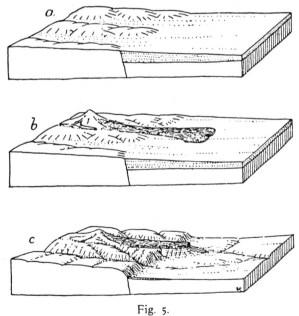

Fig. 5.

Brief reference should be made to the rectilinear valleys which frequently develop along the weakened, sheared zones of single faults or rifts. These have been called fault-line valleys, and their burial by debris and subsequent exhumation would give resurrected fault-line valleys. Examples of both types are known, but as they are not common features along the bounding faults of rift valleys they need not further be considered in the present discussion.

I have taken some time to review the morphological aspects of single fault phenomena because once these are clearly in mind the problems of double and multiple faulting present no special difficulties. The principles reviewed are not new, and one may find references to resequent fault-line scarps, obsequent fault-line scarps, and resurrected scarps in the literature of geomorpholy. There is therefore nothing radical in recognising the morphologic and other differences peculiar to each type, nor in applying them to the discussion of rift valleys and others forms due to multiple faulting.

In treating this latter phase of our subject I shall consider only four major types of blocks included between bounding faults: I. Tilted or monoclinal blocks; II. Rift blocks; or those relatively raised or lowered between normal faults; III. Thrust blocks; those relatively raised or lowered between reverse or thrust faults; and IV. Step block; those raised or lowered progressively between successive faults of concordant displacement. For purposes of this classification I define "normal fault" as any fault which prevailingly hades toward the downthrow; "reverse fault" as any fault which prevailingly hades toward the upthrow.

I. Tilted blocks are the familiar forms commonly described simply as block mountains, or fault block mountains, although this term has also been extended to include horsts which have no obvious monoclina tilting. Tilt block mountains are separated by tilt block basins, the combination being sometimes described as basin and range topography, or basin and range structure. We are concerned here only with such aspects of these forms as may be of significance in connection with rift valleys.

Emphasis has been placed upon the triangular facets separated by V-gorges which frequently characterize the fault face of a tilt block mountain, and it is evident that similar forms may mark the bounding scarps of rift valleys. One point of importance, frequently overlooked in the discussion of tilt block mountains, is that this type of triangular facets and V-gorges indicates continued, progressive displacement along the fault plane in all cases where dissection of the raised block is accompanied by deposition of alluvium on the depressed block. The V-gorges will give place to flat-bottomed valleys as the mounting alluvium enters the gorges and partially buries them. Only when faulting is periodically renewed will the gorges be lifted above the alluvial plain, any accumulated debris

47

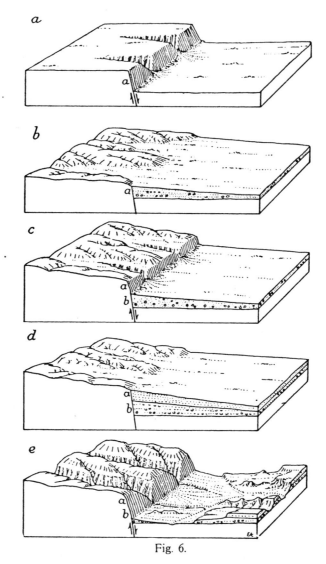

Fig. 6.

be washed out of them, and their V-form restored by renewed erosion. In
this way a mountain block, bounding a block basin or a rift valley, may be
relatively raised and subjected to re-invigorated erosion many times (Fig.
7), and yet retain a mature topography (although with increasing relief),
and equally retain triangular facets separated by V-gorges, as is shown by
the classic case of the Wasatch Mountains in Utah, where both maturity
of block dissection and youth of facets and gorges are evident.

Differential tilting of a faulted block may be indicated not only by
varying height of the fault scarp, but also by various disturbances of the
previously established drianage regime. Uplift may even be followed by
aggradation of valleys in the raised block; or by aggradation in one part

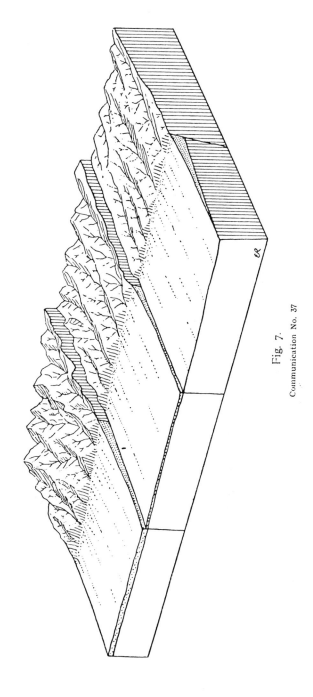

Fig. 7.

Communication No. 37

49

and rejuvenation of erosion in another part of the same block. One such possibility is illustrated in the second diagram of Figure 8, where re-invigorated erosion of the further tributary, which flows with the tilt and is therefore steepened by it, contributes so much debris to the main stream on the depressed block that the latter is aggraded. This aggradation may affect the less uplifted tributary in the foreground of the raised block, especially if the latter flows against the direction of tilt and therefore has its gradient decreased by the differential movement. We may take this as merely one illustration of the kind of drainage disturbances which must be expected in a region of faulting accompanied by tilting, disturbances which will give alluvial deposits on different streams similar in appearance but possibly formed at widely different times and by different tectonic disturbances.

It would not be profitable in the present connection to discuss the varied types of landforms belonging in the tilt block family. But it is perhaps worth while stressing the point that resequent tilt block valleys, that is valleys formed by excavation of infaulted wedges of weak rocks, are of very common occurrence. I have examined several beautiful examples in the Adirondack mountains of New York, while better illustrations have been described from southern Sweden. Such resequent tilt block valleys, as their name implies, reproduce the asymmetrical form due to the earlier faulting, and have therefore in both the localities mentioned been mistaken for original tilt block basins. Obsequent forms, or those involving a renewal of the original topography, are presumably less numerous; while resurrected tilt block basins and resurrected tilt block erosion valleys (both resequent and obsequent) should be included in any complete discussion of this family of land forms.

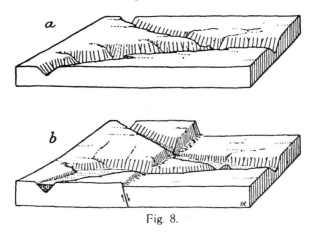

Fig. 8.

II. We come now to the second group, the rift blocks. For convenience I include here blocks relatively raised, as well as those relatively dropped, between more or less parallel normal faults. The term rift is

widely used for a single major fracture or fault, without implication as to
direction of displacement. Any block between two parallel rifts may thus
be termed a rift block, a block separated from adjacent masses by rifts.
If the rift block is dropped between normal faults, as shown in the first
diagram of Figure 9, we have an original rift block basin, commonly
called a graben or rift valley. Since the term rift is employed, as previously noted, for single faults or fault zones like the so-called earthquake
rifts; and since true valleys are frequently formed by stream erosion along
such single rifts, giving rift valleys which are not grabens; and since,
further, we shall have to recognize by some appropriate term a very
common type of valley produced when stream erosion excavates a block
of weak rocks bounded by parallel normal faults, I should prefer to call
the initial form represented by the first diagram a rift block basin, from
analogy with the tilt block basins of the basin and range topography of
the Western United States; or else to adopt the simple term graben, long
naturalized in the English language.

If the erosion which wears back the graben walls, as shown in the
second diagram of Figure 9, continues until the country is levelled, the
stage will be set for the development of a new form. In case the down
dropped block brings weak rock in the centre in contact with more resistant
rock on either side, a new cycle of erosion will sweep out the weak rock
belt, leaving a broad valley bounded by rectilinear walls coincident with
the fault planes. We may even find the same rock capping the bordering
uplands and covering the valley floor. In form and structure the new
valley may appear precisely like the original basin of the first diagram,
and I have not taken the trouble to reproduce it in the series. But its
immediate origin is erosional, not tectonic; its history is long and somewhat
complicated, not short and simple; and the bounding walls are resequent
fault-line scarps, not true fault scarps. The new form might be called
a resequent rift block valley.

If, on the other hand, we imagine that the down dropped rift block
brought more resistant beds in contact with weaker rocks on either side,
as shown in the third diagram (Fig. 9), subsequent erosion will lower the
weaker formations and leave the infaulted resistant mass standing in relief.
As indicated by the fourth diagram, the resulting form resembles an
original rift block mountain or horst; but it is an erosional, not a tectonic
feature, and the bounding walls are obsequent fault-line scarps, facing in
opposite direction from the original fault scarps. We have here to deal
with an obsequent rift block mountain.

It is obvious that if we began the history with a true rift block
mountain or horst, and if its uplift brought weak rocks in the centre in
contact with resistant rocks on either side, a later erosion cycle would find
an erosional rift block valley, bounded by obsequent fault-line scarps,
occupying the site of the uplifted block. The Burnet region of Texas is

51

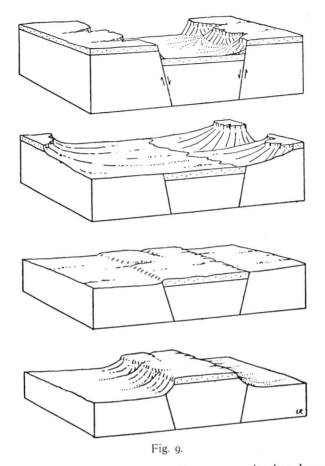

Fig. 9.

described as one in which limestone resting on granite has been so faulted as to give a succession of horsts and grabens. Subsequent erosion destroyed the original topography due to faulting, but left down-dropped blocks of limestone in contact with upraised blocks of granite. In this case the granite proved an easy victim to weathering and erosion in a new cycle, while the limestone was left standing in relief. Thus we have in this region remarkably good examples of both obsequent rift block valleys and obsequent rift block mountains.

Any of these rift block forms may of course be buried and later resurrected. I will cite one beautiful example from the central plateau of France: The Limagne Basin, a resurrected rift block basin where remnants of the alluvial filling, especially were preserved under post-alluvium lava flows, reveal the true history with unusual clarity.

Before leaving the subject of rift block form, I would like to call attention to one feature of original rift block basins worthy of mention. This is the series of fault slivers or splinters which not infrequently characterize the bounding scarps, where the major displacement takes place not along a single clean-cut fracture, but by means of a series of fractures arranged en echelon. In such a case, as will be apparent from the first diagram of Figure 10, one may start on the down dropped block, ascend the warped surface of one of the fault splinters, and so reach the surface of the raised block without crossing any fracture or fault. If the same number of faults produce the same displacement, but with the faults more oblique to the axis of the basin and hence more widely spaced as in the second diagram, shall we call the resulting form a rift block basin? And if this case passes muster, what shall we say of the third example, where warping of the broadened fault splinters predominates over fracturing to such an extent that a cross section of the basin, as at the front of the diagram, may not show any fault? In the Klamath Lakes Basin in Oregon (Fig. 11), some of the fault splinters are small, while others attain the magnitude of imposing blocks oblique to the axis of the major depression. This basin presents a second problem in that the principal fault bounding the depression on the west has every appearance, when examined in the field, of being much more ancient than the faults on the east. When the fault scarp on the west was fully developed and even badly eroded, there apparently was still no basin. Not until faulting of very recent date occurred was the basin structure completed. Is a relatively depressed block between two faults of widely different ages a graben? Perhaps not; but we should certainly have to include it in the class of rift block basins.

I am tempted to say a word further about this interesting region, because on visiting it in 1915, I found there three phenomena not often encountered by the geologist: true fault scarps of considerable magnitude so young as to be practically untouched by stream erosion; what appears to be the actual fault plane exposed on the face of a range with fault breccia and slickensided surfaces beautifully preserved; and an antecedent stream valley left hanging in the raised eastern block when the rate of uplift exceeded the rate of stream erosion. The latter form is just visible in the northern part of the birdseye view of the basin (Fig. 11) drawn by one of my former students, Dr. Erwin J. Raisz. The fault phenomena were so unusual that I sent my notes to the late G. K. Gilbert, then engaged on a special study of block faulting. Dr. Gilbert later made a detailed examination of the region, from the published results of which is reproduced the cross section of one of the large fault splinters or blocks, cut by the hypothetical trench in the right foreground of Dr. Raisz's drawing.

III. In the family of thrust block forms a central block is bounded by two oppositely inclined surfaces (diverging downward), up which the bounding blocks are pushed or thrust; or the central block may itself be forced downward; or a central block may be pushed first up one, then up the other, of two converging planes. To this family of forms belong the ramp valley of Professor Bailey Willis.

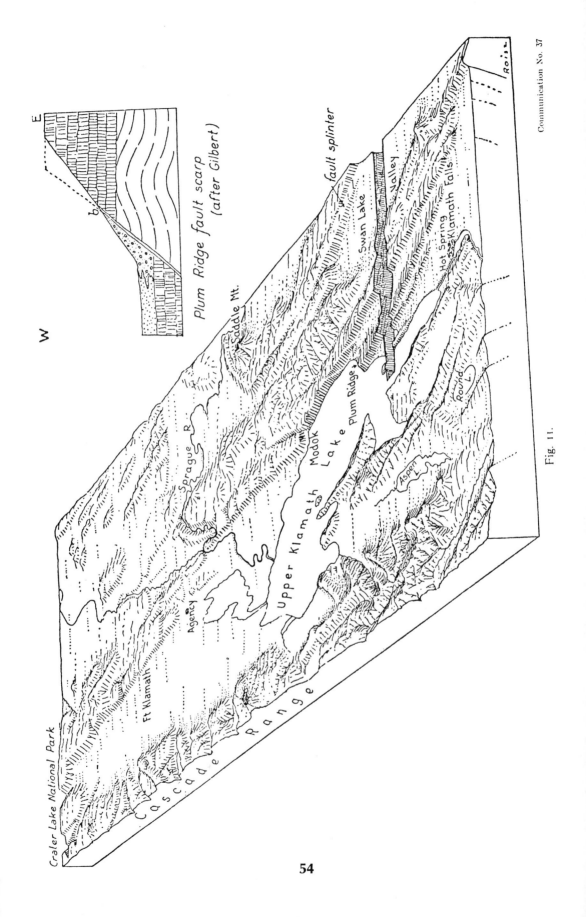

Plum Ridge fault scarp
(after Gilbert)

Fig. 11.

Communication No. 37

54

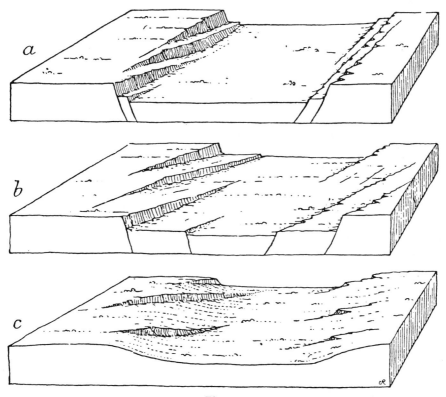

Fig. 10.

Various types of depression may be developed on thrust blocks. It seems desirable to follow the system of nomenclature previously employed, and designate the original depression, produced when the bounding masses are thrust upward and toward each other (Fig. 12), a thrust block basin. After the initial form shown in the first diagram has been softened by weathering and erosion, the base of the scarps may be sharpened and steepened by renewed movement along the thrust planes. Scarps with double or multiple slopes, gravel benches, rock terraces, and rejuvenated valleys may thus in time be observed along the bounding walls.

One peculiar aspect of the thrust walls deserves special consideration. As the mass advances, its forward edge may be unsupported or imperfectly supported on talus or alluvium, with the result that conditions are especially favourable for the development of landslides, shown in the enlargement of a portion of the first diagram in Figure 13. In time weathering and erosion, and deposition of alluvium, will obliterate the slide topography, and the form shown in the second diagram will then be indistinguishable from a weathered and dissected normal fault scarp. Renewed movement will recreate conditions favourable for slipping, and it is conceivable that large landslip blocks, separated by local normal faults, may break from the parent mass thrust forward over imperfectly compacted talus or alluvial deposits. Such blocks whether fresh as shown in the third diagram, or weathered as represented in the fourth, might be mistaken for ordinary fault splinters associated with normal block faulting. It is a fair question whether fault splinters and normal fault blocks, as well as slickensided fault surfaces, attributed to rifting of the normal type, may not in some cases at least be the result of large scale landslips on the imperfectly supported front of a thrust block. It seems probable, however, that slipping would occur before any great portion of the thrust block had advanced into an unstable position. For this reason one should expect normal landslide topography, rather than block mountain topography, to characterize the borderlands of thrust block basins. In this connection it should be noted that the diagrams are drawn with the fault planes relatively flat, in order to represent conditions favourable to extensive overthrusting and slipping. In nature we should expect adjacent thrust faults dipping in opposite directions to be more steeply inclined, and hence less favourable to local slipping of broad blocks.

It is hardly necessary to discuss in detail the other members of the thrust block family. A little consideration will suffice to show that we must recognize the possible existence of resurrected thrust block basins, resequent and obsequent thrust block valleys, and resurrected resequent and obsequent block valleys. Thrust block mountains of various types may also exist. Their essential characteristics may readily be deduced from considerations already presented.

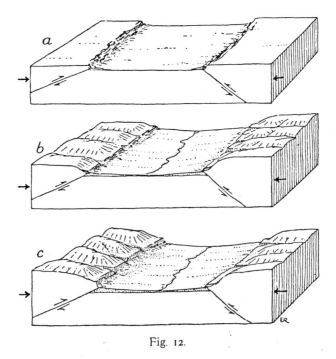

Fig. 12.

IV.   It might appear, at first thought, that the fourth group of fault blocks, step blocks, should not concern us, since such blocks, horizontal and each relatively displaced in the same sense, give rise to a series of ascending or descending platforms or steps, but neither to basins nor to mountain blocks.   We must remember, however, that both normal and reverse fault blocks may be so displaced that resistant and non-resistant rocks are brought into contact.   Later diffential erosion of the faulted mass may then give valleys bounded by parallel rectilinear fault-line scarps, and resembling other forms previously described so closely as to be almost indistinguishable from them.   Thus is the case of three step fault blocks, if after peneplanation the central block happens to have exposed at the surface weak shales, let us say, in contact with resistant quartzite on one hand and resistant limestone on the other, erosion in a new cycle will sweep out the shales leaving a step block valley between quartzite and limestone uplands.   A little consideration will show that this valley will be bounded on one side by a resequent fault-line scarp, on the other side by an obsequent fault-line scarp.   We may therefore speak of this form as a resequent-obsequent step block valley.

Another case is illustrated in the series of diagrams forming Figure 14. A mass is faulted, peneplaned, and further eroded in a new cycle to give an obsequent fault-line scarp, facing in the opposite direction from the scarp consequent on the initial faulting, as may be seen in the third diagram.   Now a new block is uplifted, in the same sense and in front of the obsequent scarp.   This produces, as the fourth diagram shows, a true fault scarp, consequent on the new faulting, facing toward the other scarp.   The depression between may be called an obsequent-consequent step block valley.   I use valley in preference to basin, because one wall and the floor are erosion products, even if the remaining wall be a tectonic product.   In a similar manner we may have produced resequent-consequent step block valleys.

Repeated movement along a single plane may, with the intervention of normal erosion, give a special type of obsequent-consequent step block valley.   The mass is first faulted (Fig. 15), then peneplaned, and through long-continued later erosion the obsequent fault-line scarp is worn far back from the fracture plane, as shown in the third diagram.   Then renewed displacement along the original plane and in the original sense, gives a true fault scarp for the opposite boundary of the depression, which might be called a single fracture obsequent-consequent step block valley.   The name is long, but self explanatory.   The history involved is merely that which we know to have occurred repeatedly in nature, later movement along an old fracture following peneplanation, uplift and renewed erosion.

We need only add that each of these four forms of step block valleys may be buried and resurrected; and then pass to one or two concluding observations.

57

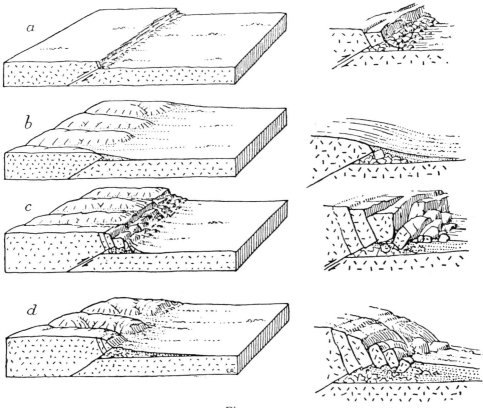

Fig. 13.

My object has been to show that forms sometimes identical in internal structure, and often so similar in outward appearance as to deceive the experienced geologist, may have histories differing widely one from the other. It may be objected that the picture I have painted is too confusing as to terminology, too theoretical as to substance, and too complicated as to successions and combinations of events to be of practical value.

To the first indictment I would make two answers. No less than forty forms (the number would be greater were all types of faulting considered), each having a geological history different in some essential respect from all the others, require to be named. This can be done with the aid of only 12 technical terms, most of them already common property of geological science, and all but one of them found in one or more of our college text books on geology. Fault scarp has been distinguished from fault-line erosion scarp, tectonic basin from erosional valley; rift, thrust, tilt and step blocks have been discriminated; and the consequent, obsequent, resequent and resurrected forms associated with each described. This does not seem to me an unduly elaborate equipment with which to designate in explanatory terms, forty important types of land forms, many of which,

58

perhaps all of which, occur repeatedly in Nature. Will the mineralogist or the palaeontologist name forty species of minerals or forty species of fossils, with so modest an equipage as 12 terms, most of them previously in common use? Of the terms I have employed, five alone will require even the most moderate effort to understand their significance. Indeed, since rift and thrust are now so well known, I may reduce the number to three: consequent, resequent, obsequent, all of them used in at least one standard textbook of geology for more than a decade, and all suggesting in some measure their own significance. I do not see how a more simple or less confusing terminology *could* be devised to define and at the same time in some measure describe and explain concisely forty distinct landforms.

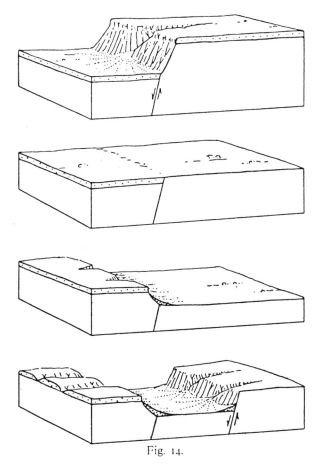

Fig. 14.

But I would not quarrel with anyone who may entertain a contrary opinion. After all, the precise terms to be used are of secondary import-ance. Our concern is first of all with the landforms themselves, and the geological histories back of them. If anyone can suggest some other terminology which will more easily keep in the mind of the field investigator of faulted regions the forty or more forms liable to be encountered there, their respective histories, and the means of discriminating them, I will wholeheartedly accept the substitute. If the discrimination between basins as tectonic forms and valleys as stream erosion forms proves unacceptable, because the term valley is often loosely applied to depressions of tectonic origin, then call the basins consequent rift block valleys, consequent thrust block valleys, and so on. Or if the terms consequent, resequent and obsequent confuse rather than clarify things, even for those who have expended the moderate effort required to understand their meaning, then discard them and invent other terms which may prove more acceptable. Any terminology which, after fair trial, fails to simplify and clarify things for the geological investigator, should give place to something better.

That the classification of fault forms outlined above is too theoretical, I can not admit. It is based on relatively simple geological histories, known to have occurred repeatedly in nature. I have given examples of many of the types mentioned, and had time permitted could have added

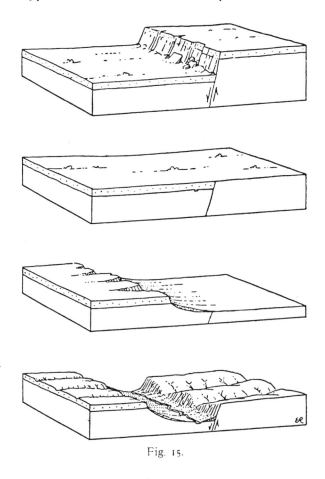

Fig. 15.

others. If a few of the types are as yet unreported from the field, it seems to me more reasonable to suppose that this is because of the deficiency of our knowledge, and especially because these forms have not been recognized even when observed, rather than to suppose that they do not exist in all the wealth of fault phenomena exhibited by the uneasy crust of our planet.

Finally, to the charge that the classification of fault forms presented is too complicated as to successions and combinations of events, I would answer that different observers have already reported fault histories more complicated than any I have described. And if nature has faulted, eroded, buried and resurrected blocks of the earth's crust in a great variety of ways, should not the geologist go into the field with a clear mental picture of all the possibilities, and make a conscious effort to find the critical evidence, often preserved in a few obscure localities, which may enable him to read the record aright?

I am, as the reader will observe, urging the importance of deliberately employing in field studies of faulted regions what CHAMBERLIN has called the method of multiple working hypotheses. The rift-valley region of east central Africa has been described as one in which cycles of diastrophism, vulcanism, and erosion have succeeded each other more than once. It is in such regions that fault phenomena are apt to be most complex; and hence I have thought it not impossible that those fortunate enough to work in this remarkable district might find CHAMBERLIN's method useful in directing attention to elements of the topography which, however obscure, possess critical value for deciphering the geological history of faulted and peneplaned masses.

Part III

EASTERN AFRICA—GENERAL

INTRODUCTION

Eastern Africa lies between the Red Sea and the Gulf of Aden on the north and the Zambezi River on the south, and between the Congo Basin on the west and the Indian Ocean on the east. Within and bordering this region are the great systems of continental rift valleys and accumulations of alkaline volcanics.

This region is dominated by the East African swell, a shield area of Precambrian crystalline rocks. Krenkel (1925, Ch. 3) designated this, the most elevated of the African swells, as "high Africa." Superimposed is the Kenya dome. The separate Arabo-Ethiopian swell is of Precambrian metamorphics, largely masked by tabular terrestrial sediments and Neogene volcanics. The two swells are separated by a crustal sag. These elevated features would have lain to the north of center of reconstructed Gondwanaland. The crustal setting was favorable to rifting processes. Furthermore, it is this region, with its repeated epeirogenic uplift episodes, that has preserved the clearest evidence of its geomorphic history. Because of these factors the study of continental rift-valley phenomena has been given greatest attention in Africa.

The rift valleys of Eastern Africa may be described under four systems.

1. The southern rift valleys and faulted troughs, containing partly infaulted Karroo and early Cretaceous terrestrial sediments, are oriented either NE-SW, extending from Zambia across Tanzania into Kenya, or SE-NW along one part of the Western Rift System.

2. The Western Rift System, having a general N-S trend but curving around the margin of the swell, extends from the Zambezi River to the Aswa mylonite zone in northern Uganda. The three zones are offset sinistrally and there is no structural continuity.

3. The Eastern (Gregory) Rift System, which extends north as the Main Ethiopian Rift, commences with block faulting in the central part of the swell and continues northward, to bisect the Kenya

63

dome and cross the crustal sag into the Arabo-Ethiopian swell as the dextrally offset Ethiopian rift. This, in turn, merges into the Afar triangular depression.

4. In Eastern Tanzania a discontinuous zone of faulting and troughs begins at the coast and extends to the NW to meet the Kenya Rift near the Kenya border.

The key map (Paper 20A) illustrates this description of the Eastern Africa rift valley systems.

FOSSIL SITES

The climatic variations, ecological conditions, and geomorphic and structural development of the rift valleys of eastern Africa created a favorable habitat for vertebrates including man; these conditions also were favorable to the burial, preservation, and eventual exhumation of vertebrate fossil remains. The sites so far investigated lie principally within the rift-valley zones, but others are situated where warping and faulting, tectonically associated with rift valley movements, provided favorable conditions.

Articles of interest and the more important references will be found in Cooke (1957), Howell et al. (1963), Bishop and Clark (1967), and Bishop (1978).

GENERAL LITERATURE

Systematic surveying and publication of geological maps and reports began in the early 1920s and continued until the late 1960s. Most of this work was done by governmental geological surveys, university and learned society expeditions from overseas, and the university colleges of the countries concerned. Bibliographies of this work can be found in Anonymous (1965), which covers all of eastern Africa including Ethiopia. Geological maps on scales 1:100,000 to 1:250,000 cover most of the length of the rift valleys. For maps of Zaire, Burundi, and Ruanda, refer to Cahen (1954).

A recent contribution is the report of the expedition of the Soviet Geophysical Committee of the Academy of Sciences of the USSR (Beloussov et al, 1974).

An international symposium on "Geodynamic Evolution of the Afro-Arabian Rift System" was held in Rome in April of 1979. The range of subjects discussed was wide and attention was primarily focused on the Eastern Rift System, including the Afar and the southern Red Sea. The resulting publication *Atti dei Convegni Lincei, No. 47*, was published by the Accademia Nazionale dei Lincei, Rome, 1980.

REFERENCES

Anonymous, 1965, Report on the geology and geophysics of the East African Rift System, *I.U.M.P. Scientific Report No. 6,* Nairobi University Press, Nairobi, 265p.

Beloussov, V. V., V. I. Gerasimovsky, A. V. Goriachev, V. V. Dobrovolsky, A. P. Kapitsa, N. A. Logachev, E. E. Milanovsky, A. I. Poliakov, L. N. Rykunov, and V. V. Sedov, 1974, *East African Rift System, Major Features of Structure, Stratigraphy,* vol. 1; *Hypergenic Formations, Geomorphology, Neotectonics,*vol. 2; *Geochemistry of volcanics, seismological investigations, main results,* vol. 3, [In Russian], Soviet Geophysical Committee, Nauka, Moscow, 775p.

Bishop, W. W., and D. Clark, eds., 1967, *Background to Evolution in Africa,* University of Chicago Press, Chicago, 935p.

Bishop, W. W. ed., 1978, *Geological Background to Fossil Man—Recent Research in the Gregory Rift Valley, East Africa,* Scottish Academic Press, Edinburgh, 600p.

Cahen, L., 1954, *Géologie du Congo Belge,* Vaillant-Carmanne, Liège, 577p.

Cooke, H. B. S., 1957, Observations relating to Quaternary environments in East and Southern Africa, *Geol. Soc. S. Africa Bull.* 60 (annex): 1-73.

Howell, F. Clark, and F. Bourliere, eds., 1963, *African Ecology and Human Evolution,* Methuen, London, 666p.

Krenkel, E., 1925, *Geologie Afrikas,* vol. 1, Borntraeger, Berlin, 1918p.

Editor's Comments
on Papers 3 Through 7

In the papers that follow, the relative influence on Mesozoic and Cenozoic rift faulting of Precambrian linear and planar structures and the trends of orogenic belts are briefly discussed. There is, however, another hypothesis regarding the trends of rift faults that deserves mention. This is that they are influenced by a megafracture pattern revealed in alignments of diverse features—structural, topographic, volcanic—for much of Africa. Furon (1963, pp. 76-83) sets out the evidence empirically but not the justification in tectonic terms for a fracture system in which "the large faults were produced and rejuvenated in two directions at right angles: SW—NE and NW—SE" (P.78). Acceptance on a continentwide basis is difficult. In any case, as can be seen on map (Paper 20A) there are only two examples of such trends—northwest along Nyasa and Rukwa and northeast along the southern Karroo troughs—that have plausibility. In Papers 3 and 5, these are attributed to other causes—the known trends of shear zones and foliation in the Precambrian orogenic belts and shear zones.

Paper 3 by King provides a concise, comprehensive, and noncontroversial description of the rift valleys and systems of eastern Africa. These rift valleys are often described, along with the enclosed Lake Victoria, as tectonically related landforms. In fact, a closer relationship exists between the Gregory and Ethiopian systems than between the Gregory and Western systems. This is recognized in Paper 3. King's greater familiarity with the rift valleys of Kenya and Uganda leads to some imbalance. For the Western Rift System south of Lake Kivu and for the earlier Mesozoic faulting in the southern rift zone, he relies largely on Dixey's descriptions and conclusions (see Papers 4 and 10). He accepts that there is parallelism between rift faults and Precambrian structure and an en-echelon pattern where this is not so. He recognizes the record provided by erosion surfaces. The prerift westerly drainage across the subsequent locations of western rift valleys, first noted by Wayland but denied by Dixey, is in his view established.

King accepts that the fault (or fault-line) scarps seen today in East Africa are, in general, the consequence of late Cenozoic normal dilatational faulting. He contrasts continental rift valleys with those of the ocean floor and relates their dimensions, widths, and vertical displacements to the thickness of the crust. Tectonic and volcanic activities are seen as neither being the result of the other but as two effects of a single cause. Volcanic activity tends to migrate toward the culminations of the arches or domes along which lie the rift valleys of the Gregory and Ethiopian zones. The accompanying map is useful but not completely accurate.

Paper 4 is an excerpt from the last comprehensive paper written by Dixey. His publications on the subject date from 1923 (see Paper 4 references). Based on his field work on the geology of Nyasaland and the Luangwa Valley of Zambia, he constructed a history of the formation of the rift valleys of the southern Western Rift System. His views have so influenced subsequent research that, although now in some important aspects modified by later work, they cannot be disregarded in any review of the subject. In the excerpt his conclusions expressed in an earlier paper (Dixey, 1946) are repeated and strengthened. He makes additions from the results of fresh information. Although there has been implicit and explicit refutation of some of Dixey's most important conclusions by later workers, there is no publication in which these later results are consolidated and where comparisons are made.

Briefly, Dixey regarded the history of the Luangwa trough (Dixey, 1937) and its extension, the upper Zambezi, which is recorded in folding and faulting, sedimentation, partial removal of sediments, and resurrection by erosion and later fault movements, as the prototype for the formation of the Nyasa-Rukwa and the Tanganyika rift valleys. He regards nearly all visible fault scarps, whether or not they have been modified by erosion, as fault-line or exhumed fault scarps. His hypothesis requires the infilling of the Nyasa, Rukwa, and Tanganyika troughs by Karroo and Cretaceous terrestrial sediments (as in the manner of the Luangwa and mid-Zambezi troughs) and their virtually complete removal by the Shire River in the case of Nyasa and the Lukuga River in the case of Tanganyika. There are arguments against this hypothesis. The floors of these lakes are now below the levels of their outlets. Furthermore, there are no recorded occurrences of residual sediments of this age in the Tanganyika trough and in the Rukwa-Nyasa trough their presence in the north has an alternative explanation which will be given later.

However, it must be stated that the contributions made by Dixey to the solution of rift problems and to the geomorphic history and its relationship to rift valleys have had great significance. In Paper 4 and the earlier papers, he extends his conclusions to rift-valley phenomena in general, but the reader must judge if such speculations are soundly based.

Paper 5 is a consolidated statement by McConnell of his views on the relation of modern rift valleys of Africa to Precambrian structure. Although most workers in this field recognize that structure in older rocks can provide pre-existing planes of weakness, few go as far as McConnell.

The Western Rift System is set in Precambrian rocks—cratonized nuclei (shields) and orogenic belts—and rift faults there can actually be seen to displace continental basement rocks. There are few areas blanketed by younger formations. McConnell's conclusions were first advanced in 1951 and expanded and given wider application in 1972 (which, incidentally, is a valuable source paper). Other recently published papers discuss the downward extension into the mantle of phenomena that he had formerly treated as belonging to the continental crust (McConnell, 1972). Other writers (see Harpum, Paper 6) agree that pre-existing linear structural features of post-Archaen age have been followed locally by late rift faulting, but that similar features of unequivical Archaean age are not in evidence. McConnell's original model (1951) required that planes of weakness were created during the Archaean by right-lateral wrench movements along the orogenic belts between cratons, subsequent movements on such planes being dominantly vertical displacement on the rift faults. However, he later moderated his view on the wrench movements.

Dixey directed his attention chiefly to exogenetic association of rift valleys, and McConnell concentrated on endogenesis and structure in depth. Neither has submitted his hypothesis to rigorous testing but because of the influence of these men on subsequent research, statement of their views is required.

Paper 6 by Harpum is an excerpt from a comprehensive account of the structure and geotectonics of the Precambrian of Tanzania, conclusions drawn from fifteen years of study in the field. The excerpt is included because it provides a lucid and concise statement of the problems involved. The structure and history of the Archaean and Proterozoic rocks, and the manner in which these could have influenced younger rift faulting, form the substance of this excerpt. Harpum dispels what are regarded by many as misconceptions and clarifies the issues. He is primarily concerned with structural relationships, and the tectonic or geodynamic aspects are only touched on. For Harpum's treatment of the part played by rifting within the ideal geological cycle, his paper on this subject should be consulted (Harpum, 1960).

Paper 7 is by Mohr whose earlier accounts of the geology of Ethopia and the Afar were followed by a series of papers on the structural geology of the African rift valleys from ERTS-1 imagery. They are a valuable contribution, especially regarding the Ethiopian and Gregory (Kenya) rift valleys. For reasons stated, their treatment of the Western Rift System, especially in the south, has less definition. Mohr's interpretation of the features seen on the ERTS-1 imagery can be studied as a model for application of this technique to rift valley study. He describes and discusses the strong influence of Precambrian structure, especially along zones of cataclastic deformation but stresses the ultimate independence of the African rift system in origin and nature from Precambrian tectonism. The Pliocene-Quaternary rift faulting is well expressed in the topography and thus easily discerned on the ERTS imagery. It is easily separated from manifestations of Precambrian structure. The discussion on the relation of rift faulting to Precambrian structures over such an extensive area is probably the most objective and therefore the most valuable contribution on this subject to date.

In some matters Mohr reinforces Harpum's views. Some misconceptions (but not all) that have come to be repeated in the new literature are discussed and their fallacies exposed.

VOLCANISM OF EASTERN AFRICA RIFTS

For general observations on petrogenesis and volcanic activity as related to the tectonics of rift faulting in both the Western and Eastern systems, papers by Bowen (1938), Harris (1969), Gilluly (1971), Le Bas

(1971), Gass (1972), Bailey (1974, 1978), Baker (1978), and the book by Tuttle and Gittens (1966) are among those that can be consulted. More detailed and localized treatment of volcanic activity can be found in Part V for the Western Rift System, that is, for south Malawi, Rungwe, south Kiva, Virunga and Bufambiro, Ruwenzori, and Fort Portal; and in Part VI for the Eastern Rift System, that is, for north Tanzania and Gregory, Kavirondo, Ethiopia, and Afar. However, it is emphasized that much additional field and laboratory work is proceeding or is as yet unpublished.

GEOPHYSICS OF EASTERN AFRICA

The seismicity of the region is well documented and described by Wohlenberg (1969) and discussed by him (1970). For the Uganda section of the Western System, Maasha (1975) supplements Wohlenberg; Bath (1975) contains information on the Western Rift System as well as the Tanzanian zones. Only modern works are referred to here. For gravity, Masson Smith and Andrew (pers. comm.) observed 1200 stations in Tanzania, and this was supplemented with additional observations by a Newcastle University team and B. P. Exploration. Interesting profiles were observed over the Western Rift valleys, Fairhead (1973) compiled a map for the Gregory zone of the Eastern Rift. Early work was done by Bullard (1936) and others.

REFERENCES

Bath, M., 1975, Seismicity of the Tanzania region, *Tectonophysics* **27:**353-379.

Bailey, D. K., 1974, Continental rifting and alkaline magmatism, in *The Alkaline rocks,* , H. Sorenson, ed., Wiley, New York, pp. 148-159.

Bailey, D. K., 1978, Open system magmatism through continental lithosphere: the East African rift as a case example (Abstract), in *Rio Grande Rift, Tectonics and Magmatism,* American Geophysical Union, Washington, .D.C., pp. 11-14.

Baker, B. H., 1978, Relations between tectonism and magmatism in the Kenya rift valley (Abstract, in *Rio Grande rift, Tectonics and Magmatism,* American Geophysical Union, Washington, D.C., pp. 14-15.

Bowen, N. L., 1938, Lavas of the African rift valleys and their tectonic setting, *Amer. Jour. Sci.* **35A:**19-34.

Bullard, E. C., 1936, Gravity measurements in East Africa, *Royal Soc. London Philos. Trans.* **A235:**445-531.

Dixey, F., 1937, The pre-Karroo landscape of the Lake Nyasa region and a comparison of the Karroo structural directions with those of the Rift Valley, *Geol. Soc. London Quart. Jour.* **93:**77-91.

Dixey, F., 1946, Erosion and tectonics in the East African Rift System, *Geol. Soc. London Quart. Jour.* **102:**339-388.

Fairhead, J. D., 1973, The regional gravity field of the Eastern Rift, East Africa, *Inst. African Geol. Annual Report* **17**:48-52.

Furon, R., 1963, *Geology of Africa,* Oliver & Boyd, Edinburgh, 377p.

Gass, I. G., 1972, The role of magmatic processes in continental rifting and sea-floor spreading, *Fourth Tomkeieff Memorial Lecture* University Newcastle. 16p.

Gilluly, J., 1971, Plate tectonics and magmatic evolution, *Geol. Soc. America Bull.* **82**:2383-2396.

Harpum, J. R., 1960, The concept of the geological cycle and its application to problems of Precambrian geology, *21st Intern. Geol. Congr. Proc.* **9**:201-206.

Harris, P. G., 1969, Basalt type and African rift valley tectonism, *Tectonophysics* **8**:427-436.

Le Bas, M. J., 1971, Per-alkaline volcanism, crustal swelling and rifting, *Nature* **230**:85-87.

Maasha, N., 1975, The seismicity and tectonics of Uganda, *Tectonophysics* **27**:381-393.

McConnell, R. B., 1951, Rift and shield structure in East Africa, *18th Intern. Geol. Congr. Proc.* **14**:199-207.

McConnell, R. B., 1972, Geological development of the rift system of Eastern Africa, *Geol. Soc. America Bull.* **83**:2549-2572.

Tuttle, O. F., and J. Gittens, eds., 1966, *Carbonatites,* Wiley-Interscience, New York, 591p.

Wohlenberg, J., 1969, Remarks on the seismicity of East Africa between 4 N-12 S and 23 E-40 E, *Tectonophysics* **8**:567-577.

Wohlenbert, J., 1970, On the seismicity of the East African Rift System, in *Graben Problems,* J. H. Illies and S. Mueller, eds., Schweizerbart'sche, Stuttgart, pp. 290-295.

3

Reprinted from pp. 263–267, 269–275, and 281–283 of *African Magmatism and Tectonics,* T. N. Clifford and I. G. Gass, eds., Oliver & Boyd, Edinburgh, 1970, 460p.

Vulcanicity and rift tectonics in East Africa

B. C. KING

ABSTRACT. *The East African Rift System and its associated vulcanism are among the most remarkable geological phenomena in the world. While it appears inescapable that they are related to a pattern of earth behaviour of a major order, the details of rift faulting are closely determined by older, mostly Precambrian, structures.*

In southern Africa 'rifts' in part controlled, but more largely followed the deposition of the Karroo, and while in some cases they were rejuvenated in later times, the evidence in East Africa points clearly to mid-Tertiary and later dates for the rift faulting there. Perhaps the most notable feature of the rifts is the extraordinary intensity of faulting (ranging from Tertiary to Recent) within their limits and the almost complete absence of faults outside.

There seems to be a correlation between vulcanism and topographic culminations in the 'rift system' of which the Kenya 'dome' is the most obvious example. A pattern has often been sought in the vulcanism of the rifts, both in its relation to tectonics and in terms of composition. Neither is clear, although alkaline, sometimes strongly alkaline, compositions are characteristic. Notable in Tanzania and Kenya are the early but sporadic manifestations of highly undersaturated basic volcanics, but even more are the later (mid-Tertiary) vast 'plateau' outpourings of phonolites and the great development of trachytes, and lavas and pyroclastics, in the form of huge broad volcanoes of late Tertiary, or early Pleistocene age. Recent vulcanism is marked by numerous central volcanoes along the median zone of the rift, mostly trachytic, but with basic associates, and often characterised by spectacular calderas.

Introduction

Within eastern Africa the rift system extends some 2300 miles, from the neighbourhood of the Limpopo River in the south to the Red Sea in the north. Here, despite considerable variation in pattern, the typical Rifts are graben about 30–50 miles in width. Northwards the Red Sea Rift extends the system a further 1400 miles but, like the branch rift into the Gulf of Aden, it is a much wider structure (up to 300 miles in width). The Levantine rift, which continues from the Gulf of Akaba northwards some 700 miles to the Taurus Mountains, again consists of narrower troughs resembling those of eastern Africa.

Although the most characteristic rifting is of Tertiary and later age, a similar pattern of faulting is recognisable in several sectors dating back to late Karroo or early Jurassic times. The fact that there are no older crustal structures or lineaments in this great belt of such magnitude and persistence suggests that the system has been determined by a major deep-seated crustal or sub-crustal mechanism. The occurrence of distinctive volcanic associations along many parts of the rift system is also in accord with this conclusion. Nevertheless, when the detailed pattern of dislocations is studied there is a notable correlation between fault directions and older Precambrian structural trends, a correspondence that was first clearly appreciated by Dixey (1956). The coincidence of rift faulting with Precambrian trends and the extended history of rift movements suggested to Dixey and also to McConnell (1951) that the fundamental cause was a stress system that had operated from Precambrian times in eastern Africa. This, however, is a questionable inference, for the older trends that are followed by the rift faults belong to Precambrian orogenic belts of widely differing ages and to structures of differing kinds, such as fold trends, foliations, faults or mylonite zones. It is rather that the structures of the old orogenic belts constituted a crustal 'grain' of greatly varying regularity and trend, by which the course of later faults was largely directed.

The profound dislocations that affected the Karroo in southern Africa are well documented, the vertical displacements being particularly obvious from the varying altitudes of the Stormberg lavas (Cox, this volume, p. 223). Throws of 10,000 ft or more are demonstrable on some faults, but there is also evidence that effective bevelling by erosion occurred simultaneously (Dixey, 1956, p. 9). The general uplift within the south-eastern and eastern margins of the continent was intermittently active from Karroo times on-wards and the complementary coastal and offshore downwarp determined the pattern of Cretaceous and Tertiary marine sedimentation. The uplift is represented by the highlands and plateaus of eastern Africa and with it the ancestral watershed between the Indian Ocean and Atlantic drainages; Dixey has appropriately referred to this major feature as the 'rising rim of Africa'.

Northwards from the Limpopo the Karroo is largely confined to a number of structural troughs (Cox, this volume, p. 214) which have much in common with the later rifts; the present morphological resemblance is, however, largely due to erosion of the less resistant infill of Karroo sediments; the troughs have alignments which were evidently largely determined by Precambrian trends (see De Swardt, 1965a). The Zambesi contrived to main-tain its course to the Indian Ocean against the rising rim of Africa which lifted the Batoka basalts from 3500 ft above sea-level at the Victoria Falls to 6500 ft above sea-level on the Rhodesian plateau.

Profound excavation and planation within the Karroo troughs and across their margins was achieved by the beginning of the Cretaceous and produced what Dixey (1939) terms 'valley-floor peneplanes'; extensive deposits of early Cretaceous 'dinosaur beds' on these surfaces point to their former much more widespread occurrence in the troughs. Early Cretaceous sediments, generally resting on eroded Karroo, occur in the Rukwa and other troughs of southern Tanzania and in the Nyasa-Shire trough, thus testifying not only to the age of the earlier movements in these troughs, but also to a period of stability that intervened before the Tertiary movements.

The recognition of the nature and importance of the earlier troughs was one of Dixey's major contributions to an understanding of the evolution of the East African Rift System. He showed that the southern rift valleys were late- or post-Karroo troughs rejuvenated by Quaternary faulting; indeed, in the case of the Malawi Rift the greater part of the eastern escarpment was determined by the Mesozoic displacements.

Alkaline igneous centres occur in association with the Mesozoic rifts and troughs. The most notable of these are the Chilwa Series of Malawi, comprising a group of relatively large alkali granite-syenite ring-complexes, together with numerous much smaller carbonatite centres (Garson, 1965; Woolley and Garson, this volume, p. 238). It is probable that volcanic activity was associated with the Chilwa centres, but it may be doubted if they were surmounted by major volcanoes. Similar carbonatite complexes occur around the junction of the rifts in northern Malawi, north-eastern Zambia and southern Tanzania.

The main rift system of East Africa

Throughout its length the rift system is characterised by Tertiary to Recent tectonic and, in many sectors, associated volcanic activity. On a continental scale it is essentially a north-south feature, yet its structural elements only exceptionally show this trend. The coincidence of detailed structures with older Precambrian trends is, however, very striking. Equally, it seems likely that the 'rejuvenation' of earlier troughs by later rifting merely reflects the fact that both sets of dislocations were determined by the ancient Precambrian grain.

Major faults showing strong parallelism and regularity are generally found where the Precambrian grain is also regular and persistent and trends approximately north-south. This is true of large parts of the Eastern Rift in Kenya. Similar persistence and continuity is also shown where a pronounced Precambrian grain is markedly oblique to the north-south trend, as in the case of the Rukwa and other troughs of south Tanzania and of the Lake Albert Rift.

Where the Precambrian structures depart considerably from north-south the rift valley may be defined by a pattern of *en echelon* faults, following the older grain, which relay the effective displacement along a more nearly north-south course. Examples are shown by the Albert Nile towards the Sudan border and to the south of Ruwenzori. Again where the Precambrian grain is complex and variable in direction the rift fault pattern is correspondingly irregular, as in eastern Tanzania.

It seems clear that the bifurcation of the system into an Eastern and a Western Rift is the result of deflection of the fracture systems around the Tanganyika 'Shield' or Craton (Fig. 1); this consists of a gneissic complex and meta-sedimentary or meta-volcanic systems (Dodoman, Nyanzian, Kavirondian) with prevalently east-west trends. The south-west of the craton is sharply limited by the Ubendian and associated formations, with their parallel and strongly marked folds, mylonite belts and lines of intrusions;

these have determined the series of north-west to south-east faults bounding troughs and horsts that constitute the rift structures of northern Lakes Nyasa, Rukwa and southern Lake Tanganyika (Fig. 1). Towards Lakes Kivu and Edward the rift continues to follow the Ubendian Belt but northwards it swings to the north-east following the grain of the Toro System and the Uganda 'basement'. Conforming also to this grain is the well-defined Lake Albert Rift, which is sited *en echelon* to the Lake Edward Trough, separated from it by the horst of Ruwenzori. Northwards again the relaying series of *en echelon* faults of the Albert Nile continue to follow the 'basement' grain. The Western Rift terminates abruptly against the strongly marked north-west to south-east trends of the Madi Series of the Sudan border and the parallel structures of the Aswa Shear Belt of northern Uganda.

The Eastern or Gregory Rift Valley in Kenya approximates in many places to a classic graben (Fig. 1); its north-south trend and general regularity are evidently related to the equally regular structures of the Mozambique Orogenic Belt. Northwards, towards the Ethiopian border, however, it splays out, again apparently largely controlled by older structures, to form a broad zone up to about 200 miles in width (Fig. 1), within which only a narrow and relatively insignificant trough passes from the southern end of Lake Rudolph and by way of Lake Stephanie to link with the great rift valley of Ethiopia (see Gass, this volume, p. 286).

Southwards in Tanzania the Eastern Rift rapidly 'degenerates' into a broad zone of faults of varying persistence and trend which define a series of tilted blocks (James, 1956; Dundas, 1965; Pallister, 1965). Control by older structures is, however, still evident. The zone of faulting, over 200 miles in northern Tanzania, is commensurate with the width of the Mozambique Belt and the fault trends, though widely varying, conform nevertheless with the complexities of structural trends in the Precambrian. More persistent and strongly marked older structures at the margin of the Mozambique Belt determine, for example, the Usungu Trough in the south, while the horst blocks of the Pare and Usambara Mountains and associated troughs reflect more regular 'Mozambiquian' trends in north-eastern Tanzania (Fig. 1). The subsidiary rifts or troughs that branch westwards from the main rift, namely the troughs of Kavirondo, Musoma, Speke Gulf and Lake Eyasi, cut across the margin of the Mozambique Belt and appear to be controlled by the transverse trends within the Tanganyika 'Shield'.

Two aspects of the morphology of the rifts that evidently have genetic significance are the rise from the surrounding land surface to the rift shoulders and the variations in altitude of floor and margins. Gregory (1921), in his classic description of the Kenya Rift Valley, recognised that it consists essentially of a downfaulted belt or graben along the crest or axis of an elongated arch; he thereby developed the notion that the rift was the consequence of the tensional stresses set up by the process of arching and that the downfaulted strip is thus analogous mechanically to the keystone of an architectural arch. Wayland (1930), however, concluded that the arch was more satis-

factorily interpreted as a compressional structure, the central graben representing in effect an inverted keystone held down by the same compressive
forces, explaining thereby, among other things, the negative gravity anomalies normally characteristic of the rift floors. This hypothesis is now largely
discredited in view of the almost ubiquitous evidence for tensional, normal
faulting associated with the rift valleys.

Traced along their length the rift valleys show great variations in altitude,
but the fact that elevations and depressions of the floor are matched by
corresponding elevations and depressions of the shoulders shows that the
overall displacements are everywhere of the same general order. The
tendency for the rifts to assume similar proportions, in terms both of width
and height, is presumably related to such common factors as crustal thickness
and strength as well as to the general magnitude of the primary stresses.

In many places the present altitudes of rift floors and shoulders do not
indicate the actual displacements, since the floors locally are infilled by
hundreds or thousands of feet or sediments and/or volcanics, and the rift
shoulders similarly have great thicknesses of volcanics surmounting the
'basement'. Nevertheless, in general terms, the floor of the Eastern Rift
ranges from below sea-level in the Red Sea to 7000 ft in central Ethiopia,
below 1000 ft to the south of Lake Rudolf, 7000 ft around Nakuru in central
Kenya and again to below 2000 ft at Lake Magadi near the Kenya–Tanzania
border (Figs. 1 and 2). Elevations of the shoulders and flanks of the rift, corresponding to those of the floor, constitute the Ethiopian and Kenya 'domes'.
Significantly these are both regions of volcanic activity in the Eastern Rift.
Elsewhere similar culminations in the rift system occur around Lakes
Edward and Kivu in the Western Rift and in southern Tanzania and northern
Malawi (Fig. 1). Again these are centres of volcanic activity. The association
of vulcanism with regions of uplift is not confined to the rift zone, but also
characterises, for example, the volcanic areas of Hoggar and Tibesti in the
Sahara (this volume: Black and Girod, p. 185; Vincent, p. 303).

Erosion surfaces

Erosion surfaces assume considerable significance in establishing the nature
and sequence of movements within the greater part of the region of the rift
valley system.

The pattern of the river systems in Uganda and north-western Tanzania
shows clearly that the original drainage was towards the Atlantic, across the
site of the Western Rift (Wayland, 1930). The watershed was approximately
along the line of the Eastern Rift, although, owing to disturbance by tectonic
and volcanic activity, the earlier drainage pattern here has been largely
obscured. The difference of situation of the two rift valleys in relation to the
drainage history accounts for the much greater sedimentation in the Western
Rift and its more extensive occupation by lakes. The sediments of the Western
Rift accumulated almost continuously from its inception during the Miocene
to the reversal of drainage in the Pleistocene. By contrast, the Eastern Rift is

[*Editor's Note:* Figures 1, 2, and 3 have been omitted.]

an internal drainage system with limited inflow and a number of shallow saline lakes.

The confusion in the recognition and correlation of erosion surfaces has largely arisen from the assumption that they were everywhere almost flat, with little or no relief, and that correlation is possible by projections based on concordance of summit levels. In the tectonically active regions, the dislocation of the surfaces by warping and faulting has been underestimated. De Swardt (1965b) affirms that there are in Uganda only two surfaces of major significance, marked by deep weathering and lateritisation. The *upper surface* (Buganda surface; P II of Wayland; or mid-Tertiary of Dixey) forms closely concordant flat-topped relics on the schists and gneisses of central Uganda, where it approximates to a peneplane, but shows broad elevations over more resistant formations with a relief up to 1000 ft or more in southwestern and western Uganda. The *lower surface* (P III of Wayland; end-Tertiary of Dixey) is widely expressed over northern and eastern Uganda. In central Uganda it is 500-1000 ft below the upper surface and extends southwards as broad valley shoulders set below the Buganda surface; it ascends with the valleys themselves until, in watershed regions between the river systems, it almost converges with the upper surface, as for example around the north-western shores of Lake Victoria. Both upper and lower surfaces rise towards the Western Rift; the rise is, however, much greater towards Lake Edward, where the upper surface reaches more than 8000 ft, than towards Lake Albert, while it is still less towards the Albert Nile (see Bishop and Trendall, 1967). The rise to the rift is more marked for the upper surface, so that the interval between the two becomes progressively greater towards the west; tilting was therefore initiated after the formation of the earlier upper surface and continued after the development of the later lower surface. Nevertheless, the drainage maintained its westward course and reversal did not occur until valleys had been cut below the lower surface. Lake Victoria, which was a consequence of the reversal, has submerged not only part of the lower surface but also the later valley incisions.

In western Kenya also only two lateritised surfaces are represented; but, since uplift was towards the Eastern Rift rejuvenating the upper courses of the original drainage and so accelerating erosion, the older surface is preserved only as remnants standing now at widely divergent levels, notably the Cherangani Hills (up to 11,000 ft), the Chemerongi Hills and the Kisii Hills (Fig. 2). The last named show flat-topped summits, morphologically similar to the outliers of the Buganda surface, which slope westwards from over 8000 ft to 5000 ft within 30 miles or so. In addition, numerous hills which rise above volcanics of the Uasin Gishu Plateau (Fig. 2) and above the general level of the basement plateau to the north and north-east of Kitale, are eroded residuals of the older surface. In eastern Uganda, erosion consequent upon upwarping towards the Eastern Rift has removed the older surface beyond regions underlain by the more resistant formations of the Buganda Series.

The lower lateritised surface can be traced continuously from eastern Uganda into western Kenya; not only does it truncate the lowest volcanics and penetrate as broad valleys into the volcanics of Mounts Elgon and Kisingiri, but it extends above the Nandi Scarp to form a lateritised planation, truncating alike the volcanics of the Uasin Gishu Plateau to the edge of the Elgeyo Escarpment and the basement of the Kitale area. On the Kericho Plateau it is similarly a modification of the surface of the lavas and rises into the Kisii Hills as shoulders to the valleys.

Sediments or tuffs, containing Lower Miocene faunas, are found at many localities below or within the lowest volcanics in eastern Uganda and western Kenya, while the earliest sediments of the Western Rift are of similar age. An extensive sub-Miocene erosion surface has been extrapolated by contouring from these various occurrences (see Pulfrey, 1960). In fact, however, the visible surface is a landscape of greatly diverse and often high relief that reflected the effects of current tectonic and paravolcanic activity resulting in rapid erosion of monoclinal upwarps and fault lines to form scarps, sections of which are preserved under Elgon and Kadam (Fig. 2), and complementary sedimentary accumulation in troughs, valleys and down-warps. Relatively planar elements in the landscape approximate to dislocated sectors of the older surface which have suffered little modification, as under the western flanks of Elgon and Kadam or around Napak. Even the Kitale plain, regarded as part of the sub-Miocene bevel (*ibid.*, p. 7), shows consider-able irregularity beneath the lavas (see Sanders, 1964). Only in eastern Kenya is it possible to recognise a sub-Miocene surface of planar character (Sagger-son and Baker, 1965).

These features suggest that the upper surface of Uganda (the Buganda surface) must, like the surfaces over the Kisii and Cherangani Hills, be older than mid-Tertiary (or Miocene), the age that has commonly been ascribed to it; indeed, it may well date back to the Cretaceous (see Bishop and Trendall, 1967, p. 407).

The lower surface cannot as yet be dated with any precision, although it is broadly 'late Tertiary', but it is likely to prove an important 'horizon' in a regional demarcation of tectonic and volcanic events in and around the Eastern Rift.

The Western Rift

The Western Rift System has been reviewed recently by Macdonald (1965). Hopwood and Lepersonne (1953) showed that in the Albert–Lower Semliki part of the rift the oldest sediments are Lower Miocene. The initial down-warp of the rift was thus at least as early, depressing the upper (Buganda) erosion surface and, presumably, severing the direct connection of the Uganda drainage with the Congo. The floor of the rift sank continuously, for the sediments are all of shallow water type. In Lake Albert drilling and geophysical work have established a maximum thickness of some 8000 ft of sediments. Older faulting can be recognised, as near the Murchison Falls, where faults are overlapped by sediments, but the major, visible faulting is Pleistocene, and later than the formation of the main erosion surfaces. The total amount of displacement increases from hundreds of feet near the Victoria Nile to over 10,000 ft in the Semliki area, the increase being marked by a rise of the shoulders from 2800 ft to 5500 ft, while the floor declines to 6000 ft below sea-level. On the Congo side of the rift the shoulders rise to around 10,000 ft indicating an even greater displacement of the north-western boundary fault. No more than a 100 ft or so of later Pleistocene sediments occur, reflecting the effect of reversal of the drainage. The fault pattern in the Albert Rift is comparatively simple, but numerous small faults in the sediments have been recorded (see Bishop, 1965) and continuance of movement to the present day is shown by the small scarp produced by the 1967 earthquake.

In the Albert Nile sector to the north the sediments are thinner, but 1300 ft have been proved by drilling and up to 3000 ft inferred geophysically. The faults here are also mostly north-east to south-west, but have an *en echelon* pattern which relays the movements northwards. Pleistocene faulting is evident where sediments are thrown against basement.

The continuation of the Lake Albert graben south-westwards along the Semliki River is confirmed by the form of the belt of high negative anomalies. To the south-east of the Semliki, Ruwenzori forms a parallel horst block rising to 17,000 ft. The upper surface is arched from north-east to south-west and tilted to the south-east, and despite its altitude it may well represent the uplifted Buganda surface.

Further south is the offset Lake Edward Trough, which is shallow at its north-eastern end, but deepens towards the Rwanda border; from Lake George it is defined by a series of faults which form an *en echelon* pattern.

In the Birunga–Bufumbira region eight major volcanoes have been built up in the rift along a trend which is almost at right angles to that of the trough itself (Holmes and Harwood, 1937). They are steep-sided cones, rising to a maximum of well over 14,500 ft and are composed of lavas ranging from leucitites or nephelinites to trachytes, but are characterised by a high potash/soda ratio. The two westernmost volcanoes are still active. Nyamlagira (10,010 ft) erupted in 1938–42 and its double caldera, $1\frac{1}{2}$ miles wide and 1000 ft deep, is unique among African volcanoes in containing a lava lake; Niragongo (11,385 ft) erupted in 1948. The volcanoes may have commenced activity in Pliocene times, but the bulk of the lavas are ascribed to the later part of the Pleistocene. There are also several hundred smaller cones, mostly formed of scoria. The culmination of the floor of the Western Rift is Lake Kivu (4790 ft) which, as a result of a volcanic accident, drains into Lake Tanganyika and thence to the Congo. South of Lake Kivu (Fig. 1) there are lava fields largely composed of alkali (olivine) basalts and basanites, but trachytes and trachyandesites are also represented.

In western and south-western Toro, Uganda, within and marginal to the rift zone, is a unique region of numerous explosion craters, often so closely spaced that they are contiguous or even overlap (Combe, 1930). They are mostly surrounded by insignificant rings of ejectamenta, but issuing from the Katunga crater there is a flow composed of a highly undersaturated basic lava with kalsilite. This mineral is also recorded in the lavas of the main Birunga field (Fig. 1) and, indeed, has proved to be rather widespread in the pyroclastics of Toro and Ankole (Holmes, 1942). Formations originally recorded by Combe as carbonated pyroclastics have more recently been interpreted as carbonatite lavas (Von Knorring and Du Bois, 1961).

Southwards from Lake Kivu the Western Rift swings into the 400 mile north-north-west to south-south-east belt occupied by Lake Tanganyika (Fig. 1). The structures here have not as yet been closely studied, but appear to be relatively simple. The lake is not comparable to Lake Albert, for it has a negligible sedimentary infill and its floor is well below sea-level. The lake itself occupies sub-parallel graben defined by various pairs of faults and, towards its southern end, the rift structures extend over a width of about 100 miles, involving parallel graben and horsts, into the rejuvenated post-Karroo Rukwa Trough (Fig. 1).

The Livingstone Fault bounding the eastern side of the Nyasa Trough continues the line of the fault to the east of the Rukwa Trough and at Mbeya Mountain and the Livingstone Mountains the shoulders rise to over 8000 ft, some 2000 ft higher than the Mbozi block on the opposite side of the rift. Between Mbeya Mountain and the Livingstone Mountains the transverse Usangu Trough joins the Rukwa–Nyasa Trough and it is here that the Rungwe volcanics form extensive highlands. This complex volcanic region shows the association of moderately alkaline basalts, phonolites and trachytes with more strongly alkaline nephelinites; lavas are accompanied by pyroclastics and although activity was at its maximum during the Pleistocene, it may have begun during the Pliocene, and there is evidence of tuff eruptions during the historic period (Harkin, 1959).

The Eastern Rift

In the northern part of eastern Africa lies the great Abyssinian Plateau, an enormous mass of volcanics rising to over 13,000 ft and extending eastwards into Somalia (Gass, this volume, p. 286). It is traversed by the rift graben widening to the north-east as the Afar depression, the bounding faults of which diverge to the Red Sea and Gulf of Aden. The first movements in the early Tertiary involved uplifts of the region by a maximum of 8000 ft accompanied by faulting. The volcanics mostly overlie various Mesozoic sediments, which rest, in turn, on the Precambrian. The older basalts were succeeded by over 3000 ft of later basalts before the Middle Tertiary. Major faulting and warping form the main graben and the Afar depression, and were contemporaneous with general uplift of the plateau. This was followed in the Quaternary by further vulcanism which continued with the development of basaltic volcanoes (Mohr and Rogers, 1965), including the numerous craters of the Red Sea. Although this is by far the greatest volcanic region of eastern Africa, it has been very inadequately investigated.

By contrast the Eastern Rift of Kenya and northern Tanzania has been examined in comparative detail, largely by the systematic work of the Geological Surveys in those territories (Figs. 2 and 3).

The older alkaline complexes of south-eastern Uganda may be considered as the earliest manifestations of igneous activity associated with the Eastern Rift although they lie well to the west of its main course. Extending over about 40 miles in a north-north-east to south-south-west line they comprise the centres of Sukulu (about 22 miles diameter), Tororo and Bukusu, formed of carbonatites with associated alkali syenites (including ijolites) and fenites, and the centres of Sekululu and Budeda, which have been revealed by partial removal of the Elgon volcanics. It is evident from the configuration of the surface under the Elgon volcanics (Fig. 1) that these complexes produced relatively abrupt domes, originally 1000 ft or more high. Although the radiometric ages obtained so far show a wide scatter they do suggest that the complexes may be no older than early Tertiary, rather than Mesozoic as formerly assumed.

In general terms the Eastern Rift is a north-south trough which roughly bisects the oval dome of the Kenya Highlands (Fig. 2). Corresponding to the greatest height of the rift floor, over 6000 ft between Nakuru and Naivasha at the centre of the dome, the lava plateaus of the rift shoulders rise to 8000 or 9000 ft in the west, culminating at over 10,000 ft in the Mau area, and from 5000 to 6000 ft in the east, but stepping up successively to around 8500 ft on the Kinangop Plateau and 13,000 ft at the crest of the Aberdare Range (Fig. 2).

The rift appears to have been initiated in the early Miocene as a downwarp with only minor faults; thereafter faulting, vulcanism and episodes of fluviatile or lacustrine sedimentation have continued to the present day. The most characteristic feature of the rift here is that it is a sharply-defined zone of more or less intense faulting, for the adjacent regions are largely or totally devoid of such dislocations. In general, faulting has developed inwards from the rift margins, so that in places there is a later, narrow central graben. The major faults have displacements of several thousands of feet and often provide evidence of repeated or long-continued movement. Uplift of the shoulders has occurred, as well as subsidence of the floor. The latest episodes of faulting have consisted of close-set, sub-parallel 'grid' faults, which generally displace the surface by no more than hundreds and often only tens of feet, producing a succession of narrower or wider horsts and troughs (Baker, 1965a and b).

There are great structural variations from one section of the rift to another. In southern Kenya and northern Tanzania the western margin is marked by the Nguruman Fault (Fig. 1) with a throw of at least 5000 ft in basement rocks. Later Tertiary basalts, banked against the fault scarp, were faulted to form a lower scarp-bound terrace above the floor of the rift. Northwards, the margin of the rift is lower and is essentially an eroded downwarp in Pleistocene trachytic tuffs and lavas. East of the Mau, the scarp, although again of greater height, is largely modified by a mantling of late tuffs, and descends to a subdued feature where the Kavirondo Trough joins the main rift. Thence to the north the western margin is downwarped, step-faulted and strongly dissected, but, with the decline in the floor to less than 4000 ft towards Lake Baringo (Fig. 2), it becomes an increasingly prominent feature and passes into the Kamasia Range rising to 8000-9000 ft. Here, however, the side of the rift is relayed farther to the west by the Elgeyo Fault with a throw of upwards of 5000 ft; this great displacement develops within a few miles from a series of minor step-faults. The Kamasia Range is essentially a block tilted to the west to the Kerio Valley at the foot of the Elgeyo escarpment (Fig. 2), and with a step-faulted eastern slope in which rather steep eastward dips are largely offset by repeated faults which northwards, however, become dominated by a single major fault, the Kamasia Fault. Thence, the Kamasia fault passes into several faults of smaller throw, while the Elgeyo Escarpment continues as an impressive feature surmounted by the Cherangani Hills which rise to 11,000 ft.

Farther to the north the rift becomes less well defined and from a width of 30–40 miles, widens out to almost 200 miles. The western margin is relayed by a splay fault, branching at a large angle from the Elgeyo Fault and carrying most of the displacement in passing the Chemerongi Hills; a similar splay structure following the Turkana Escarpment (Fig. 1), and marking the Kenya–Uganda border, is probably an eroded downwarp rather than a fault. Of these structures the Turkana downwarp was evidently formed at an early stage, for it was eroded and locally buried by the Miocene lavas from Moroto Mountain and Yelele in eastern Uganda, whereas the main movement of the Elgeyo Fault was in late Tertiary times.

The structures of the eastern margin of the rift are generally different from those of the western, so that the rift itself is commonly asymmetric both in structure and morphology. From northern Tanzania almost to the latitude of Nairobi (Fig. 2), the eastern side of the rift is a step-faulted downwarp, marked by fault strips which are tilted away from the floor. West of Nairobi and for some distance northwards the eastern margin is largely defined by a single fault scarp and the rift as a whole approximates to a simple graben. Northwards the faults which define the Kinangop Plateau die out and the main wall of the rift is relayed to the east by the fault marking the western side of the Aberdares which continues into the faults of the Laikipia Escarpment (Fig. 2).

The Laikipia sector is marked by erosion scarps etched along or across a pattern of north–south faults which are relayed successively eastwards in a northerly direction; the fault-bound strips in general step down into the rift but are variously tilted towards or away from it. Farther to the north, from Maralal to the Ethiopian border, the eastern margin is a broad downwarp, with comparatively minor faulting, towards the central trough.

The Kavirondo Trough (Shackleton, 1951) appears to have originated as a transverse downwarp or lag, faulted along part of its northern margin, probably in association with early arching along the site of the main rift, and contemporaneous with the formation of the west-facing downwarp or lag under Elgon and Kadam and the east-facing downwarp of the Turkana Escarpment to the north (Fig. 2). These movements dislocated the older (? late Mesozoic) lateritised erosion surface which subsquent erosion modified to form the sub-volcanic (sub-Miocene) surface.

[*Editor's Note:* Material has been omitted at this point.]

Conclusions

The East African Rift System is a structure of a major order in the earth's crust. Its history extends from at least early Miocene to the present day, while in the south it has partly rejuvenated similar rifts of early Jurassic age. The trends of rift faults are closely dependent on the older Precambrian 'grain'. The faulting is mostly normal and tensional; only to this extent are the rifts dilational phenomena. They are therefore unlike the so-called 'rifts' of the ocean floors.

Arching over the sites of the rifts is characteristic, and intermittent uplift of the rift shoulders has occurred concomitant with subsidence of rift floors. Typical rifts show rather regular widths between 30 and 50 miles and vertical displacements of 5000-10,000 ft, which are independent of their absolute altitude and probably reflect relations between thickness and strength of the crust and imposed stresses. Activity, both tectonic and volcanic, has tended to migrate towards the central graben during rift history and the latest movements commonly took place along numerous close-set faults of small displacement. These features suggest a sequence of secondary adjustments to strains induced by primary dislocations.

Vulcanism is not a direct consequence of rifting, for extensive sectors of the rift are devoid of volcanic activity and volcanic centres rarely, if ever, show a close association with faults. Nevertheless, the coincidence of vulcanism in East Africa with parts of the rift system shows that there is a relation, ship of some kind. Indeed, volcanic regions are associated with culminations or broad topographic domes in the rift system.

The volcanic rocks range from mildly to strongly alkaline and include highly undersaturated types. There is no evidence of simple sequences or cycles, but in the Eastern Rift the first appearance of particular compositions on a major scale may be of genetic significance. The earliest (Miocene) volcanics are highly undersaturated nephelinites and/or alkali basalts. Widespread phonolites followed in later Miocene. Trachytes were abundantly represented in the Pliocene, whilst alkali rhyolites are Pleistocene to Recent in age and of limited distribution. Alkali basalts are, indeed, of frequent occurrence throughout the volcanic sequences, while central volcanoes ranging in age from Miocene to Recent, and correspondingly more or less dissected, testify to the association between nephelinite lavas and intrusive ijolite-carbonatite.

Highly characteristic of the volcanic province is the great development of phonolite and trachyte, comparable in extent with basalts; the proportions are strikingly different from those of similar rock types in the volcanics of the ocean basins. The total volume of volcanic rocks associated with the Eastern Rift is probably at least 150,000 cubic miles (600,000 km^3).

Acknowledgements

Adequate recognition of source material for a largely appreciative and synthetic account is not possible, but reference is conveniently made to the very comprehensive bibliographies that were prepared for the UMC/ UNESCO Symposium held in Nairobi in 1965. I am also much indebted to the work, as yet largely unpublished, of members of my research teams in East Africa and I thank B. Collins in particular for the preparation of the maps.

REFERENCES

BAKER, B. H. 1965a. The Rift System in Kenya. *Rep. UMC/UNESCO Seminar on the E. African Rift System (Nairobi)*, 82.

—— 1965b. An outline of the geology of the Kenya Rift Valley. *Rep. UMC/UNESCO Seminar on the E. African Rift System (Nairobi)*, 1.

BISHOP, W. W. 1965. Quaternary geology and geomorphology of the Albertine Rift Valley, Uganda. *Spec. Pap. geol. Soc. Am.*, **84**, 291.

—— and TRENDALL, A. F. 1967. Erosion-surfaces, tectonics and volcanic activity in Uganda. *Q. J. geol. Soc. Lond.*, **122**, 385.

COMBE, A. D. 1930. Volcanic areas of Bunyaruguru and Fort Portal. *Ann. Rep. geol. Surv. Uganda* (for 1929), 16.

DAWSON, J. B. 1962. Sodium carbonate lavas from Oldoinyo Lengai, Tanganyika. *Nature, Lond.*, **195**, 1075.

DE SWARDT, A. M. J. 1965a. Rift faulting in Zambia. *Rep. UMC/UNESCO Seminar on the E. African Rift System (Nairobi)*, 105.

—— 1965b. Lateritisation and landscape development in parts of equatorial Africa. *Rep. UMC/UNESCO Seminar on the E. African Rift System (Nairobi)*, 134.

DIXEY, F. 1939. The early Cretaceous valley-floor peneplain of the Lake Nyasa region and its relation to Tertiary rift structures. *Q. J. geol. Soc. Lond.*, **95**, 75.

—— 1956. The East African Rift System. *Bull. colon. Geol. Mineral Resources*, Suppl. **1**.

DUNDAS, D. L. 1965. Review of rift faulting in Tanzania. *Rep. UMC/UNESCO Seminar on the E. African Rift System (Nairobi)*, 95.

GARSON, M. S. 1965. Summary of present knowledge of the rift system in Malawi. *Rep. UMC/UNESCO Seminar on the E. African Rift System (Nairobi)*, 94.

GREGORY, J. W. 1921. *The rift valleys and geology of East Africa*. London.

HARKIN, D. A. 1959. The Rungwe volcanics. *Mem. geol. Surv. Tanganyika*, **2**.

HOLMES, A. 1942. A suite of volcanic rocks from south-west Uganda containing kalsilite. *Miner. Mag.*, **26**, 197.

—— and HARWOOD, H. F. 1937. The petrology of the volcanic area of Bufumbira. *Mem. geol. Surv. Uganda*, **2**, Pt. 2.

HOPWOOD, A. T., and LEPERSONNE, J. 1953. Présence de formations d'âge miocène inférieur dans le fossé tectonique du Lac Albert et de la Basse Semliki (Congo belge). *Annls. Soc. géol. Belge*, **77**, 83.

JAMES, T. C. 1956. The nature of rift faulting in Tanzania. *C.C.T.A. east centr. reg. Comm. Geol.*, Dar-es-Salaam, 81.

—— 1959. Carbonatites and rift valleys in E. Africa. Abstract. *Int. geol. Congr. 20, As. Serv. geol. Afr.*, 325.

KING, B. C., SUTHERLAND, D. S., COLLINS, B., DIXON, J. A., LE BAS, M. J., CLARKE, M. C. G., FLEGG, A., and FINDLAY, A. L. 1966. *In* Volcanism in eastern Africa. *Proc. geol. Soc. Lond.*, **1629**, 16.

—— and SUTHERLAND, D. S. 1960. Alkaline rocks of eastern and southern Africa. *Sci. Prog.*, **48**, 298, 504, 709.

MACDONALD, R. 1965. The status of rift valley studies in Uganda. *Rep. UMC/UNESCO Seminar on the E. African Rift System* (Nairobi) 52.

MARTYN, J. E. 1967. Pleistocene deposits and new fossil localities in Kenya. *Nature, Lond.* **215**, 476.

McCONNELL, R. B. 1951. Rift and shield structures in E. Africa. *Int. geol. Congr.*, **18**(14), 199.

MOHR, P. A., and ROGERS, A. S. 1965. Status of geological and geophysical studies and resumé of the geology of Ethiopia. *Rep. UMC/UNESCO Seminar on the E. African Rift System* (Nairobi), 47.

NIXON, P. H., and CLARK, L. 1967. The alkaline centre of Yelele and its bearing on the petrogenesis of other eastern Uganda volcanoes. *Geol. Mag.*, **104**, 455.

PALLISTER, J. W. 1965. The rift system in Tanzania. *Rep. UMC/UNESCO Seminar on the E. African Rift System* (Nairobi), 86.

PULFREY, W. 1960. Shape of the sub-Miocene erosion bevel in Kenya. *Bull. geol. Surv. Kenya*, **3**.

SAGGERSON, E. P., and BAKER, B. H. 1965. Post-Jurassic erosion-surfaces in eastern Kenya and their deformation in relation to rift structure. *Q. J. geol. Soc. Lond.*, **121**, 51.

—— and WILLIAMS, L. A. J. 1964. Ngurumanite from southern Kenya and its bearing on the origin of rocks in the northern Tanganyika alkaline district. *J. Petrology*, **5**, 40.

SANDERS, L. D. 1964. Geology of the Eldoret area. *Rep. geol. Surv. Kenya*, **64**.

SHACKLETON, R. M. 1951. A contribution to the geology of the Kavirondo Rift Valley. *Q. J. geol. Soc. Lond.*, **106**, 345.

VON KNORRING, O., and DU BOIS, C. G. B., 1961. Carbonatite lava from Fort Portal area in western Uganda. *Nature, Lond.*, **192**, 1064.

WAYLAND, E. J. 1930. Rift valleys and Lake Victoria. *C. R. Int. geol. Congr.*, **15**(6), 323.

WILLIAMS, L. A. J. 1965. Petrology of volcanic rocks associated with the Rift System in Kenya. *Rep. UMC/UNESCO Seminar on the E. African Rift System* (Nairobi), 33.

—— 1967. In *Nairobi: City and Region*, p. 1. (Ed. W. T. W. Morgan).

4

Crown copyright © 1956 by Her Majesty's Stationery Office
Reprinted from pp. 1–7 of *Colonial Geol. Mineral. Resour. Bull. Supp.* 1, 1956, 71p.

THE EAST AFRICAN RIFT SYSTEM

By F. Dixey, C.M.G., O.B.E., D.Sc., F.G.S., M.I.M.M.
*Director of Colonial Geological Surveys and Geological Adviser to
the Secretary of State for the Colonies*

ABSTRACT

RECENT WORK IN VARIOUS PARTS OF THE EAST AFRICAN RIFT ZONE, AND ITS EXTENSION
into Arabia, is reviewed. Post-war geological mapping carried out in this region
has added greatly to our knowledge of the age and character of the rift movements. In the Shire rift the close association of post-Tertiary rifting with pre-
Cambrian and Jurassic structures has been confirmed; moreover, the Jurassic
trough is shown to be of similar character to the post-Tertiary rift, in that the
margins of the trough were upthrown, thereby producing a Jurassic " rise to the
rift " similar to that of the later rift. In the Rukwa trough, the close association
of pre-Cambrian, Jurassic and post-Tertiary rifts has been confirmed, as also has
the former widespread occurrence of early Cretaceous Dinosaur Beds, thus
proving the great age and resurrected character of much of the present topography, which indeed had achieved much of its present form prior to the post-
Tertiary rifting. The eastern and western rifts, however, despite the importance
in them of post-Tertiary features, are now known to have had a history as ancient
and complex as that of the southern rift. New studies of the coastal sediments of
East Africa and of western Madagascar, and of their structures, indicate a history
comparable with that of the rifts, but more continuous. The west Arabian, or
Levantine, rift shows many important features in common with the East African
Rift System, particularly in relation to its ancient and complex history, but,
owing to its proximity to the Tertiary orogeny, it shows in addition an increasing
degree of folding and thrusting of the formations concerned as it is followed
towards the north. The rift itself, none the less, maintains its individuality across,
and even far beyond, the Taurus orogeny.

The early description of the obvious post-Tertiary rifting in Kenya and
Uganda has long caused undue emphasis to be placed on this aspect of the rift
history and has given rise to the impression that the East African rift as a whole is
essentially a late- and post-Tertiary phenomenon. That this view is untenable,
however, seems apparent when consideration is given especially to the southern
and Arabian rifts in which the Jurassic and, in places, the pre-Cambrian ages
of the system are now well established. Here it appears that the rift system should
be regarded as a feature of great importance in the earlier as well as in the later
history of the region, and that its origins extend back to crustal weaknesses
developed in ancient or primordial times. Moreover, it seems evident that
stresses due to the long continued and intermittent rise of the East African swell
and the concomitant sinking of the Mozambique geosyncline have caused the
ancient structures from time to time, and particularly in the Jurassic, to be rejuvenated, and that the post-Tertiary rift is merely a late phase—not necessarily
the most important or final phase—of a long and complex history.

The thousand-mile overlap of Jurassic and post-Tertiary structures in the
Rukwa-Nyasa-Shire-Urema rift, the contemporaneous development of the
Lebombo monocline with these structures, and the close association of both with
the N-S pre-Cambrian trends and faults, suggest that the Lebombo structure
should be included within the concept of a greater rift system. This monocline
and the East African swell, both rising intermittently, are regarded as complementary to the intermittent subsidence of the Mozambique geosyncline, which is
envisaged as a N-S extension of the Mozambique channel, known to have been

sinking since Permian times. The Mozambique geosyncline, and the rifts, thus regarded as branches of the geosyncline, have occurred as lag areas and have failed from time to time to keep pace with the rising East African and Madagascar swells. Expressed in another way, the stresses due to the development of the swell of the geosyncline have given rise at successive periods to rejuvenation of fractures first established as primordial lines of weakness.

The problem of the rifts is accordingly but part of the greater problem of the swell and the geosyncline, which in turn is only a part of that connected with the ancient basin-and-swell pattern of Africa as a whole. This pattern is regarded as due to some form of crustal adjustment to a shrinking earth, resulting in part in the subsidence of larger and smaller blocks, of ancient origin, relative to others. If a basin be regarded as a subsiding block, with the central part sinking relatively to the margins, the interaction of the margins of such blocks would provide a mechanism for much of the observed fact of rift development. Lag areas could occur between the margins of major blocks, magmatic action would tend to be at a maximum along the block boundaries, and the movements and varying pressures of the magmatic substratum could account for relative vertical displacement of lesser blocks along the rifts, for local compression and folding, and for volcanic activity.

AUTHOR'S NOTE

THIS PAPER formed the basis of a lecture delivered before the Geological Society of London on 11th January, 1956. The lecture was followed by a discussion recorded in *Proc. geol. Soc. Lond.*, 1956, No. 1533, pp. 33-40, and it is reproduced by kind permission of the Geological Society as an Appendix to the present work. A complementary lecture on Some Aspects of the Geomorphology of Central and Southern Africa was given by the author as an Alex. du Toit Memorial Lecture in Johannesburg in September, 1955, the text of which will be found in *Trans. geol. Soc. S. Afr.*, LXVIII, Annexure.

Introduction

THE VERY CONSIDERABLE expansion of Government and non-Government Geological Surveys in north-east and central Africa and in Arabia since the War has led to the accumulation of much new information on the East African rift system and its extensions to the north and south. This work has thrown fresh light on the complex surface features of the rift, and has perhaps given us a glimpse of its nature and origin. The present time is therefore opportune in which to review some of the problems presented by this immense and fascinating feature, whose origin probably dates back to the earliest period of the earth's history.

Although the rift system is a more or less continuous geographical feature throughout its length, it varies greatly from place to place in its geological and geomorphological aspects, and particularly in the extent to which its ancient structures have been revealed by deep erosion. This fact has long been obscured by the accident that the system was first brought to scientific notice by descriptions of the Kenya or Gregory rift, where, over a limited range, it consists wholly of lavas of later Tertiary and post-Tertiary age that were rifted in the post-Tertiary. From this and from the obvious geographical continuity of the rifts, the view gained early currency that the East African rift system as a whole was essentially a feature of the post-Tertiary, and that the main structures were " rifts ",

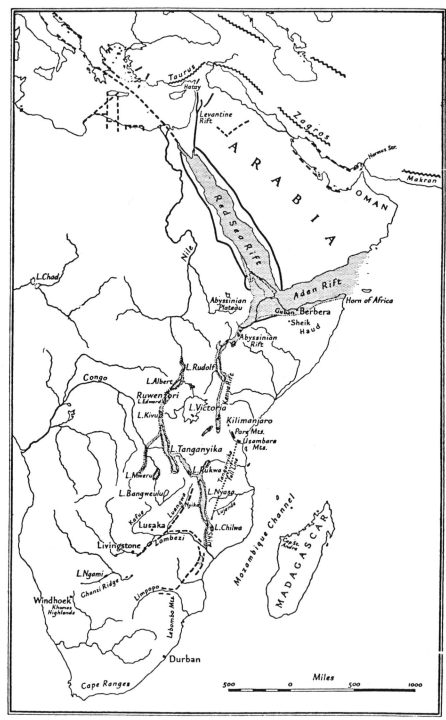

TEXT-FIG. 1. MAP OF THE GREAT RIFT VALLEY

in the sense that, as in the Kenya rift, the floors represent relatively down-faulted strips of the bordering plateau surfaces.

In some parts of the rift system the later Tertiary and the post-Tertiary movements account for all, or practically all, of the visible effects, while in other parts they give rise to only a negligible part of the current relief; there was a fairly general considerable renewal of activity in this later period, but this renewal can only be regarded as the latest phase of a complicated history extending back into very ancient times.

The first clue that the story was, in fact, immensely more complicated came from the recognition of the Dinosaur Beds in the Nyasa trough (Dixey, 1939b), which showed that many of the visible features of this part of the rift dated from the early Cretaceous, and that the only true rifting there was that of the mid-Pleistocene, which carried the floor of the ancient valley-plain downwards to a maximum of 2,300 feet to form the actual basin now occupied by Lake Nyasa; i.e. the Pleistocene rifting is now mainly hidden beneath the waters of the lake.

The term " rift " can only be applied to the system as a whole to indicate a structure due to parallel faulting, usually of the normal type, whereby the enclosed blocks have been at different times moved up or down relative to the sides, and not necessarily always in the same relative direction; moreover, owing to the operation of several cycles of erosion in the rift zone in Mesozoic and later times, the present surface of the disturbed blocks has usually long since lost any cyclic connection with that of the bordering country. Furthermore, the older scarps, long subject to erosion, are often fault-line scarps rather than simple fault scarps. For these, and for other reasons connected with a complicated geological history, in the older parts of the rift at least, the existing scarps and floors have served as a very insecure basis for the development of theories of rift structure of the kind so far presented.

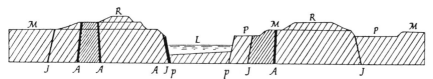

TEXT-FIG. 2. FORMATION OF AN EAST AFRICAN RIFT VALLEY

A typical East African " rift valley " formed along pre-Cambrian structures by (1) Jurassic faulting, (2) a succession of erosion cycles, and (3) post-Tertiary rifting.

The Jurassic faults (J) are older than the residual plateaux (R) on the Miocene peneplain (M) below which has been eroded the Pliocene surface (P). The Pleistocene rift faults (p), one of them rejuvenating the old fault (J), have let down the Pliocene surface to form the rift lake (L). Pre-Cambrian faults (A) are locally rejuvenated.

There is reason to believe that the rift is aligned along a very early, practically primordial, weakness of the crust, and that a long succession of stresses in the region traversed by the rift has enabled it to retain its individuality even to the present day; moreover, on the broader view now presented, the rift is regarded as extending southwards down the whole of the eastern side of Africa, and northwards, not only across western Arabia, but also through the Taurus ranges and beyond far into

Anatolia; it thus extends through 70 degrees of latitude, or nearly one-fifth of the earth's circumference, and is one of the major geological features of the world.

A few years ago the writer presented a paper on erosion and tectonics in the African Rift System (Dixey, 1946), with special reference to the Nyasa, Rukwa and Tanganyika rifts, and summarised his views as follows:

A considerable part of the rift system originated in two main series of fractures following the same lines but separated by a prolonged period of intermittent continental uplift and regional planation. In the first place, early post-Karroo movements, following in many cases still earlier lines of weakness, gave rise to widespread large-scale undulations associated with pronounced trough-faulting and block-faulting. The anticlinal movements carried up ancient resistant pre-Karroo masses that have remained as larger and smaller residuals throughout all succeeding cycles, while the down-folded areas are often still recognisable as relatively low-lying basins, in which, in many cases, patches of Karroo sediments still survive. These widespread disturbances were succeeded by a great period of planation which culminated in the late Jurassic period; subsequently, the surface was deeply dissected in the north, centre, and south, in early Cretaceous times. This planation was highly effective even on resistant rocks, but a number of high residuals remained on the ancient surface. Following widespread uplift, a lower erosion surface was developed, that of the late Cretaceous or early Tertiary (Dixey, 1943a), on which remnants of the earlier peneplain remained. The process was repeated, and ultimately led to the erosion of the " main peneplain " of this part of Africa, which was terminated by further uplift in the Miocene. During these different cycles, lowlands and troughs were developed respectively on the less resistant Karroo and pre-Karroo rocks and in the down-faulted Karroo blocks, which were usually bordered by high-level residuals of former erosion surfaces. In this way were formed the ancestral Tanganyika, Rukwa, Luangwa, Zambezi, and other troughs. (See Plate VII, fig. 2, facing p. 26.)

On further uplift, bringing in a new erosion period which continued throughout the Pliocene, the drainage followed the established pattern and still further deepened the lowlands and troughs in the less resistant rocks.

At this stage, or a little later, the second great series of fractures developed, comprising the " Rift System ". This series followed very closely the lines of the earlier fractures, which were by this time clearly picked out by erosion into a system of fault-line troughs and fault-line scarps, somewhat blurred in many cases by the long-continued denudation. The new fractures thus merely deepened the existing topographic troughs, which marked the sites of earlier structural troughs.

The second main period of faulting in the rift zone has continued locally even into modern times. It was accompanied by gentle warping, and in this way lakes Victoria, Rudolf, Mweru, Chilwa, and others were formed in the Rift Zone, while beyond the Rift Zone contemporaneous movements of similar character gave rise to the Ngami-Bangweulu line of lakes and swamps in Northern Rhodesia and comparable depressions in southern Africa, as well as to the Chad basin of West Africa.

It was accordingly suggested that the rift system developed mainly from an ancient series of fractures, and that it was due in part only—and in some sections only a minor part—to the post-Tertiary fracturing.

The sequence of post-Karroo or later Jurassic faulting, planation, pronounced regional uplift, erosion of the weaker sediments to the new base-level, and early Cretaceous sedimentation on the lowlands, already recorded from the Nyasa Rift Valley and from the Cape Ranges, was recognised in the vicinity of the great eastern scarp of the Abyssinian Plateau. Planation of Jurassic age and vigorous early Cretaceous and subsequent erosion were accordingly regarded as important features in the history of the Rift Zone, and many of the high-level residual plateaux were believed to represent a " late Jurassic peneplain ".

Since this interpretation was written much additional geological work has been carried out; the suggestions regarding the importance of pre-Cambrian structures to rift tectonics, and the probable early Cretaceous age of the Rukwa, Tanganyika and Zambezi troughs, have been confirmed and strengthened. The presence in central Africa of a late Cretaceous surface has been confirmed on palaeontological and stratigraphical grounds, and much additional evidence regarding the tilting of surfaces in the vicinity of rifts has been gained. Furthermore, the high-level plateau remnants, formerly believed to represent a late Jurassic peneplain, are now regarded as remnants of bevels developed on regional upwarps of Jurassic age, and associated, somewhat lower, " surfaces " are regarded as bevels due to intermittent upwarping along similar lines. Finally, on the basis of evidence from the Lower Zambezi, there is reason to believe that the older bevels were eroded on uplands formed by a " rise to the rift " in early Jurassic times, when the Karroo fault-troughs were being formed, and comparable with the arching that normally accompanied the Tertiary and post-Tertiary rifts.

Wayland (1930) and others have long stressed the importance of compression in at least the earliest (pre-Cambrian) phase of the rift structures. Further work has tended to confirm this view, but, as will be seen later, there still remains a considerable difference of opinion as to the relation between these structures and the obvious normal faults of the rifts. There can now be no doubt as to the importance of the pre-Cambrian association and of the frequent rejuvenation of old faults, but in the Mesozoic, the Tertiary and the post-Tertiary, the visible faults—with possibly rare exceptions—appear to have been normal, and the relations of these to the structures of deep-seated origin are still not fully understood.

In the Rukwa-Nyasa-Shire-Urema rift, over a distance of more than 1,000 miles, there is a complete overlap of Tertiary and Jurassic faulting, and, as the Jurassic structures of at least the southern part of this zone form part of the Lebombo monocline, it is reasonable to regard this structure also as contemporary with, and marking a southern extension of, an important phase of the rift movements.

Furthermore, the long-continued intermittent uplift of the co-linear rift-zone swell and the Lebombo monocline is regarded as complementary to the intermittent subsidence of a Mozambique geosyncline, which is envisaged as a major feature in north-south extension of the Mozambique channel, known to have been sinking since Permian times.

Some recent work in critical parts of the rift system, as followed from south to north, is reviewed below and some general questions of the age and origin of the system are discussed.

[*Editor's Note:* In the original, material follows this excerpt. Only references cited in this excerpt are reprinted here.]

REFERENCES

Dixey, F., 1939b, The Early Cretaceous valley-floor peneplain of the Lake Nyasa Region, and its relation to Tertiary Rift structures, *Quart. J. geol. Soc. Lond.,* Vol. 95, Pt. 1, pp. 75–108, figs., maps.

Dixey, F., 1943a, Erosion cycles in central and southern Africa, *Trans. geol. Soc. S. Afr.,* 1942, Vol. 45, pp. 151–181, refs.

Dixey, F., 1946, Erosion and tectonics in the East African Rift System, *Quart. J. geol. Soc. Lond.,* Vol. 102, Pt. 3, pp. 339–379, figs., refs., map.

Wayland, E. J., 1930, Rift valleys and Lake Victoria, *C.R. 11e Sess., Congr. géol. internat., S. Afr.,* 1929, Vol 2, pp. 323–353, figs., refs., map.

Copyright ©1967 by Macmillan Journals Ltd

Reprinted from *Nature* **215**:578–581 (1967)

The East African Rift System

R. B. McCONNELL

THE study of rift valleys has assumed a new importance in the past 10 years because of the discovery by Heezen[1] of comparable features on the world-wide system of mid-ocean ridges, and the International Upper Mantle Project[2,3] has shown considerable interest in the East African rift system. Previous reviews of this system have been published[4-7], but it is now possible to give a clearer picture of the entire field because of the recent accumulation of detailed mapping since the post-war expansion of the Overseas Geological Surveys[8] and the establishment of the Research Institute of African Geology at the University of Leeds[9], and it may be of value to outline the geological characteristics of the system in view of the latest work.

A rift valley may be defined as an elongated sunken area, with a characteristic width[6] of 30–60 km, which has descended between parallel faults. The term was introduced by Gregory[10] to designate the grabens (down-faulted strips) of the great fracture (*Brüche*) system of eastern Africa, first described by Suess[11]. In the East African rift system, as now described, true rift valleys are numerous, but the rift faults, although forming a continuous system, are not always in pairs determining a graben, but may take the form of block faults or be replaced by flexures. The rift system was believed by Suess and Gregory to include the Red Sea and the Dead Sea–Jordan Valley complex, but only the portion within the African continent will be considered here.

The classical East African rift system, as shown in Fig. 1, is more than 4,000 km in length from Asmara, near the Red Sea, to the Zambezi River, and reaches a maximum

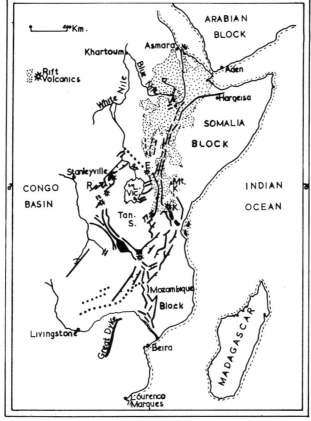

Fig. 1. Structural map of the East African rift system showing major rift scarps (schematized) and allied features. Solid black denotes raised blocks. Heavy dots denote dislocation zones. *A*, Addis Ababa; *R*, Ruwenzori Mountains; *E*, Mt. Elgon; *Mt. K*, Mt. Kenya; *Tan. S.*, Tanganyika Shield; *K*. Kilimanjaro.

width of 1,000 km in Tanzania. This immense system is much quoted in the literature of world tectonics, but its complexity is little understood because of the difficulty of synthesizing the many detailed investigations which are necessarily restricted to limited map sheets. Enough geological mapping is now available, however, to give a preliminary view of many important aspects of the rift system which concern: (1) the geological formations involved; (2) the pattern of faulting; and (3) the age of faulting.

Geological evidence[5,12,13] now appears to confirm that the pattern of the rift system originated in Pre-Cambrian times[7,14,15] as a line of deep faults centring on a N.N.E.–S.S.W. axis, diverted locally around pre-existing cratons, and that movements have occurred during major orogenic cycles down to the present. Dixey[16] has given evidence for the existence of actual rift valleys as early as the Jurassic, but during periods of stability the fault scarps and grabens were generally planed down by erosion[17,18] and even crossed by river systems.

It is a fact of outstanding geological importance that the formations principally affected are of Pre-Cambrian age; the pattern of rift faulting has been found to reflect the Pre-Cambrian structures of the basement and formations of later date are merely relatively thin cover rocks which conceal these structures. To attain an objective view of the history and geological structure of the rift system as a whole it is therefore necessary to give special consideration to those few areas, particularly in Malawi, Tanzania and Uganda, in which the remarkable Pre-Cambrian structures within the rifts are clearly visible, and not, as is the general rule, concealed by young volcanics and sediments, or by lake waters.

In view of the inherent difficulties of Pre-Cambrian geology, however, the study of the very significant association between rift faulting and Tertiary–Recent volcanism[19-23] which is so well exposed in the Kenyan and Ethiopian sectors is also most important for the unravelling of the history of the rift system. This volcanism[24,25] is characteristically alkaline and of great interest petrologically; the associated post-Miocene rift structures are also particularly clearly seen, but it must be remembered that these phenomena are typical only of the northern part of the system and that elsewhere Cainozoic volcanism is rare, being either completely absent or limited to intersections in the fault system.

A map, Fig. 1, showing the pattern of faulting compiled from the latest information[26-31] illustrates the difficulty, which results from the curvature of the component faults and grabens, of explaining the formation of the rift system by invoking either tension[6,19] or compression[32-34] alone. The strike of the components of the rift system varies between N.W.–S.E. and E.N.E.–W.S.W., but the map clearly indicates an overall N.N.E.–S.S.W. axis which is represented in the north by the strike of the Gregory and Ethiopian rift valleys. This axis is split into eastern and western branches by the resistance of a central craton termed the Tanganyika Shield[7], thus giving rise to the great width of the system in its equatorial sector.

In the Ubende area of Tanzania, adjacent to Lake Tanganyika, a portion of the Pre-Cambrian floor of the western branch of the rift system is exceptionally well exposed. In 1943–44 I mapped the region and showed[14] that the N.W.–S.E. rift fault scarps in this area follow Pre-Cambrian movement zones identified by blastomylonites, phyllonites, migmatites and elongated anatectic granites whose origin was assigned to an Ubendian Diastrophism[7]. These metamorphic zones were later studied by Sutton and Watson[35,36] and the presence of Pre-Cambrian movement zones was confirmed. Similar associations and parallelism of Pre-Cambrian mylonites and migmatites with rift faulting have also been described from many different sectors of the rift system[7], irrespective of strike.

At the southern termination of the rift system, in the Zambezi River sector, the fault pattern appears to split again around the Rhodesian Shield, and Vail[37] has described elongated troughs of Karroo sediments, to the east and west of this shield, which he believes to be the southern continuation of the rift system manifested by faulting in late Karroo and Cretaceous times. It has been pointed out elsewhere[7] that the line of the Great Dyke of Rhodesia continues exactly the strike of the Gregory rift system, and this line also bisects the Bushveld Igneous Complex. Cousins[38] has since shown that a chain of igneous complexes of Bushveld type follows this same line as far south as Trompsburg[39] (lat. 30° S.), a nearly straight line between the Zambezi and Orange rivers of some 1,500 km. The main axis of the rift system (Fig. 2) appears therefore, from present geological indications, to be a major N.N.E.–S.S.W. lineament (deep fault) of great age extending over a strike of some 5,500 km, diverted and split where penetrating pre-existing cratons representing the ancient core of Africa, and outlined by intrusive bodies which can be dated. This feature may be compared in strike with the great N.N.E.–S.S.W. fracture zones in the floor of the Indian Ocean, independent of the mid-ocean ridges, which have recently been described by Heezen and Tharp[40,41] and are schematically shown in Fig. 2.

The question of the age of the rift system can now be approached with more confidence in view of the recent synthesis of the geochronology of equatorial Africa by Cahen and Snelling[42] in which five geological cycles comprising syntectonic and post-tectonic episodes are described. An early cycle, sparsely represented by dates older than $3,000 \times 10^6$ yr, is followed by the Shamvaian Cycle (compare Superior Cycle[43,44]) dated $2,700-2,900 \times 10^6$ yr and recognized in East Africa. The term Uberdian is now firmly attached to a widely recorded cycle dated $2,150-1,650 \times 10^6$ yr (compare the Svecofennid-Hudsonian

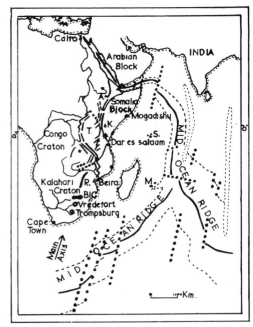

Fig. 2. Structural map illustrating main axis of rift system features of eastern Africa. Structure of Indian Ocean outlined after Heezen and Tharp. Black lines denote major fault scarps. Heavy dots denote dislocation zones. *A*, Addis Ababa; *K*, Mt. Kenya; *T*, Tanganyika Shield; *S*, Seychelles; *R*, Rhodesian Shield; *M*, Mauritius; *BIC*, Bushveld Igneous Complex.

cycle) in Tanzania (compare also ref. 54) and Uganda, where it coincides with the Buganda-Toro-Kibalian movements and granitization in the Western Rift (for example the Ruwenzori Mountains[45,46]). Table 1 shows that the chain of Bushveld type igneous complexes has given ages comparable with these early cycles. The latest investigation[50] of the Great Dyke of Rhodesia gives an age of $2,530-2,800 \times 10^6$ yr, corresponding to the Shamvaian Cycle, and two independent and very thorough studies[47,48] of the Bushveld Complex have given Ubendian ages. Evidence from the Ruwenzori Mountains (ref. 45, p. 263) placed the age of rift faulting tentatively between $2,500 \times 10^6$ and $3,000 \times 10^6$ yr, and so, relating this result with the age of the Great Dyke, it is here suggested that an extensive dislocation of Africa, originating the East African rift system, took place in the Shamvaian Cycle. This original disruption was followed by the event now termed Ubendian, which imposed its isotopic age pattern on many of the migmatites and granites associated with the older event.

I have proposed[7,45] that the rift system arose in these early Pre-Cambrian times from a right-lateral wrench as a result of a tendency for the African continent to drift to the north-north-east while its eastern margin (Somalia, Mozambique) either remained fixed or drifted to the south-south-west. This hypothesis was based on the movement pattern observed around the Tanganyika Shield, which was postulated to be a pre-Ubendian craton around which the rift system had split. Later work has produced supporting evidence for early movements of this nature, but it must be emphasized that recent mapping has confirmed that no appreciable horizontal displacement has occurred along the rift faults since Ubendian times. If indeed the original movements were of lateral slip nature, then it must be supposed either that the strain varied during the many later revivals, or that strike slip was impeded by the complication of the fault system along N.W.–S.E. and N.E.–S.W. lines, following the regmatic pattern[55,56], and also by widespread granitization leading to a sort of "grouting" along fault zones. Renewed activity was, in any event, manifested by vertical[12,21,57] tectonics, perhaps accompanied in depth by metamorphic or magmatic activity as a result of the conversion of dynamic to thermal energy.

Later Pre-Cambrian cycles (including syntectonic and post-tectonic phases) are defined by Cahen and Snelling[42] as Kibaran (Karagwe–Ankolean) from $1,290 \times 10^6$ to 850×10^6 yr, and Katangan–Mozambiquan from $1,100$ to 450×10^6 yr. These episodes began with extensive deposition in basins which locally followed the pattern of the rift system[9,42], although details are still under consideration, and were accompanied by extensive migmatization and granitization along the rift zones which are reflected in the radiometric updating of previously consolidated areas. Evidence of sedimentation in basins along the rift system is clear in Karroo times[4,16,37] and this cycle was closed by extensive vertical movements in the Jurassic which appear to have been manifested by arching on an immense scale (compare cymatogeny[58]) followed by the formation of grabens, perhaps on the lines proposed by Cloos[59]. After a period of planation the same pattern of events appears to have been followed in time and in space during the Alpine cycle, which culminated in the formation of the present rift valleys and single scarps in post-Miocene times[23].

At the present time the East African rift system is represented by rift valleys and fault scarps on a wide arch, elongated meridionally, forming the eastern high plateaus of Africa. This structure has been compared with the world-wide system of mid-ocean ridges with their central rift valleys[60,61], but King[22] has pointed out fundamental geophysical and geological differences between these two systems, and it must be emphasized that the mid-ocean ridges are exclusively in oceanic crust and associated with geologically rapid spreading accompanied by the formation of new oceanic crust by the upwelling of basic magma[62-65]. The East African system, on the contrary, is confined to continental crust, and shows no spreading or formation of new crust. Any distension arising from graben faulting is slight because of the steepness and commonly antithetic nature of the fault planes: although many faults dip 50°–70° towards the rift, vertical faulting is more widespread than has been generally assumed following Dixey[5], but reverse faults are exceptional. There is a sharp contrast between the dyke-filled extension fractures of Iceland[66], regarded as characteristic of a mid-ocean ridge, and the structure of a typical continental rift valley. The marked decoupling between structures in the oceanic crust and on the adjacent continent which is seen on the Pacific coast of North America[68] is also significant.

The physiographic and seismic characters of the East African rift system and the Red Sea–Gulf of Aden sectors of the world-wide mid-ocean ridges[63,67] are remarkably similar, however, and the close conjunction of the two systems in Ethiopia is a point of great importance and invites further detailed investigation.

It is intended to exclude theoretical considerations from this article so far as possible, but it may be suggested that the repeated rejuvenation along similar lines of movements which originated in early Pre-Cambrian times is of some significance and suggests that Africa has remained frozen to the higher layers of the upper mantle, and that the main N.N.E.–S.S.W. axis of the rift system reflects a deep-seated mantle lineament of great age. Beloussov[69,70] has pointed out that the survival of regional structural characteristics over many geological periods reflects a permanent relationship between crust and upper mantle. In view of

Table 1. RADIOMETRIC DATES ALONG THE EAST AFRICAN RIFT SYSTEM

Formation	Reference	Rock and stratigraphic position	Locality	Sample	Method	Age (10^6 yr)
Rusizi-Ubendian Belt	42	Kate Granite emplaced in Ubendian gneisses	Abercorn Dist., Zambia. 9° S.; 31·5° E.	Biotite	K : Ar	$1,725 \pm 70$
	42	Quartz-muscovite vein cutting metamorphosed sediments	Ditto	Muscovite	K : Ar	$1,800 \pm 70$
	42	Greisen associated with post-orogenic granite intruding Ubendian gneisses	Chunya Dist., Tanzania. 8° S.; 33° E.	Muscovite	K : Ar	$1,800 \pm 70$
	42	Post-Ubendian mineralization. Average of three galenas	Mpanda Dist., Tanzania. 6·5° S.; 31° E.	Galena	Common lead	$1,658$
	42	Pegmatite cutting Rusizian	Uvira, Kivu Prov., Rep. of Congo. 3° S.; 29° E.	Microcline	Rb : Sr	$1,820 \pm 240$
Buganda-Toro Belt	42	Andalusite-mica schist, Stuhlmann Pass Series, Buganda-Toro-Kibalian belt	Luzilubu R. Valley, Ruwenzori Mtns. Uganda. 0°; 30° E.	Biotite	K : Ar	$1,820 \pm 60$
Tanganyika Shield	54	Dolerite from N.E.–S.W. dyke swarm intruding gneisses of Tanganyika Shield near Gregory Rift Valley	Near Lake Evasi, Tanzania. 3° 58′ S.; 35° 10′ E.	Pyroxene / Plagioclase	K : Ar / K : Ar	$1,630 \pm 80$ / $1,730 \pm 80$
Bushveld Igneous Complex	48	Age assigned on basis of four independent investigations on minerals from granites, mafic rocks and alluvials	Bushveld Igneous Complex, South Africa, between 25–29° S. and 23–25° E.	Micas / Monazite / Zircon / Galena	Rb : Sr / U : Pb / Th : Pb / U : Pb / Common lead	$1,950 \pm 150$
	47	Red granite intruding mafic rocks of Bushveld Igneous Complex. Average of four samples	Four localities north of Pretoria, South Africa	Whole rock	Rb : Sr	$1,800$*
Great Dyke of Rhodesia	50	Picrite from Wedza, Selukwe and Hartley igneous complexes	Three localities on Great Dyke, Rhodesia	Whole rock / Biotite / Plagioclase	K : Ar / Rb : Sr	Min. $2,530 \pm 30$ / Max. $2,800$

* Value adjusted following Faure et al., ref. 49.

the recent confirmation of spreading from the centre of the mid-ocean ridges[64,65], the reality of continental drift becomes more and more widely accepted, but the historical persistence of structural features in the crust would support the opinion of Heezen[71] that this continental separation takes place by expansion of the earth[6,44], combined with the formation of new oceanic crust, rather than by the drift of the crust over the mantle. The extent of Cainozoic volcanism in the northern sector of the system may be allied to the development of the Gulf of Aden–Red Sea oceanic-type rifts during the Alpine orogenic cycle[67].

I thank P. E. Kent for critical comments, N. J. Snelling for advice on radiometric ages, and J. V. Hepworth and J. B. Kennerley for unpublished information on the rift system.

[1] Heezen, B. C., *Sci. Amer.*, **205**, 99 (1960).
[2] Intern. *Upper Mantle Project, Rep. No.* 1 (1964). *Rep. No.* 2 (1965), *Rep. No.* 3 (1965) (Secretariat Upper Mantle Committee, University of California, Los Angeles).
[3] *East African Rift System*, Upper Mantle Committee, Unesco Seminar, University College, Nairobi (1965).
[4] Cahen, L., *Géologie du Congo Belge* (Imprimerie H. Vaillant-Carmanne, Liège, 1954).
[5] Dixey, F., *The East African Rift System*, *Overseas Geol. Min. Res. Bull.*, suppl. 1 (HMSO, London, 1956).
[6] Holmes, A., *Principles of Physical Geology* (Thomas Nelson, London, 1965).
[7] McConnell, R. B., *Rep. Eighteenth Sess. Intern. Geol. Congr.*, *London*, 1948, Pt. 14, 199 (1951).
[8] Dixey, F., *Colonial Geological Surveys*, 1947–56, *Overseas Geol. Min. Res. Bull.*, suppl. 2 (HMSO, London, 1957).
[9] *Research Institute of African Geology*, *Ann. Reps. Dept. Earth Sci.*, Univ. Leeds (1957–66).
[10] Gregory, J. W., *The Great Rift Valley* (John Murray, London, 1896).
[11] Suess, E., *Die Brüche des östlichen Afrika, in Beitr. Geol. Kennt. östl. Afrika, Denk. Akad. Wiss. Wien*, **58**, 555 (1891).
[12] Dixey, F., *Rep.* 20 *Intern. Geol. Congr. Mexico*, 1956, *Assoc. Afr. Geol. Surveys*, 359 (1959).
[13] Harpum, J. R., *C.R. Réun. Assoc. Serv. Geol. Africains*, Nairobi, 1954, 169 (1955).
[14] McConnell, R. B., *Geol. Surv. Tanganyika, Bull.*, 19 (1950).
[15] McConnell, R. B., *Geol. Surv. Tanganyika, Short Pap.*, 27 (1947).
[16] Dixey, F., *Quart. J. Geol. Soc. Lond.*, **102**, 339 (1946).
[17] Cooke, H. B. S., *Bull. Geol. Soc. S. Afr.*, **60**, Annex. (1957).
[18] Bishop, W. W., in *Essays in Geomorphology* (edit. by Dury, G. H.), 139 (Heinemann, London, 1966).
[19] Gregory, J. W., *The Rift Valleys and Geology of East Africa* (Seeley, Service, London, 1921).
[20] Pulfrey, W., *Geol. Surv. Kenya, Bull.*, 2 (1960).
[21] Shackleton, R. M., *Quart. J. Geol. Soc. London.*, **106**, 345 (1951).
[22] King, B. C., *Proc. Geol. Soc. Lond.*, No. 1629, 16, March 23, 1966.
[23] Baker, B. H., in *East African Rift System, Rept. UNESCO Sem.*, *Univ. Coll.*, *Nairobi*, Pt. 2, 1 (1965).
[24] Williams, L. A. J., in *East African Rift System, Rep. UNESCO Sem.*, *Univ. Coll.*, *Nairobi*, Pt. 2, 33 (1965).
[25] Wilcockson, W. H., *Adv. Sci.*, **21**, 400 (1964).
[26] Hepworth, J. V., *Geol. Surv. Uganda, Rep.* 10 (1964).
[27] Swardt, A. M. J. de, *et al.*, *Geol. Soc. Amer. Bull.*, **76**, 89 (1965).

[28] Baker, B. H., Macdonald, R., Mohr, P. A., and Pallister, J. W., in *East African Rift System, Rep. UNESCO Sem.*, *Univ. Coll.*, *Nairobi*, Pt. 2, 115 (1965).
[29] Hepworth, J. V., and Macdonald, R., *Nature*, **210**, 726 (1966).
[30] Bloomfield, K., *Nature*, **211**, 612 (1966).
[31] Kennerley, J. B., *Fourth Symp. on African Geology*, Sheffield, 1967 (unpublished).
[32] Wayland, E. J., *Rep.* 15 *Sess. Intern. Geol. Cong.*, *S. Africa*, 1929, **2**, 323 (1930).
[33] Willis, B., *East African Plateaus and Rift Valleys*, *Carnegie Inst. Washington, Publ.* 470 (1936).
[34] Bullard, E. C., *Phil. Trans. Roy. Soc.*, A, **235**, 445 (1936).
[35] Sutton, J., Watson, J., and James, T. C., *Geol. Surv. Tanganyika, Bull.*, 22 (1954).
[36] Sutton, J., and Watson, J., *Journ. Geol.*, **67**, 1 (1959).
[37] Vail, J. R., *Fourth Symp. on African Geology*, Sheffield, 1967 (unpublished).
[38] Cousins, C. A., *Trans. Geol. Soc.*, *S. Afr.*, **62**, 179 (1959).
[39] Ortlepp, R. J., *Trans. Geol. Soc.*, *S. Afr.*, **62**, 33 (1959).
[40] Heezen, B. C., and Tharp, M., *Physiographic Diagram of the Indian Ocean*, *Geol. Soc. Amer.*, *Spec. Publ.* (1965).
[41] Heezen, B. C., and Tharp, M., *Phil. Trans. Roy. Soc.*, A, **258**. 90 (1965).
[42] Cahen, L., and Snelling, N. J., *The Geochronology of Equatorial Africa* (North Holland Publishing Co., Ltd., Amsterdam, 1966).
[43] Sutton, J., *Nature*, **198**, 731 (1963).
[44] Dearnley, R., in *Physics and Chemistry of the Earth* (edit. by Ahrens, L. H., *et al.*), **7**, 3 (Pergamon Press, Ltd., Oxford, 1966).
[45] McConnell, R. B., *Overseas Geol. and Min. Res.*, **7**, 245 (1959).
[46] Freeman, R. P., *Third Symp. on African Geology*, Tervuren, Belgium, 1966 (unpublished abstract).
[47] Schreiner, G. D. L., *Proc. Roy. Soc.*, A, **245**, 112 (1958).
[48] Nicolaysen, L. O., *et al.*, *Trans. Geol. Soc. S. Afr.*, **59**, 137 (1958).
[49] Faure, G., *et al.*, *Nature*, **200**, 769 (1963).
[50] Allsopp, H. L., *J. Geophys. Res.*, **70**, 977 (1965).
[51] Harpum, J. R., *Map Sheet Kipengere*, Sheet No. 260, 1 : 125,000 Series, Geol. Surv. Tanganyika (1958).
[52] Bagnall, P. S., *Map Sheet North Pare*, Sheet No. 73, 1 : 125,000 Series, Geol. Surv. Tanganyika (1960).
[53] Bagnall, P. S., *Rec. Geol. Surv. Tanganyika*, **10**, 7 (1963).
[54] Snelling, N. J., in *Inst. Geol. Sci.*, *London*, *Ann. Rep.* 1966 (in the press).
[55] Moody, J. D., *Tectonophysics*, **3**, 479 (1966).
[56] Brock, B. B., in *The World Rift System, Rep. Int. Upper Mantle Project Symp. Ottawa*, 1965, *Geol. Surv. Canada, Paper* 66–14, 99 (1966).
[57] Brock, B. B., *Rep.* 19 *Sess. Intern. Geol. Congr.*, *Algiers*, 1952 (1953).
[58] King, L. C., *The Morphology of the Earth* (Hapner Publ. Co., New York, 1962).
[59] Cloos, Hans, *Geol. Rund.*, **30**, 405 (1939).
[60] Heezen, B. C., in *Continental Drift* (edit. by Runcorn, S. K.), 235 (Academic Press, New York and London, 1962).
[61] Wilson, J. T., *Nature*, **198**, 925 (1963).
[62] Dietz, R. S., *Nature*, **190**, 854 (1961).
[63] Drake, C. L., and Girdler, R. W., *Geophys. J. Roy. Astron. Soc.*, **8**, 473 (1964).
[64] Vine, F. J., and Matthews, D. H., *Nature*, **199**, 947 (1963).
[65] Heirtzler, J. R., and Pitman III, W. C., *Science*, **154**, 1164 (1966).
[66] Walker, G. P. L., *Phil. Trans. Roy. Soc. Lond.*, A, **258**, 199 (1966).
[67] Laughton, A. S., in *The World Rift System, Rep. Intern. Upper Mantle Project Symp. Ottawa*, 1965, *Geol. Surv. Canada, Paper* 66–14, 78 (1966).
[68] Menard, H. W., *Marine Geology of the Pacific* (McGraw-Hill Book Co., 1964).
[69] Beloussov, V. V., *Quart. J. Geol. Soc.*, *Lond*, **122**, 293 (1966).
[70] Beloussov, V. V., *Sci. Jour.*, *Lond.*, **3**, 56 (1967).
[71] Heezen, B. C., in *Continental Drift* (edit. by Runcorn, S. K.), 235 (Academic Press, New York and London, 1962).

6

Reprinted from pp. 51–52 of *Summary of the Geology of Tanzania. V: Structure and Geotectonics of the Precambrian*, Mineral Resources Division, Dodoma, Tanzania, 1970, 58p.

SUMMARY OF THE GEOLOGY OF TANZANIA. V: STRUCTURE AND GEOTECTONICS OF THE PRECAMBRIAN

J. R. Harpum

[*Editor's Note:* In the original, material precedes this excerpt.]

THE RELATIONSHIP BETWEEN PRECAMBRIAN AND RIFT STRUCTURES

Although the present study is directly concerned with Precambrian (or more correctly pre-Karroo) structures, the accompanying structural overlay shows the major pattern of faults associated with the Rift system. It is at once obvious that there is marked parallelism between the Rift fault and certain of the Precambrian trends. This is only a general observation but is a phenomenon that has been commented upon by several writers (McConnell, 1951; Shackleton, 1951; Dixey, 1956; Pallister and Hepworth, 1956; Henderson, 1960a, 1960b). The Rift system, therefore, apparently lies above a fundamental geofracture or zone of prolonged activity in the mantle.

The East African system is a cratonic structural element and completely non-orogenic in nature. It does not, therefore, correspond to those graben (such as Lake Titicaca in the Andes and the Midland Valley of Scotland in the Caledonian fold-belt) that are produced in a mountain chain during its late-stage morphogenic uplift. It is true that it may represent renewed movement within such graben (as does the Midland Valley of Scotland; Kennedy, 1958), but such movement is unrelated to orogenic mountain-building. The East African Rift system is assumed to be a direct consequence of the uplift of the East African cratonic swell (Quennell, 1960). During pre-Karroo times this type of epeirogenic activity probably became at all important only towards the close of the Bukoban Period. Early forces were predominantly tangential. True rift conditions, therefore, only came into being after the main episode of Bukoban tectonism but before Karroo deposition.

The parallelism of Rift and Precambrian structures does not prove that rifting (in the cratonic non-orogenic sense) began in the Precambrian; it merely indicates that rifting has sometimes taken advantage of and followed zones of weakness established by Precambrian tangential forces. These forces were mainly orogenic in nature and produced by somewhat different conditions in the mantle from those giving rise to epeirogenic uplift on the scale of the East African swell. The *pattern* of the Rift was established in Precambrian times, but the *style* of its activity was established only towards the end of the Bukoban Period. The effects of Rift faults on the style of Precambrian sedimentation, volcanism and folding are almost negligible. These effects are confined to tilting and *immediately adjacent* to the faults there may be sharp up- or down-turning of structures. The overall style of these events was established, however, before rifting began, and to some extent they controlled and directed the zones along which Rift stresses were relieved.

The second major misconception concerns the parallelism of Rift faulting with Precambrian trends. In general, Rift trends are parallel to structures established in Early-Proterozoic times and possibly reactivated in Late-Proterozoic times. Occasionally the structures may be primarily Late-Proterozoic or even younger in origin. In the main, however, the main Rift pattern can be said to have been established in Early-Proterozoic times. The northern end of the Nyasa Rift, for example, is closely linked to Ukingan structures, and it is not improbable that the present Nyasa Rift Valley is located in an old orogenic graben (of the same type as Lake Titicaca) developed during the Ukingan mountain-building episode.

When parallelism of Rift structures to Archaean structures is mentioned this can be misleading. If there is parallelism it is probably because Proterozoic activity is also parallel to Archaean
trends. Archaean structures, particularly in the Usagaran zone, tend to be recumbent and sub-
horizontal, while Rift structures are often vertical. Any parallelism of trend here may be fortuitous,
or may be related to *ac*-joint patterns rather than to the disposition of penetrative *s*-planes.

[*Editor's Note:* Only references cited in this excerpt are reprinted here.]

REFERENCES

Dixey, F., 1956, The East African rift system, *Colon. Geol. min. Resour.,
Bul. Suppl. I,* 71 pages.

Henderson, G., 1960a, Air-photo lineaments in Mpanda area, Western
Province, Tanganyika, Africa, *Bull. Amer. Assoc. Pet. Geol.,* Vol. 44,
pp. 53–71.

Henderson, G., 1960b, The geology of the Bukoban-type rocks of the Kigoma
and Mpanda Districts, Western Province, Tanganyika, *Trans. geol.
Soc. S. Afr.,* Vol. 63, pp. 11–44.

Kennedy, W. Q., 1958, The tectonic evolution of the Midland Valley of
Scotland, *Trans. geol. Soc. Glasgow,* Vol. 23, pp. 106–133.

McConnell, R. B., 1951, Rift and shield structure in East Africa, *Rep. Int.
geol. Congr.,* 18th Sess., Gt. Brit., Pt. XIV, pp. 199–207.

Pallister, J. W., and Hepworth, J. V., 1956, Notes on mylonite and rift faulting
in Uganda, *Proc. East-Central reg. Comm. Geol. C.C.T.A.,* 1st Meeting,
Dar es Salaam, pp. 95–97.

Quennell, A. M., 1960, The Rift System and the East African Swell, *Proc.
geol. Soc. Lond.,* No. 1579, pp. 78–86.

Shackleton, R. M., 1951, A contribution to the geology of the Kavirondo Rift
Valley, *Quart. J. geol. Soc. Lond.,* Vol. CVI, pp. 345–392.

7A

Reprinted from *Third Earth Resources Technology Satellite-1 Symposium, Vol. I: Technical Presentations*, S. C. Freden, E. P. Mercanti, and M. A. Becker, eds., National Aeronautics and Space Administration, Washington, D. C., 1974, pp. 767-782

STRUCTURAL GEOLOGY OF THE AFRICAN RIFT SYSTEM: SUMMARY OF NEW DATA FROM ERTS-1 IMAGERY

P. A. Mohr, *Smithsonian Astrophysical Observatory, Cambridge, Massachusetts*

INTRODUCTION

Unified mapping of regional structures of the African rift system (Figure 1) on a scale of 1:1 million has been completed except for a few persistently cloud-covered areas. The mapping has been done by direct tracing from 18-cm square black-and-white prints (spectral bands 5 and 7). Incorrect indication of coordinates on some images has been adjusted for using the USAF Operational Navigation Charts, and distortion from a precise 1:1 million scale has been allowed for by arithmetical scaling.

The results of the mapping have been charted on fifteen 100 x 50cm sheets, and they include Precambrian strike, Cainozoic faults, and lineaments of an uncertain nature. In some areas, young lava shields, calderas, cones and craters, and lithological boundaries have additionally been recorded. The area mapped covers nearly 5 million sq.km, and obviously it has been impossible to give full attention to fine detail, fascinating though this can be on the best ERTS images.

River courses and lake shorelines in eastern Africa are accurately seen for the first time, though due to uncertainties in satellite position the precise coordinates of mapped features can be subject to as much as 2-3 minutes of arc error in some areas.

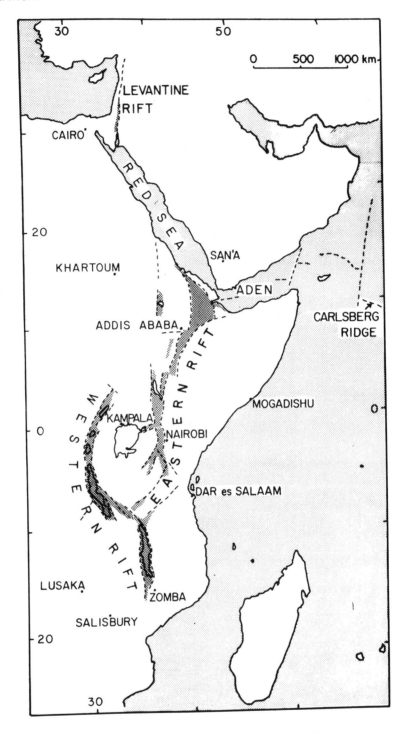

Figure 1.   The East African rift system (rift valleys stippled).

This paper presents a summary of information interpreted from the mapped sheets.   The sheets themselves will be published by Smithsonian Astrophysical Observatory in the final report on this work.   In lieu of these sheets here, the reader should seek geographical assistance from the maps in Baker et al. (1972) and McConnell (1972).

METHOD OF MAPPING AND GROUND CONTROL

Although detailed mapping and study of the Precambrian fold-belts of eastern Africa falls outside the scope of the present work, an important question concerning the African rift valleys is the degree of parallelism of rift faults to pre-existing structures. Therefore, trends of strike-lines in exposed Precambrian terrain have been mapped from the ERTS imagery where these lines are clearly expressed.   The problem is complicated by the fact that strike-lines are best observed in regions of appreciable topographic relief where the degree of lateritic palaeosoil cover is minimised.   Instances of an apparent transition from strongly to weakly expressed Precambrian strike features, for example in Katanga and northern Ethiopia, can be related to the degree of soil cover and not to any change in structural style.

Precambrian strike-lines identified on ERTS imagery usually denote lithological strike of tilted or folded strata, particularly for younger Precambrian rocks such as comprise the Bukoban System of northwestern Tanzania, the Karagwe-Ankolean of Burundi, the Lufilian and Irumide belts of northern Zambia, or the Precambrian sequence of northern Ethiopia.   This type of strike-line is characteristically curvilinear and frequently reveals broad synclinoria/anticlinoria.   Where deformation and metamorphism have been intense, lithological strike is obliterated or at least obscured to the extent that the identified strike-lines can be said to represent a new, refoliated grain.   This is notably the case in zones of deep-seated cataclasis, known from ground surveys, as for example in the Rukwa rift and the central sector of the Lake Tanganyika rift, and in the Aswa Mylonite Zone of southern Sudan against which the Western Rift terminates.   This type of strike-line is characteristically precisely linear over distances of the order of 100km, and when followed by Cainozoic rift faulting the two types of structure can become virtually indistinguishable on the ERTS imagery.

Boundaries between different, regional lithological types in Precambrian terrain, though in some places clearly expressed on the ERTS imagery, have not been mapped in the present work.   Some circular structures, including ring-intrusions, have however been included as possibly marking otherwise masked lines of ancient crustal weakness.

The faulting which gives form to the present-day rift valleys of eastern Africa is generally recognised to be of Pliocene-Quaternary age (Baker et al., 1972).   The topographical expression of the faults is therefore usually sharp, and because of this they are easily detected on the ERTS imagery. This is especially so for east-upthrown faults, which cast a sun-shadow at the time of imaging. Northwest-upthrown faults can sometimes be detected by illumination of their scarp, showing as a bright strip on the imagery, but where the scarp is somewhat denuded then the modest increase in brightness can be lost amongst background variations. It cannot be denied, therefore, that in regions where ground-truth is lacking there may be a bias favouring representation of the eastern boundary faults of graben (and western boundary faults of horst) on the author's maps.

The direction of throw on rift faults can be detected not only from sun-shadow or illumination, but frequently also from identi-fication of the drainage pattern. The darker tones of afforest-ation are not a reliable guide: in some semi-desert areas, forest occurs along the foot of fault-scarps, presumably where there is near-surface groundwater. In some hilly terrain, forests are denser above the scarps because prevalent mountain mists foster a cold, rain forest. Complete afforestation makes the detection of faults very difficult, even where strong faults are seen to enter such an area.

Evidence for large-scale transcurrent faulting in the African rift system has not been found in this study, and it is probably best that this phenomenon be sought initially from ground and aerial photographic surveys. Likewise, no thrust-faulting has been identified in the rift system; the Precambrian Nandi fault of western Kenya (Sanders, 1965) is remarked on the ERTS imagery, but its nature, controverted even on the basis of ground studies, is not revealéd thereon. ERTS imagery is a suitable tool for determining the nature of regionål faults only insofar as the linearity of the fault trace can be related to topography and lithological strike. Even here some ambiguity can arise, as shown by the fact that regional rift faults with steep hade can show an almost perfectly linear trace whereas small, short faults of the graben floor, also with near-vertical hade, yet have a characteristically sinuous trace.

The third class of structural feature identified in this work is covered by the non-generic, descriptive term of lineament. Lineaments usually have the form of a linear arrangement of topo-graphic features, either continuous or interrupted, without clear evidence of crustal displacement that defines a fault. Many of them trend at a large angle to the main rift faulting. Where Precambrian structures are largely masked by superficial cover, their vague and intermittent surface expression can form a type of lineament; but lineaments as mapped in this study can also cut across the strike of well-exposed Precambrian terrain, and could represent regional jointing or an overprinted or secondary structural trend. Lineaments are distinctly rare in the Mesozoic-Eocene sedimentary basins of the Horn of Africa, perhaps because of the weakly consolidated nature and low rigidity of the rocks concerned, but in the Cainozoic volcanic fields can be common.   Where a volcanic field is only a thin pile of lavas, the structural trend from adjacent Precambrian terrain can, as in central-north Ethiopia, be traced as lineaments into the volcanic field;   but other lineaments in such fields can be discordant both to the nearest exposed Precambrian structures and rift faults. Indeed, the latter form the most characteristic type of lineament, whose trace can extend, though discontinuous or offset, for several hundred kilometres.

A not-yet-understood property of the ERTS imagery is its differential emphasis of structural features according to the overall tone of an image.   Some darker images emphasise linear faults and lineaments, whereas lighter, more contrasted images bring out faults with large topographic displacements.   Differential emphasis also comes, of course, from the different spectral bands employed by the ERTS-1 satellite, and in the present work it has been consistently noted that band 5 (6000-7000Å) best brings out geological structures, except where the overall tone is dark when band 7 (8000-11000Å) yields a lighter and more contrasted image. Band 7 images are also preferred for regions of extensive vegetation cover, and for regions subject to partial cloud-cover.

Ground survey data have been referred to throughout the author's mapping as a check against doubtful features expressed on the ERTS imagery.   Whilst a few structural lines were recognised on this imagery only subsequent to a search originating from knowledge of ground data, it must be emphatically pointed out that in all cases the mapping has proceeded first from the images and then been checked against ground data.   In this process, for example, the direction of fault-throw, or the manner of bifurcation or intersection of faults, has occasionally required correction from closer re-scrutiny of the images.   In no case whatsoever has the author indicated on his maps structures that cannot be discerned on the ERTS imagery, even though there be an important feature (according to ground surveys) which is actually present.   Therefore it will be discovered that the new ERTS-based maps lack some structural data shown on geological maps, and add new data previously unsuspected.   The ERTS-based maps add to, and do not supersede existing structural maps.

Owing to the present lack of cloud-free imagery, no mapping has yet been possible of the Western Rift between latitudes 1°N and 3°S, nor the southern end of the main Ethiopian rift, nor of the Kilimanjaro area.

RESULTS

Relation of rift faulting to Precambrian structures

This knotty and persistently controversial topic basically concerns two alternative interpretations: 1. the pattern of the

rift faults is a faithful reflection of major Precambrian structures, owing (to an extent which is disputed) to continuity of a
single tectonic style from the Precambrian up till the present.
The principal proponent of this viewpoint has been McConnell
(1972, and numerous other papers referred to therein).
2. the rift faults have taken advantage of the pre-existing Precambrian tectonic 'grain', but the overall pattern of the rift
system has not been determined by this grain but by a fortuitously
superimposed, completely different structural event.   This viewpoint has been advanced by, amongst others, King (1970) and Baker
et al. (1972).

The ERTS imagery provides, for the first time, an overview of the
entire problem, and indeed this is the sort of task to which
satellite imagery is well-suited.   Precambrian rocks are extens-
ively exposed in proximity to the African rift system, the centres
of the Ethiopian and Kenyan swells excepted, and along much of the
Western Rift the rift faults themselves occur in exposed Precambrian
terrain.

It immediately has to be admitted that, on a regional scale, the
frequent coincidence of Precambrian structures and rift faults is
uncanny.   The Eastern Rift, from Tanzania to Ethiopia, essentially
follows the meridional Mozambique Belt, whose final activity is
dated at 550 ± 100 m.y. (Clifford, 1970; Shackleton, 1967).   The
Western Rift commences in the south (L. Malawi rift) by following
the Mozambique Belt, but then turns northwestwards along the Ubendian (1850 ± 250 m.y.) deformation zone before bending round to the
north-northeast along the Kibaran (1100 ± 200 m.y.) deformation
zone (Clifford, 1970).   The Western Rift abruptly terminates in
the north against a cross-cutting Precambrian structure, the NW-
trending Aswa Mylonite Zone (Whiteman, 1971).   The essentially
upfaulted crustal blocks between the rift zones in eastern Africa
are formed of cratonic nuclei, typically older than about 2500
m.y.

Some specific regions mapped from the ERTS imagery can now be
discussed.   In Yemen, major ENE-NE trending fracture zones in
the Precambrian appear to have been reactiviated by tectonism
accompanying the opening of the Red Sea and Gulf of Aden, and some
were sites for Cainozoic basaltic volcanism (Mohr, 1972b).   However, typical Precambrian terrain in eastern and northern Yemen
takes the form of a curving mesh of strike-lines with interspersed
granitic intrusive bodies: young faults in this terrain are
rarely identifiable, but where they are they usually run markedly
oblique to the Preambrian strike.   In fact, non-parallelism of
Precambrian and rift structures in Yemen is hardly surprising in
view of the proximity of an RRR-triple junction at Afar, unless
this coincides with a Precambrian nexus.   South of the Gulf of
Aden, in northern Somali, non-parallelism is general (Mohr, 1962;
Beydoun, 1970).

In Ethiopia, Precambrian zones of possible cataclastic deformation

(Mohr, 1972b) trend NNE from western Tigray province into Eritrea, where they turn to a meridional trend with several bifurcations and reunions.   It is difficult to allow these structures as being parallel to the adjacent Red Sea basin.   Central Tigray province exposes broad synclinoria and anticlinoria trending NNE-NE (Beyth, 1972):   this regional Precambrian trend is slightly but definitely oblique to the Cainozoic margin structures of western Afar further east (Mohr, 1972a, 1973), and to the Tana graben further south. In northern Ethiopia and eastern Sudan, strongly expressed E-ESE Precambrian lineaments are not related to any known rift structures.

In central Ethiopia the Precambrian is masked by thick volcanic cover, but east of the main Ethiopian rift, along meridian 39°E, persistent and rather linear Precambrian structures (Chater and Gilboy, 1970) trend N-S, oblique to the NNE strike of Ethiopian rift faulting (But N-S faulting determines the Amaro horst and Galana graben at the southern end of the rift - see Levitte et al. 1974).   Strongly discordant angles separate the meridional Precambrian strike of the Kenya-Ethiopia border region from superimposed NW-SE Cainozoic faulting, though numerous small basaltic centres appear to be aligned at intersections along the Precambrian trend.

East of the Gregory rift, in north-central Kenya between latitudes 2°N and 0°, the detailed ground-mapping of Baker (1963) is fully confirmed.   The Precambrian strike here trends N-NNW, whereas the rift faulting immediately to the west trends NNE.   This might seem to be a clear case of non-parallelism, yet the Precambrian structures project northwards into the precise alignment of the Mt Kulal volcanic range and related rift structures, east of Lake Rudolf.   To the south, the same ancient structures pass via the Pleistocene volcanic centre of Mt Kenya (Baker, 1967) to the Cainozoic volcanics and faulting of the Yatta plateau.   If the Gregory rift itself is following a 'discordant' NNE Precambrian trend, this is of course hidden by young volcanics and sediments. However, this possibility is rendered less likely by the fact that, west of Lake Rudolf, the western margin structures of the rift are precisely parallel to again NNW Precambrian strike-lines, though the rift structures are notably more linear than are the Precambrian ones (Mohr, 1972b).

The Western Rift is often regarded as the example par excellence of a rift valley determined by pre-existing structures, and a description of the polyphase Precambrian tectonism of the region and its influence on the Cainozoic rift faulting has been given by McConnell (1972).   The Western Rift meets important NE-SW faulting at a structural node at latitude 9°S, longitude 33½°E; here the Rift trends NW-SE and its unusually linear faulted margins are strongly developed in the southern L. Tanganyika-L. Rukwa -northern L. Malawi sector.   This strong expression and linearity are matched in the precisely parallel Ubendide structures the cataclastic character of which emphasises a zone of intense crustal deformation separating two cratonic nuclei (Sutton and Watson, 1959; Brown, 1962; McConnell, 1972). By contrast, the NE-trending Cainozoic faulting, expressed both northeast of the node (Usangu and Fufu faults) and to the southwest (Luangwa valley faults), are seen on the ERTS imagery to be superimposed on mild, NE-trending Irumide structures that are obliterated by the ostensibly older Ubendide structures at the node itself. At any rate, the pre- dominant Precambrian trend is also the predominant Cainozoic fault trend here. Directly to the west, NE-SW faulting of the Upemba and Mweru half-graben is generally close to parallelism with struc- tures of the relatively mildly deformed Kibaride belt (Cahen, 1970); however, no important Cainozoic faults are revealed along the impos- ing structural arc of the end-Precambrian Katangides (Cahen, 1970).

The L. Tanganyika rift, roughly 70km wide, is at first glance strongly controlled by the NW-SE Ubendide structures. In fact this only holds for the central sector of the rift; at latitude 6°S the rift trough is offset dextrally and thus, proceeding north- wards, the rift escapes from the influence of the Ubendide struc- tures continuing west of the lake, and follows the meridional trend of the related Rusizian structures though with some influence also from the Kibaride belt. Again, this younger Precambrian belt is largely obliterated where it crosses the older Rusizides (see McConnell, 1972, for a discussion on this problem). Similarly, proceeding southwards from the central sector of the L. Tanganyika rift, the Ubendide structures run straight, southeast across moun- tainous terrain to the L. Rukwa rift. But the L. Tanganyika rift itself, maintaining its normal width, extends SSE-wards until ter- minating in a region of Cainozoic cross-faulting.

ERTS imagery of the L. Tanganyika region therefore suggests two key points for an understanding of the Precambrian-Cainozoic struc- tural relations problem. First, where Precambrian crustal defor- mation has been most intense, rift faulting is also more strongly developed. It is as though the Precambrian cataclastic zones provided trans-crustal sutures which became planes of weakness when tensional strain accumulated during the Cainozoic. Second, the L. Tanganyika trough persists in its curving arc, even where forced to make a lateral offset, in defiance of the linear Ubendide struc- tures: the rift here is not a straight NW-trending valley super- imposed on the length of the Ubendide belt.

In conclusion of this section, Precambrian structures have a power- ful local influence on the Cainozoic rift structures. On the regional scale revealed by the ERTS imagery, however, the rift valleys are seen to persist in trends that transgress the Precam- brian structures, suggesting that a new stress-field has been imposed with the generation of the rift valleys. Whether, on the sub-continental scale, the near parallelism of the African rift system and the Mozambique Belt is of fundamental structural signi- ficance cannot be answered here, but the question is of importance to possible relations of thermal plumes and plate motions.

The regional fault pattern

The fault pattern of the African rift system is known from ground and aerial surveys in varying degrees of detail for different sectors of the system. Ground surveys are fairly complete for the southern and central sectors of the Gregory rift, for the southern and western regions of the L. Malawi rift, and for Afar; they are only of reconnaissance type for much of the Western Rift and the Ethiopian rift, and are virtually non-existent for the graben occupied by lakes Tana, Stefanie and Mweru, and the Luangwa valley.

The great merit of the ERTS imagery is that, for the first time, it provides the means for a unified mapping of the major structures of the whole African rift system, formidable task though this is. As emphasised previously, this mapping is not a substitute for, but an adjunct to ground mapping. Nevertheless the satellite imagery enables a first interpretation to be made of unsurveyed regions in the light of what it reveals about well-surveyed regions. Furthermore, some significant additions and revisions are provided from ERTS imagery of even well-surveyed regions.

The fault pattern in Yemen is revealed by ERTS imagery to show a coastal zone of major warping and associated antithetic faults, both for the Red Sea and the Gulf of Aden (Gass et al., 1965). In the interior of the Yemen plateau, NE-ENE trending fracture zones have been identified. Other important faulting runs southeast across northern and eastern Yemen, parallel to the Red Sea structural trend north of its Yemeni, SSE-trend narrowing. These and other features have been discussed by Mohr (1972b).

The African rift system meets the Red Sea and Gulf of Aden at the Afar triple junction (Tazieff, 1970, 1972; Mohr, 1970, 1972a). Mapping of Afar margin structures from ERTS imagery has been described elsewhere (Mohr, 1973). Essentially, the western margin of Afar shows a gently sinuous plan which, in detail, reveals the powerful influence of NNW-trending 'Red Sea' structures. As the western margin runs close to meridian 40°E throughout its extent, this requires that the NNW-trending structures be offset dextrally. These offsets occur in sectors where the dominating structural trend in NNE; this trend parallels the Precambrian 'grain' west of northern Afar, and in the south occurs as a forceful extension of main Ethiopian rift faulting projecting right across southern Afar. Marginal warp-zones, and marginal graben associated with belts of antithetic faulting can be identified on the ERTS imagery.

The southern margin of Afar can be mapped accurately for the first time from the ERTS imagery. The western sector of the margin shows the influence of dextrally offset Ethiopian rift structures that turn off northwards into Afar (Mohr, 1973); this pattern changes abruptly near longitude 41°E where a Gulf of Aden trend is imposed, though this fades out south of the Aisha horst (Canuti

et al., 1972; Mohr, 1967, 1972a).   Further east, along the northern margin of the Somalian plateau, the dominant fault trend is
WNW and is especially strongly developed in the Asseh graben and
on the Cape Guardafui peninsula - the latter is revealed for the
first time on the ERTS imagery.   Inland, WNW faulting again
determines the Nogal and Darror graben (Azzaroli and Merla, 1957).

The SSW-trending faulting of southern Afar continues directly as
the main Ethiopian rift to almost latitude 5°N (Baker et al.,
1972; DiPaola, 1973).   The aerial photographic mapping of DiPaola
is largely confirmed and amplified by the ERTS imagery, which shows
however that the importance of dextral en-echelon offsets in the
rift has been exaggerated (eg. Mohr, 1967).   Only the axial Wonji
fault belt of the rift floor shows such offsets (Mohr, 1962), and
as in places this belt can show two parallel developments, and
elsewhere can be absent, the term en-echelon is a misnomer on the
regional scale.   Immediately west of the main Ethiopian rift, the
parallel structures of the Omo valley can be traced northwards via
the Guder valley to possibly as far as the Tana graben.   The
asymmetrically developed Tana graben occurs within the Ethiopian
plateau block, and its fault pattern has been accurately mapped
for the first time, using the ERTS imagery.

The regional link between the Ethiopian and Gregory (Kenyan) rifts
was poorly known until the advent of the ERTS imagery.   This
reveals that the faulting at the southern end of the Ethiopian rift,
and the Kino Sogo fault belt east of L. Rudolf, lie on a common
NNE-oriented alignment.   An unfaulted 'gap' of about 100km length
separates the terminal faults of the two rifts (note: to the west
of this region the Omovalley faulting passes through a broad zone
of horst-graben structures into the Lake Rudolf basin).   The
transition of the rift system across this gap was thought by Mohr
(1967) to be effected through the L. Stefanie graben via a dextral
offset;   but the imagery shows that this graben lies west of the
alignment, and is formed of NNW-trending faults crossed by secondary but important ENE-faulting.

The transverse faulting of the L. Stefanie graben projects as lineaments across the structural gap between the Ethiopian and Gregory
rifts.   In the opposite direction it projects across Turkana,
where lineaments line up with WSW-SW faulting on the eastern side
of the L. Albert rift.   Here is tentative evidence that the
Western Rift, though indubitably terminating against the Aswa
Mylonite Zone north of L. Albert, may persist through a minor
tectonic offshoot across to the Eastern Rift, though not, it must
be emphasised, as a graben.   The structural gap between the Ethiopian and Gregory rifts is likewise a gap to some NW-SE tectonism,
superimposed on a prominent, parallel Precambrian structure,
extending from Moyali to the Omo valley.

The regional structures of the northern half of the Gregory rift
(Baker et al., 1972) are shown in their unity on the ERTS imagery
as a magnificent, regular fanning-out northwards.   The rift widens
from 60km at latitude 0° (at the junction with the transverse

Kavirondo rift) to 130km at latitude 2°N, and to about 300km for
less clearly defined structural margins at latitude 4°N. Within
this structural widening, the weakly faulted L. Rudolf basin shows
a northward fanning-out into the southern Ethiopian plateau.
Two major points require emphasising here: firstly, that although
the northward widening of the Gregory rift structures coincides
with the topographic decline from the centre to the margin of the
Kenyan swell, it also coincides with a widening of the Precambrian
structures.   This raises a fundamental question: would swell
uplift in the Cainozoic have formed a continuous, quasi-midoceanic
ridge from Ethiopia to Tanzania if dissipation had not been enfor-
ced by a pre-existing structural pattern?   Second point: strongly
expressed rift structures do cross from the Kenyan to the Ethiopian
swell without apparent diminution in strength (N.B. the gap noted
above lies upon the fringe of the Ethiopian swell, and not between
the two swells), though there does seem to be a greater complexity
and less regularity to the structural pattern in the boundary
region.

At the conjunction of the ENE-trending Kavirondo rift with the
main Gregory rift, the latter undergoes an abrupt change from a
N-NNE structural trend in the north, to a NNW-NW trend in the
south.   The southern half of the Gregory rift comprises a majes-
tic curve, concave to the west, from the equator to latitude 4°S
where the structural trend has finally turned to NE-SW.   Characteristic platforms, not found in the Ethiopian or Western rifts,
occur along the eastern side of the Gregory rift (Baker et al.,
1972), and from the western margin several turn-off structures
are revealed on the ERTS imagery.   West of the rift, within the
L. Victoria block, several important structures trend ENE-NE:
these include the Kavirondo rift, the Uitimbara-Siria faults,
the Speke Bay graben, and the L. Eyasi faults at the southern
end of the Gregory rift.   If these tensional features are contemporaneous with the Gregory rift, what homogeneous stress-field
could have given rise to the overall fault pattern?   Has there
been a slight anti-clockwise rotation of the L. Victoria block?

Although the Gregory rift terminates southwards in a fanning-out
zone, this phenomenon is shown on the ERTS imagery to be much more
abrupt and less symmetrical than at the northern end of the rift
(Baker et al., 1972).   The main faulting, as mentioned above,
turns to a SW direction, and lineaments of this trend continue
across the Dodoman nucleus to the Western Rift.   Faults of S and
SE trend are relatively minor in the fanning zone, and the ERTS
imagery shows that there is no significant connexion of the Gregory rift with the Usangu-Fufu faults farther south (see also
Hepworth, 1972).   Therefore it seems misleading to speak of the
Eastern Rift as meeting the Western Rift at Mbeya: the Eastern
Rift has already terminated further to the north.

East of the Gregory rift, a NNW-trending belt of strongly deformed
Precambrian rocks extends from southern L. Rudolf, passes close to
Mt. Kenya and along the Yatta plateau, and reaches northeastern

111

Tanzania at the Pare-Usambara graben-horst structures (McConnell, 1972). Once again, the control of Precambrian 'grain' on in this case extra-rift tectonism is emphasised: the corollary of course is that there must be important structural sub-divisions to be made within the Mozambique Belt.

We now turn to the Western Rift. Near the northern end of the Western Rift, the powerful NE-trending faulting of the Lake Albert sector curves, proceeding northwards, to a NNE-trend and a weak graben is juxtaposed west from the northern end of the lake. The faults of this weak graben impinge upon, and are abruptly terminated by the Aswa Mylonite Zone, on the Uganda-Sudan border (Whiteman, 1971). No NE-trending faults are revealed by ERTS imagery to continue northeast of the Aswa Mylonite Zone, where rather the tectonic trend is NW-SE, related to the northward widening of the Gregory rift.

The Aswa Mylonite Zone is a strongly, almost violently expressed structure for which there is as yet little ground-survey information. It forms a narrow, linear zone of cataclasites, about 150km long, whose ends are splayed with a slight clockwise bias; the southern splay is closed, indicating the presence of a broad folded structure. Cainozoic faulting is almost certainly imposed upon the Mozambiquian Aswa structures (Almond, 1969), with upthrows to the southwest if interpretation from the ERTS imagery is correct; Almond (1969) also regards the zone as being one of sinistral shear, and this is compatible with the clockwise splay. Faults continue northwestwards for at least a further 200km from the Aswa zone. Lineaments are also observed to project southeastwards from the zone, and pass via the southern fringe of Mt Elgon to the Precambrian Nandi thrust fault (Sanders, 1965). However, this southeastward continuation of the Aswa zone across northern Uganda cannot be matched with the narrow, continuous belt of cataclasites indicated by Almond (1969), perhaps owing to the effect of variably masking soil cover on the ERTS imagery. It is interesting to note that the Aswa Mylonite Zone and its continuations form a major Precambrian structure that has not been utilised in the development of the African rift system: this gives further support to the view that the rift system is a new and distinct structural unit rather than an ongoing re-activation of ancient transcrustal sutures.

ERTS imagery of the Western Rift southwards from Lake Albert is largely cloud-covered, and no useful structural maps can be constructed for as far as the northern end of the L. Tanganyika rift. The important relationships between the faulting of this rift and the Precambrian structures have been discussed briefly in the preceding section. The northern part of the L. Tanganyika rift is a graben superimposed on the Rusizian trend (N-S), but with some intersecting or adjusting faults of NNE-NE, Karagwe-Ankolean trend immediately east of the rift. The central sector of the L. Tanganyika rift is faulted parallel to the NW-NNW Ubendide structures, though the rift trough itself jumps across these structures at latitude 6 S: thus the Kiyimbi horst west of the lake lies on the precise alignment of the Kungwe horst east of the lake, such that the rift is asymmetrically developed in one sense north of latitude 6°S, and in the opposite sense south of 6°S.   At latitude 7½°S there is a further, 35km dextral offset of the graben faulting, which continues SSE as far as latitude 9°S.   On the western side of the graben, however, at latitude 8½°S the marginal faulting is largely overruled by strong NE-SW faulting that extends southwest to the Mweru Wantipa graben-horst structures.

West of the L. Tanganyika rift, NE-ENE faulting, with predominant southeasterly upthrows, forms the Kabamba-Upemba and L. Mweru half-graben as well as more doubtful features north of the lower Lukuga river.   It is of interest that the same NE-ENE trend is manifested in graben faulting west of the Eastern Rift (see above), as well as in numerous lineaments between the two Rifts.   The termination of the L. Tanganyika rift in the south may be related to the presence of such cross-lineaments.

The L. Rukwa rift develops south of latitude 7°S, and runs about 130km east of the L. Tanganyika rift and with a SE trend that is significantly different from the SSE trend of the L. Tanganyika rift at the same latitudes.   The intervening terrain is also strongly faulted, with a prominent high horst.   The ERTS imagery reveals this faulting accurately for the first time.   The faults of both margins of the L. Rukwa rift are exceptionally long and linear, and follow Precambrian cataclastic zones (Brown, 1962) of the Ubendides.   The faulting of the L. Rukwa rift continues beyond the Mbeya node and into the L. Malawi (Nyasa) rift, and is particularly strongly developed on the northeastern side of the latter, in the Precambrian migmatites of the Livingston Mts. Mapping of the L. Malawi rift (Bloomfield, 1966) is not yet completed from ERTS imagery (Mohr, 1972b), and so discussion on the southern end and termination of the Western Rift is deferred.

In summary, the overall pattern of the African rift system as revealed from ERTS imagery suggests an incipient plate boundary struggling to express itself.   The diffuse complexity of the pattern perhaps stems from both the slow rate of extensional movement in thick continental crust (McKenzie et al., 1970), and from the influence of the pre-existing Precambrian structures.   It can be remarked that the Red Sea and Gulf of Aden cut straight enough and singly in their respective traverses across the Arabo-African continent, but their spreading rates are appreciably faster and there were no cratonic nuclei to deflect their initial paths. The Tanganyika block (Dodoman nucleus) has surely played an important role in the division of the Western from the Eastern Rift.

Other features

The ERTS imagery reveals numerous other features of interest to the structural geologist, the volcanologist, the glaciologist, the sedimentologist and the economic geologist.   Not all these features can be even briefly referred to here.

Noteworthy from the ERTS mapping is the persistent occurrence of ENE-trending lineaments.   As mentioned in the previous section, there are important extra-rift normal faults of this trend, branching from or intersecting with the western margins of both the Western and Eastern Rifts.   However, these faults tend to be curvilinear or even sinuous, whereas the lineaments, as their name implies, are linear.   Well then, are these lineaments real structural elements, or are they 'hallucinosutures' (Shackleton, 1973; Tazieff, 1973)?   Whilst the regular sun-angle obtaining during ERTS imaging might be expected to favour addiction to ENE-NE trending hallucinations, in fact there is sufficient variation of this angle with latitude and season that hallucinosutures might be expected to vary more in their trend than they actually do.   In this work, given lineaments have been identified regardless of season, though their intensity of expression may vary.   Also, although the writer once grossly exaggerated the importance of ENE-trending lineaments in Ethiopia (Mohr, 1967), yet there is sufficient ground-knowledge to relate some lineaments of this trend to linear faults, lines of warping, dykes, and possible fracture lines without significant displacement, such that their existence must be faced and not pre-diagnosed.   The lineaments, now precisely located from ERTS imagery, require to be examined carefully on the ground, though their structural significance will probably only be realised from a synthetic study of the fracture pattern of the African continent as a whole.

Calderas, volcanic craters and associated young lava fields are usually prominent on the ERTS imagery of eastern Africa.   Mapping from the imagery is revealing that at least some of the major volcanic centres are situated at tectonic nodes.   That this is the case for the Rungwe volcanoes at the Mbeya node of the Western Rift was evident enough from ground surveys.   But it is now seen, for example, that Alid caldera in northern Afar lies on the intersection of the Precambrian Atsbi horst (Kazmin and Garland, 1973) lineament of the Ethiopian plateau with the Quaternary fault-belt of the floor of northernmost Afar.

Glaciated valleys can generally be recognised without difficulty on the highest mountains of eastern Africa.   In the Sagatu Mts, east of the main Ethiopian rift, such valleys cut across the margin dyke-swarm of the rift valley and have been mapped by the author and E.C. Potter (in preparation).   The largest glaciated valleys in the whole of eastern Africa have been recognised from the ERTS imagery of the Simien Mts, northern Ethiopia (Mohr, 1963). There, glaciers as long as 40km flowed south and east from the southeastward tilted crest of the Miocene Simien shield volcano.

No certain meteorite craters have been identified from the ERTS imagery of eastern Africa. Worthy of possible attention, however, are the circular feature at 10°45'S, 27°45'E, and a better preserved but less symmetrical feature at 7°25'S, 28°15'E.

REFERENCES

Almond, D.C., 1969, Structure and metamorphism of the Basement Complex of north-east Uganda. Overseas Geol. Min. Resources, 10, 146-163.

Azzaroli, A., and Merla, G., 1957, Carta geologica della Somalia e dell'Ogaden (1:500,000). Agip Mineraria-Consig. Naz. Ricerche, Firenze.

Baker, B.H., 1963, Geology of the Baragoi area. Geol. Surv. Kenya, Rep. 53, 74pp.

Baker, B.H., 1967, Geology of the Mount Kenya area. Geol. Surv. Kenya, Rep. 79, 78pp.

Baker, B.H., Mohr, P.A., and Williams, L.A.J., 1972, Geology of the Eastern rift system of Africa. Geol. Soc. Amer., Spec. Paper 136, 67pp.

Beydoun, Z.R., 1970, Southern Arabia and northern Somalia: comparative geology. Phil. Trans. R. Soc. London, A, 267, 267-292.

Beyth, M., 1972, To the geology of central-western Tigre. Rheinische Friedrich-Wilhelms Univ., Bonn. 159pp.

Bloomfield, K., 1966, Geological map of Malawi. Geol. Surv. Malawi.

Brown, P.E., 1962, The tectonic and metamorphic history of the Pre-Cambrian rocks of the Mbeya region, southwest Tanganyika. Quart. J. Geol. Soc. London, 118, 295-317.

Cahen, L., 1970, Igneous activity and mineralisation episodes in the evolution of the Kibaride and Katangide orogenic belts of central Africa. In: African Magmatism and Tectonics (eds, Clifford, T.N., and Gass, I.G.), Oliver & Boyd, Edinburgh, p. 97-117.

Canuti, P., Gregnanin, A., Piccirillo, E.M., Sagri, M., and Tacconi, P., 1972, Volcanic intercalation in the Mesozoic sediments of the Kulubi area (Harrar, Ethiopia). Boll. Geol. Soc. Ital., 91, 603-614.

Chater, A.M., and Gilboy, C.F., 1970, Stratigraphic and structural relations in the Shakisso-Arero region of southern Ethiopia. 14th Ann. Rep. Res. Inst. Afr. Geology, Univ. Leeds, 8-11.

Clifford, T.N., 1970, The structural framework of Africa. In: African Magmatism and Tectonics (eds. Clifford, T.N., and Gass, I.G.), Oliver & Boyd, Edinburgh, 1-26.

Di Paola, G.M., 1973, The Ethiopian rift valley (between 7°00' and 8°40' lat. North). Bull. Volc., 36, 517-560.

Gass, I.G., Cox, K.G., and Mallick, D.I.J., 1965, Royal Society volcanological expedition to the South Arabian Federation and the Red Sea. Nature, 205, 952-955.

Hepworth, J.V., 1972, The Mozambique orogenic belt and its foreland in northeast Tanzania: a photogeologically based study. J. Geol. Soc. London, 128, 461-500.

Kazmin, V., and Garland, C.R., 1973, Evidence of Precambrian block-faulting in the western margin of the Afar depression, Ethiopia. Geol. Mag., 110, 55-57.

King, B.C., 1970, Vulcanicity and rift tectonics in East Africa. In: African Magmatism and Tectonics (eds. Clifford, T.N. and Gass, I.G.), Oliver & Boyd, Edinburgh, p. 263-283.

Levitte, D., Columba, J., and Mohr, P.A., 1974, Reconnaissance geology of the Amaro horst, southern Ethiopian rift. Bull. Geol Soc. Amer., (in press).

McConnell, R.B., 1972, Geological development of the rift system o eastern Africa. Bull. Geol. Soc. Amer., 83, 2549-2572.

McKenzie, D.P., Davies, D., and Molnar, P., 1970, Plate tectonics o the Red Sea and East Africa. Nature, 226, 243-248.

Mohr, P.A., 1962, The geology of Ethiopia. Univ. Coll. Addis Abab Press, 268pp.

Mohr, P.A., 1963, Geological report on an expedition to the Simien Mountains. Bull. Geophys. Obs. Addis Ababa, 6, 155-167.

Mohr, P.A., 1967, The Ethiopian rift system. Bull. Geophys. Obs. Addis Ababa, 11, 1-65.

Mohr, P.A., 1970, The Afar triple junction and sea-floor spreading. J. Geophys. Res., 75, 7340-7352.

Mohr, P.A., 1972a, Surface structure and plate tectonics of Afar. Tectonophysics, 15, 3-18.

Mohr, P.A., 1972b, ERTS-1 imagery of eastern Africa: a first look a the geological structure of selected areas. Smithsonian Astro. Obs., Spec. Rep. 347, 57pp.

Mohr, P.A., 1973, Structural elements of the Afar margins: data from ERTS-1 imagery. Bull. Geophys. Obs. Addis Ababa, 15 (in Press)

Sanders, L.D., 1965, Geology of the contact between the Nyanza Shie and the Mozambique Belt in western Kenya. Geol. Surv. Kenya, Bull. 7, 45pp.

Shackleton, R.M., 1967, Complex history of the Mozambique Belt. 11th Ann. Rep. Res. Inst. Afr. Geology, Univ. Leeds, 12-13.

Shackleton, R.M., 1973, In: Implications of continental drift to the earth sciences (eds. Tarling, D.H., and Runcorn, S.K.). Academic Press, London (in press).

Sutton, J., and Watson, J., 1959, Metamorphism in deep-seated zones of transcurrent movement at Kungwe Bay, Tanganyika Territory. J. Geology, 67, 1-13.

Tazieff, H., 1970, The Afar Triangle. Scientific Amer., 222, 32-4(

Tazieff, H., 1973, About air and space photo interpretations of Afar Trans. Amer. Geophys. Un., 54, 470 (abstract only).

Tazieff, H., Varet, J., Barberi, F., and Giglia, G., 1972, Tectonic significance of the Afar (or Danakil) depression. Nature, 235, 144-147.

Whiteman, A.J., 1971, The geology of the Sudan republic. Oxford Univ. Press, 290pp.

7B

Reprinted from pp. 3–4 of *Smithsonian Astro. Obs. Spec. Rep. 361, 1974*, 86p.

MAPPING OF THE MAJOR STRUCTURES OF THE AFRICAN RIFT SYSTEM

P. A. Mohr

Figure 2 Structural map of Ethiopia, Somalia, and Yemen, based solely on mapping from ERTS-1 imagery, at 1:1 million scale. Thicker traces mark Cainozoic faults with visible displacements (the trace is dashed for fractures with uncertain or no apparent displacement; the latter include lineaments). Thinner traces mark major structural lines in the Precambrian basement. The downthrown side of faults is ticked, where known, and arrows indicate the sense of motion on transcurrent faults.

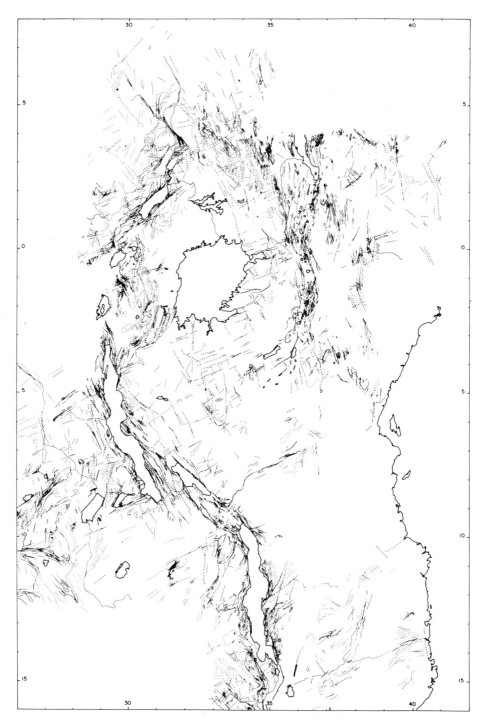

Figure 3 Structural map of East Africa and some adjacent areas, based solely on mapping from ERTS-1 imagery, at 1:1 million scale (see caption to Figure 2 for further details). Persistent cloud cover has prevented mapping of the Lake Kivu area, of southeastern Tanzania, and of parts of central Kenya, northern Mozambique, southeastern Malawi, and Ruwenzori.

Part IV

EASTERN AFRICA—KARROO

Editor's Comments
on Papers 8 and 9

8   VAIL
The Southern Extension of the East African Rift System and Related Igneous Activity

9   SWARDT
Rift Faulting in Zambia

The earliest recognizable rift-valley forms are associated with the Karroo System of Equatorial and Southern Africa (Upper Carboniferous to Lower Jurassic), which are predominantly terrestrial continental facies sediments with basic volcanics, although there are intercalations of marine Karroo into terrestrial formations in Madagascar (Haughton, 1963, pp. 199–249). Deposition was either as an extensive blanketing formation laid down in a shallow basin on a surface having some relief or in a trough bordered by monoclinal scarps and minor faults formed contemporaneously with sedimentation. Of the latter nature are the Luangwa Luano-mid-Zambezi aligned troughs of Zambia and Zimbabwe, the Ruhuhu, Mbamba Bay, and Kidodi troughs of Tanzania. They are aligned NE-SW. The lower Zambezi Karroo trough has a general E-W orientation but continues to the south as the Lower Shire and Urema troughs. Of the former nature are remnants of the originally flat-lying beds preserved in fault-angle depressions, their bedding being generally parallel with the tilted floor. They are found in northern Nyasa and Rukwa and in eastern Tanzania.

The Karroo (Lukuga Beds) of eastern Zaire appear to have infilled troughs as well as having been down-faulted (Veitch, 1935).

It is maintained by Dixey (1946, 1956) and others that the distribution and structural setting of the Karroo of Central and East Africa constitutes evidence that the late Cenozoic rift faulting of the Tanganyika, Rukwa, and Malawi zones inherits earlier trough faulting.

Paper 8, by Vail, is a useful outline of the Karroo and post-Karroo rift phenomena and extends the study southward. What is learned of the relation of the fracture to orogenic zones and parallelism of fractures with foliation trends is of general application.

Paper 9, by de Swardt, presents a picture somewhat at variance with that given by Dixey (1937). This is important because it is on comparisons with Luangwa that Dixey has based his hypotheses for the formation of Nyasa-Rukwa and Tanganyika. The structural setting of the Luangwa and related troughs is within the Irumide orogenic belt (see also Drysdall and Weller, 1966).

For the northeastern-aligned troughs of Tanzania and other Karroo occurrences in Tanzania, McKinlay (1963) contains valuable descriptions and a map. An earlier account is by Bornhardt (1900). Stockley (1932) describes the Ruhuhu. The occurrences of north Malawi are described by Andrew and Bailey (1910) and their extension between Malawi and Rukwa are detailed by Spence (1954) and Harkin (1955). McConnell (1950) describes occurrences in the Rukwa trough and on the Ufipa plateau, but many in north Rukwa ascribed by him to Karroo are now given either older or younger ages (McKinlay, 1963) and are not evidence for extension of the Rukwa trough to the Tanganyika. The concealed rift valley of western Madagascar is described by Cliquet (1957) and Haughton (1963).

King (1978, abstract only) emphasizes that the older rifts were "dead" before separation of Africa and that there is no evidence for "rejuvenation." Because no crustal extension had taken place, he rejects the popular conception that continental rifts are "aborted" oceans.

REFERENCES

Andrew, A. R., and T. E. G. Bailey, 1910, The geology of Nyasaland, *Geol. Soc. London Quart. Jour.* **66**:189–251.

Bornhardt, W., 1900, Zur Oberflachengestaltung und Geologie Deutsch-Ost-Afrikas, in *Deutsch-Ost-Afrika,* vol. 7, Reiner, Berlin, 595p.

Cliquet, P. L., 1957, La tectonique profunde du Sud du Bassin de Morondora, *CCTA Comités Regionaux Geologie,* Tananarive, 1957, pp. 199–219.

Dixey, F., 1937, The geology of part of the upper Luangwa Valley, north-eastern Rhodesia, *Geol. Soc. London Quart. Jour.* **93**:52–74.

Dixey, F., 1946, Erosion and tectonics in the East African Rift System, *Geol. Soc. London Quart. Jour.* **102**:339–388.

Dixey, F., 1956, Erosion surfaces of Africa: Some considerations of age and origin, *Geol. Soc. S. Africa Trans.* **59**:1–16.

Drysdall, A. R., and R. K. Weller, 1966, Karroo sedimentation in Northern Rhodesia, *Geol. Soc. S. Africa Trans.* **69**:39–69.

Harkin, D. A., 1955, The geology of the Songwe-Kiwira Coalfield, Rungwe District, *Tanganyika Geol. Survey Bull.* **27**:1–33.

Haughton, S. H., 1963, *The Stratigraphic History of Africa South of the Sahara,* Oliver & Boyd, Edinburgh, 365p.

King, B. C., 1978, A comparison between the older (Karroo) rifts and the younger (Cenozoic) rifts of Eastern Africa, in *Tectonics and Geophysics of Continental Rifts,* I. B. Ramberg and E. R. Neumann, eds., D. Reidel, Dordrecht, pp. 347–350.

McConnell, R. B., 1950, Outline of the geology of Ufipa and Ubende, *Tanganyika Geol. Survey Bull.* **19**:40-46.

McKinlay, A. C. M., 1963, The coalfields and the coal resources of Tanzania, *Tanganyika Geol. Survey Bull.* **38**:47.

Spence, J., 1954, The geology of the Galula Coalfield, Mbeya District, *Tanzania Geol. Survey Bull.* **25**:4-16.

Stockley, G. M., 1932, The geology of the Ruhuhu coalfields, Tanganyika Territory, *Geol. Soc. London Quart. Jour.* **88**:610-622.

Veitch, A. C., 1935, Evolution of the Congo Basin, *Geol. Soc. America Mem. 3,* pp. 91-98.

8

Reprinted from *Geol. Rundschau* **57**:601–614 (1967)

The southern extension of the East African Rift System and related igneous activity

By J. R. VAIL, Leeds [*])

With 2 figures and 1 table

Zusammenfassung

Neue Einzelheiten über das Bruchmuster in der Spät- und Post-Karroo im Osten des südlichen Afrikas sowie über die damit verbundenen Magmagesteine sind auf einer Karte angegeben. Das Bruchmuster hat eine außerordentliche Ähnlichkeit mit dem Ostafrikanischen Graben, welcher eine Fortsetzung davon ist.

Ein enger Parallelismus besteht zwischen den Brüchen des tektonischen Grabens und den intrakratonischen, orogenetischen Zonen sowie auch örtlich innerhalb der Zonen zwischen den Brüchen und dem Schieferungsstreichen der Gneise.

Der Unterschied im Relief über die Verwerfungen hin ist bis zu 2000 feet; viele dieser tektonischen Gräben sind asymmetrisch, und Verschiebungen geschehen nur entlang einer Seite.

Obgleich Erdbewegungen durch lange Zeiträume unterbrochen waren, konnten jedoch zwei Hauptepisoden von der Post-Karroo bis zur Kreide und von dem Tertiär bis zur heutigen Zeit festgestellt werden. Neunundvierzig isotopische Alter für Magmagestein sind in der Literatur zitiert und deuten auf eine eruptive Tätigkeit von über 100 M. J. hin.

[*]) Author's adress: J. R. VAIL, Research Institute of African Geology, Department of Earth Sciences, University of Leeds, England.

123

Abstract

New details of the late- and post-Karroo fracture pattern in eastern southern Africa are shown on a map together with associated igneous rocks. The pattern is remarkably like that of the East African Rift Valleys with which it is continuous.

There is a close parallelism between the Rift fractures and intracratonic orogenic belts, and also locally between the fractures and foliation trends of the gneisses in the belts. Relief differences across the faults range up to 2000 feet; many of the Rifts are asymmetrical with displacement only along one side.

Although earth movements were intermittent over a long period of time, two main episodes occurred during post-Karroo to Cretaceous and Tertiary to recent times. Forty-nine isotopic ages of igneous material are taken from the literature and indicate eruptive activity over a 100 m. y. period.

Résumé

Nouveaux détails concernant le dessin de fracture pendant le Karroo Supérieur et le Post-Karroo dans l'est de l'Afrique méridionale ainsi que les roches ignées associées sont indiqués sur une carte. On y observe une similarité très remarquable entre le dessin de fracture de l'Afrique méridionale et celui du fossé tectonique de l'Afrique-Est, du quel il est un prolongement.

Il y a un parallélisme étroit entre les fractures de fossé tectonique et les zones orogéniques ainsi que localement entre les fractures et la direction de foliation des gneisses dans les zones orogéniques. La différence en relief à travers les failles se range près de 2000 feet; beaucoup de fossés tectoniques sont asymmétriques; ils montrent un déplacement sur un côté seulement.

Les mouvements de terre étaient intermittents pendant une longue période de temps; mais deux épisodes principaux se présentaient pendant le Karroo Supérieur jusqu'au Crétacé et pendant le Tertiaire jusqu'au Quaranteneuf âges isotopiques de roches ignées sont cités de la littérature consultée et les âges indiquent une activité éruptive plus de 100 m. a.

Краткое содержание

Приложена карта, дающая новые данные о разломах формации Карру на востоке южной Африки и связанные с ней магматические породы. Картина разломов имеет большое сходство с восточно-африканским грабенем. Хотя движения масс шли непрерывно, различают два главных периода их : первый от после-Карру до мела, второй от третичного периода до сегодняшнего дня. На основании 49 определений возраста по изотопному методу установили, что эруптивная деятельность протекала в течение 100милл. лет.

Introduction

Mention of Rift Valleys in eastern Africa usually implies those systems of fault graben which are found between the Red Sea and southern Malawi. However, GREGORY (1921) who did much of the pioneer work on the Great Rift Valley System of East Africa considered a further extension southwards along two main branches: (1) A western extension along the Luangwa valley and up the Zambezi valley to the vicinity of the Botswana border; (2) an eastern branch south from Lake Nyasa (now Lake Malawi) along the Shire valley towards the Sabi river and along the coast to Cape Corrientes in Mozambique (GREGORY, 1921, fig. 42).

The southern parts of the Rift System have never attracted as much attention as those in East Africa. DIXEY (1946, 1956) did consider these areas and modified GREGORY's map to include portions of the middle reaches of the Zambezi valley in Mozambique and the Limpopo valley between Rhodesia and the Transvaal. However, the information concerning the southern area was based on the reconnaissance work of THIELE & WILSON (1915), DIXEY's work before 1940 along the Shire valley, and the publications of the Southern Rhodesia Geological Survey prior to 1946 (Provisional Geological Map, scale 1 : 1,000,000).

The information used to outline this Rift pattern is thus at least twenty years old, and some of it was collected over fifty years ago. The writer has had the opportunity of mapping post-Karroo fracture patterns along parts of the Shire and Zambezi valleys, parts of central Mozambique and in the Limpopo valley. New information also has become available recently from the various Geological Survey Departments. This data is now brought together on a single map in order to show the pattern of all the known late- or post-Karroo fractures; from it the southern extension of the East African Rift System can readily be seen (Figure 2).

GREGORY (1921) described Rift Valleys as having a characteristic history of faulting accompanied by the outbreak of volcanicity with eruption alternating with periodic earth movements. These events in the main Rift system took place in geologically quite recent times. DIXEY (1946, 1956) discussed the question of their age further and showed two main periods of Rift valley formation: one immediately post-Karroo (Mesozoic) and the other from the Pleistocene to Present. The latter is exemplified in the Great Rift Valley system of East Africa; it is not clear whether the older movements occurred there as well.

The usual methods for dating the age of the fracturing have been either palaeontological or geomorphological. Recent isotopic age determinations provide new information for the date and duration of the igneous activity, which is in many places associated with the formation of the fracture pattern.

Fracture Pattern of the Southern Rift Zone

There are three principal areas to be considered in a review of the southern Rift zone of east Africa: (1) Zambezi valley, (2) Shire valley and central Mozambique, and (3) Limpopo valley (Figure 2). In these areas the Rift valleys are formed by relative subsidence between parallel fracture zones which have produced characteristic graben and by rotational faults. In many places these tectonic valleys were filled with sediments and lavas. Subsequent exhumation, with or without rejuvenation of the early faults, has formed the Rifts troughs as they are today.

1. Zambezi valley

In that part of southern Africa with which we are concerned the Zambezi valley can be divided into three sections: (a) Lake Kariba section to the Luangwa confluence. (b) The middle reaches in which the river flows in an easterly direction, and (c) the lower stretch from the Capoche junction to the coast. For most of its length the Zambezi River occupies a faulted trough largely filled with Karroo (Upper Carboniferous-Jurassic) sediments and minor lavas.

a) The Lake Kariba section follows a remarkably straight NE trend along a trough about 80 km wide. A well-developed fault pattern affects Karroo sediments and upper basalts in Rhodesia; the south-western extension of this fracturing is obscured by Kalahari sands (Tertiary), but there is a suggestion of similar faulting at Makarikari in Botswana (GREEN, 1966). On the north-western side of Lake Kariba a complex fault pattern is shown on the maps of the Zambia Geological Survey, Karroo sedimentary rocks are repeatedly strike faulted against the Basement; secondary NNW-trending cross fracturing is also present. GAIR (1956) considered the faulting to have been contemporaneous with sedimentation and the resultant structure to be one of step faults along the margins of a synclinal trough. No volcanic rocks post dating the fracturing are evident here. The difference in relief between the valley floor and the main Rhodesian plateau is about 2000 feet on the Zambian side and slightly less on the Rhodesian side of the valley.

It is clearly seen from the map (Fig. 2) that the Kariba section of the Zambezi valley is directly in line with the Luangwa valley. Near the confluence of these two valleys a complex zone of intersecting fractures is pierced by the carbonatite igneous centres described by BAILEY (1960). Details of the Luangwa Rift faults are lacking. The valley floor is mostly covered with Karroo sediments and several igneous centres are known along its length. The west side is formed by the high Muchinga escarpment which is fault controlled. The eastern margin is not so distinct topographically but is known to be fault bound in the north. DE SWARDT (1965) has shown that two ages of movement were involved, — the earlier Karroo faulting in the south and the more recent fracturing in the north.

b) The middle Zambezi valley, in contrast to the Kariba and coastal portions, is aligned east-west. The south side of the valley is marked by a high escarpment where Karroo strata are down-faulted against ancient gneisses. Examination by the writer has disclosed a faulted contact over a length of 480 km. The relief difference between the flat colluvial covered valley floor and the gneisses of the Rhodesian plateau south of the river is as much as 2,000 feet along the central portion but dies out to almost nothing at the eastern end; this may reflect a progressive diminution of displacement along the boundary fault. The southern escarpment is due mainly to a single E—W fault; in places zones of parallel fractures occur on either side of the main break and subsidiary faults with NE and NW trends, apparently associated with the main rifting, have been observed. The line of fracture is marked by quartz breccia. The northern side of the valley is much less easy to define; the Karroo sediments are faulted against basement rocks in places but there is no major topographic break and the country slopes up through dissected hills to elevations of about 6000 feet above sea level 75 km north of the river. The Kafue river has a remarkably straight course for about 320 km and may be considered as flowing in a western continuation of the mid-Zambezi trough. The Kafue Flats, an infilled basin which it crosses are now covered with alluvium; this may well represent an early Rift valley filled by Karroo rocks. In the east, the mid Zambezi Rift terminates in a number of en-echelon SE-striking troughs.

c) That part of the Zambezi valley which trends south-eastwards joins the southern end of the Nyasa-Shire fracture system and continues across central Mozambique. Faulting controls the present outcrop of Karroo and younger strata in the Lupata area but relief differences are small. Displacement on these faults is generally not much.

Throughout the length of the Zambezi valley the rocks covering the floor of the troughs are of Karroo or younger age. At three places, Kariba gorge, Mpata gorge upstream from the Luangwa confluence, and at Cahora Bassa gorge near the Capoche junction, the river cuts through basement rocks. The Karroo comprises mainly arenaceous and argillaceous sediments, in places coal bearing, which are overlain by basaltic lava flows (Stormberg Series). Along much of the valley floor the Karroo rocks are covered by colluvial deposits and in the central part a sedimentary sequence with intercalated lavas has been classified as Cretaceous in Rhodesia and Lupata Series in Mozambique by the respective Surveys.

The precise age of the faulting on either side of the Zambezi valley has not been ascertained throughout its length; some is mid-Karroo but most of the movement is post-Karroo and pre-Tertiary. The pre-Karroo rocks are folded parallel to the troughs along much of the Rift zone, for example on the Zambezi Escarpment of Rhodesia (see fig. 1). Elsewhere the fractures cut obliquely across these earlier trend directions.

2. Central Mozambique and Shire valley of Malawi

From Lake Malawi the main Nyasa Rift faults continue southwards on either side of the Shire river to join the lower Zambezi valley. The eastern margin of the Shire Rift is formed by the Makongwa fault which cuts across the syenite complex of Chaone (for age see Table 1) and its neighbours, Mongolowe and Zomba. The relief difference may reach 2000 feet, but displacement on the fractures is difficult to measure. A subsidiary fracture pattern is developed either normal or oblique to the main break, the latter being deflected on crossing the Mongolowe Complex (Vail & Mallick, 1965).

The Shire faults continue south across the Zambezi and along the Urema Sunklands of Mozambique. One branch may continue south to mark the coastline, while the other crosses the central Manica plateau of Mozambique to form the Lucite-Buzi fracture pattern which runs into the WSW-trending Limpopo zone.

Like the Zambezi and Luangwa Rifts, the troughs are filled with Karroo and younger rocks. Across the central plateau the fractures occur in Mozambique Belt gneisses and many are marked by quartz breccia veins and low escarpments. In the Urema trough Cretaceous and Tertiary rocks are faulted. New features of the fracture pattern shown on the map (Fig. 2) are the faults and fractures now filled by dykes along the edge of Karroo rocks between the Zambezi and Buzi Rivers. Elevation differences across the faults are quite small, except in the case of the Lucite fault (VAIL,

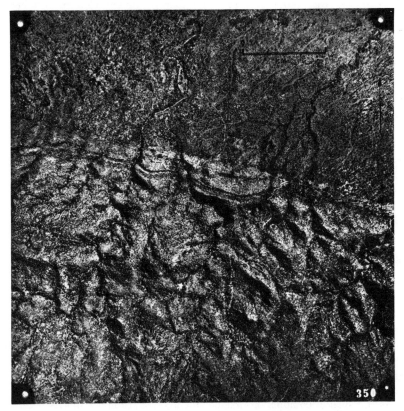

Fig. 1. Aerial photograph of southern margin of Zambezi valley, Rhodesia. Downfaulted sedimentary strata (north) contact Precambrian gneisses (south) which here strike parallel to the Rift fault (ESE). Reproduced with the permission of the Surveyor General of Rhodesia. Crown Copyright reserved.

1964) where the northern side is 1500 feet above the south. The Gorongosa massif, which rises about 6000 feet above the surrounding plains west of the Urema valley, owes its relief to differential erosion in a similar way to the southern Malawi igneous complexes (DIXEY, 1956).

In age the faulting is post-Karroo, post-Lupata, and some post dates the Malawi syenite complexes (see Table 1 for dates). Several periods of tectonic activity are to be found together; sedimentation and volcanism occur in cycles associated with earth movements (FLORES, 1967), the most recent of which represents the true East African Rift Valley type.

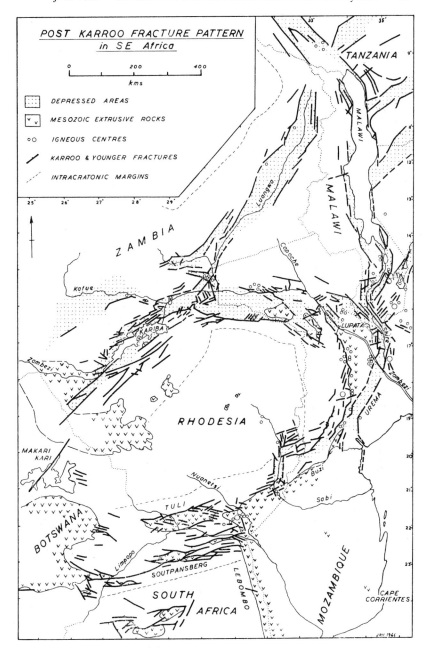

Fig. 2. The southern extension of the East African Rift System and related igneous rocks. Modified from geological survey publications.

3. Limpopo valley.

The predominently WSW fracturing south of the Rhodesian craton extends from the Lucite and Buzi rivers in the east across the Sabi river (where a N—S subsidiary trough is developed), roughly parallel to the Limpopo metamorphic belt and into Botswana. The Tuli Syncline is a partially faulted trough of sediments and lavas and is the site of a northern set of fractures. A parallel southern set extends along much of the Soutpansberg Ranges. The northern end of the Lebombo monocline intersects these Limpopo fractures at their eastern end. Further east, Cretaceous and Tertiary rocks extend to the coast.

Karroo sediments and lavas are preserved in many of the troughs. Fracturing on the whole is not as pronounced as in the Zambezi valley for example, and the topographic displacements are much less, fault scarps being relatively poorly developed. On the other hand, monclinal flexuring has been an important feature not only along the great Lebombo monocline (Du Toit, 1929) but between the Limpopo valley and the Buzi near where the Karroo rocks dip southwards beneath a younger cover. Well-developed dyke swarms, usually of dolerite but also granitic, are either oblique to the intersection with the major structures or parallel to them.

Most of the fracturing post-dates the Karroo igneous rocks (Triassic and Jurassic) and pr-dates the Cretaceous sediments (Cox, et al., 1965). There has been some rejuvenation along earlier fractures in the Soutpansberg (van Eeden, et al., 1955) and also some post-Cretaceous movement. The Lebombo flexure was partly contemporaneous with the extrusion of the lavas and the intrusion of the dykes. The close association of igneous activity with the formation of the troughs is evident in the case of the Soutpansberg both in Waterberg and in Karroo times. The alignment of the igneous centres and the thickening of both sedimentary and volcanic rocks in the Nuanetsi syncline shows dependance on structural control (Cox, et al., 1965).

Related igneous activity

Extensive igneous activity is associated with the late- and post-Karroo tectonic events. This takes the form of basaltic lava flows and dykes, rare rhyolites and igneous complexes which range in composition from basaltic to granitic, syenitic or carbonatitic.

These rocks comprise the Karroo igneous cycle as envisaged by Cox, et al., (1965). They extend from the northern part of South Africa into Botswana where two extensive basins of plateau lavas are separated by the Makarikari pan (Green, 1966). The southern basin is largely concealed beneath Kalahari sands, but the northern basalts are well exposed in the Zambezi river gorges. Both areas have obscure relations with the Rift structures. Small patches of Karroo basalts are exposed along the Zambezi valley, and several igneous centres occur along the Luangwa valley. Along the central stretch of the Zambezi valley carbonatites are reported from the north bank (R. de Barros, personal communication,

1962) and basalts and rhyolites occur on the south side. Downstream the Lupata gorge bisects a succession of sediments and basalts, rhyolites and alkaline rocks, the latter form the Lupata Group (FLORES, 1967). Numerous vents, dykes and igneous complexes of the Chilwa Alkaline Province occur south of Lake Malawi. Although Karroo basalts and dolerite dykes also are found in the lower Shire valley most of the igneous rocks of South Malawi are Cretaceous in age.

Karroo basalts and associated dykes and plugs may be traced southwards along the edge of the Manica platform to the Buzi valley and thence south-westwards into the Limpopo valley where they occupy the Nuanetsi and Tuli synclines and the Soutpansberg monocline (Cox, et al., 1965). Numerous igneous centres and dyke swarms are associated with these structures. Basalt and rhyolite flows crop out southwards for over 550 km along the Lebombo monocline to form an almost unbroken arc through Swaziland to the Lesotho (Basutoland) basin, while farther afield Karroo dolerite dykes extend to the south and west. Deep boreholes south of the Sabi river in Mozambique have proved the existance of basalt beneath the Cretaceous and Tertiary rocks at a depth of about 3000 m (ARCHAMBAULT, 1962).

In many places the presence of igneous rocks is closely related to the fracturing, in others however, no spacial relationship between Rift movements and igneous activity is apparant. Because the igneous cycle is a continuous process, emplacement of the various igneous rocks is by no means contemporaneous and consequently the faulting related to it is presumed to be a continuous process as well. It is usually difficult to date these events precisely by geological means and thus an attempt is made to supplement field observations with isotopic analyses of igneous material and to use palaeomagnetic polarity measurements to aid more precise correlations.

Dating the igneous cycle

Most of the igneous rocks under discussion post-date the Karroo sediments, some are intruded into Tertiary strata and others only occur in pre-Karroo basement rocks.

Forty-nine isotopic age determinations, mainly K-Ar, are now available on Mesozoic rocks and minerals from this part of Africa. Palaeomagnetic measurements are as yet limited to about half a dozen sites. The isotopic results are listed in Table 1. It must be noted, however, that the results do not necessarily reflect the precise age of emplacement and should therefore be treated with caution. Nevertheless, the time range can readily be seen. The table is arranged in order of increasing isotopic age but also illustrates the progressive northwards shift of activity.

Discussion

In a discussion of the southern extent of the East African Rift Valley System two points require consideration:

131

Table 1. Mesozoic age determinations from eastern southern Africa.

Locality	Latitude	Longitude	Material	Reference	Method (number of analyses)	Isotopic age in m.y.
Tertiary — Cretaceous boundary						
Musensi, Tanzania	8° 55' S	33° 05' E	biotite, pyroxenite	1		70 ± 2
Mbeya Range, Tanz.	8° 50'	33° 10'	feldspar, carbonatite	2	K-Ar	98 ± 14
Lupata, Mozambique	16° 45'	34° 10'	feldspar, rhyolite	2	K-Ar	100 ± 10
Panda, Tanzania	8° 59'	33° 14'	biotite, carbonatite	3	K-Ar (2)	108 ± 8
Panda, Tanzania			pyrochlore	4	K-Ar	113 ± 6
Lupata, Mozambique	16° 37'	34° 06'	feldspar	5	U-Th-Pb	73 ± 185
Mbeya, Tanzania	8° 50'	33° 10'	biotite, carbonatite	6	K-Ar	115 ± 10
Chaone, Malawi	15° 06'	35° 24'	biotite, syenite	4	K-Ar	115 ± 6
Chambe, Malawi	15° 56'	35° 37'	biotite, syenite	7	K-Ar	116 ± 6
Kangankunde, Malawi	15° 08'	34° 53'	biotite, carbonatite	7	K-Ar	116 ± 6
Mlanje, Malawi	16° 02'	35° 43'	biotite, syenite	4	K-Ar	125 ± 6
				7	K-Ar	128 ± 6
Cretaceous — Jurassic boundary						
Movane, Mozambique	26°	32° 20'	feldspar, basalt	1	K-Ar	135 ± 5
Chaone, Malawi	15° 08'	35° 22'	zircon, syenite	8	Pb alpha (2)	137
Alice, S. Africa	32° 39'	27° 00'	feldspar, dolerite	9	K-Ar (3)	138 ± 14
Elephants Head, S. A.	30° 09'	28° 57'	W. R., dolerite	10	K-Ar	152
Tambatsura, Rhodesia	20° 17'	32° 25'	W. R., dolerite	10	K-Ar	159
Nuanetsi, Rhodesia	21° 15'	31° 10'	various	7	Rb-Sr	160 ± 10
Lebombo, S. Africa	25° 30'	32°	rhyolite	11	Rb-Sr	160
Grootvlei, S. Africa	26° 15'	28° 10'	W. R., dolerite	12	K-Ar (2)	160
Lupata, Mozambique	16° 35'	33° 59'	feldspar, rhyolite	10	K-Ar	163
Fraserberg, S. Africa	31° 50'	21° 32'	W. R., dolerite	6	K-Ar	166 ± 10
Cookhouse, S. Africa	32° 35'	25° 24'	W. R., dolerite	10	K-Ar	168
Qora River, S. Africa	32° 26'	28° 42'	various, dolerite	10	K-Ar (6)	171

Sasolberg, S. Africa	26° 50'	27° 53'	W. R., dolerite	10	K-Ar	173
Mananga, Swaziland	26° 10'	32° 00'	hornblende, granophyre	7	K-Ar	175 ± 15
Nyarauri, Rhodesia	20° 23'	32° 25'	W. R., dolerite	7	K-Ar	178 ± 10
Jurassic — Triassic boundary				1		180 ± 5
Rooikranz, S. Africa	32° 15'	27° 20'	hornblende, dolerite	10	K-Ar	180
Jagersfontein, S. A.	29° 42'	25° 24'	W. R., dolerite	10	K-Ar (3)	187
Stegi, Swaziland	26° 25'	32° 00'	hornblende, granophyre	7	K-Ar	190 ± 10
Marangudzi, Rhodesia	22° 04'	30° 40'	various	3	K-Ar (6)	190 ± 12
Ile, Mozambique	16° 05'	37° 30'	biotite, ? schist	8	K-Ar	192
Shawa, Rhodesia	19° 21'	31° 42'	biotite, ijolite	13	Rb-Sr	195 ± 15
Start Mesozoic				1		225 ± 5

Rb-Sr results standardized to Rb ½ t = 4.7 × 10^10 yr. Other ages as stated in references: 1. HOLMES 1959 time scale. 2. MILLER and BROWN (1963), 3. GOUGH et al. (1964 b), 4. SNELLING (1965), 5. SCHURMANN et al. (1960), 6. FLORES (1967), 7. SNELLING (1966), 8. Personal Communication, Director Geological Survey of Mazambique, 9. BLOOMFIELD (1961), 10. McDOUGALL (1963), 11. MANTON (1965), 12. HALES (1960), 13. NICOLAYSEN et al. (1963).

1. S t r u c t u r e

New details of the late- and post-Karroo fracture pattern have been brought together on a map (Fig. 2). It is striking that there are no recognizable Karroo fractures in the cratonic areas. This is partly due to the difficulty in recognizing them in a Precambrian terrain, but also reflects a real difference between the cratons and the non-cratonic regions. On comparison with a tectonic map of the pre-Karroo areas (e. g. VAIL, 1965, Pl. 1), a broad parallelism between the Rift pattern and the pre-Karroo structures is apparant on two scales: (a) that of structural trends within the basement gneisses, and (b) a parallelism of the Rifts to the margins of the Precambrian cratonic nuclei and an alignment along the intracratonic orogenic belts, in some cases almost along the axes of these belts.

In many places the Rift fractures strike parallel to the foliation trend of the gneisses, elsewhere they cut obliquely across them. This suggests that the more recent fracturing locally has made use of pre-existing lines of weakness, while on a regional scale it follows the major structures to a remarkable degree, irrespective of the age or direction of the orogenic belts.

Movements along the Rift zones occording to the available published descriptions are essentially vertical, and displacement range from negligable to several thousands of feet. Many of the zones are asymmetrical due to faulting along one margin and resultant tilting to produce the topographic difference. Along the Lebombo, Soutpansberg and Tuli-Buzi ranges and in central Mozambique movements took the form of monoclinal flexuring with associated tensional fracturing and dyke emplacement. Lateral movements are rare; the horizontal displacements in central Africa recognized by DE SWARDT et al., (1965) appear to predate the Rift fractures, in Malawi BLOOMFIELD (1966) considers lateral rejuvenation to have occurred on an earlier dislocation thus displacing the Nyasa Rift.

The association of Karroo sedimentation, volcanic extrusion, igneous intrusion and crustal fracturing in many places can be closely related to the broad Rift pattern. GREGORY (1921) noticed this association of events in the East African Rifts and considered it to be a characteristic of the Rift Valley pattern. This concept can now be extended to include the southern extension where it is exemplified by the great Lebombo monoclinal flexure, by the dyke swarms filling fractures in the Tuli-Nuanetsi area and in central Mozambique, and by the location of younger igneous centres close to and within the depressed areas of the Rifts.

2. Age

The establishment of the precise age of the fracturing is difficult without accurately dated strata, especially when the fractures occur in basement rocks. However, the time of the fracturing varies and it was probably intermittent over a relatively long period. Two main episodes can be recognised from the geology: (a) post-Karroo and (b) post-Cretaceous fracturing. Geochronological measurements (Table 1) indicate that the igneous material dated is confined to Mesozoic times and that the activity had a duration of some 100 m. y.

The application of palaeomagnetic measurements to the problem of the Rift Valleys is not as yet very enlightning. Palaeomagnetic pole reversals could be used to correlate and identify the volcanic activity more precisely than by isotopic means. The available data (GOUGH, et al., 1964 a and b and others) can be summarized as follows: (a) Negative (normal) polarity: Victoria Falls basalts, Shawa Complex, Lupata rhyolites; (b) Mixed polarity: Basutoland basalts, Karroo dolerites in Transvaal and Natal, Marangudzi Complex; (c) Positive (reversed) polarity: Nuanetsi igneous complexes and Nuanetsi basalts. Detailed correlations, the dating of the pole reversals and the application of palaeomagnetic measurements to the dating of the Rifts await further data.

Both Gregory and Dixey recognized that the Rift valleys characteristic of East Africa die out southwards. It is not so clear whether or where the older Karroo tectonism dies out northwards. Karroo sedimentary rocks occur at the northern end of Lake Malawi whilst Karroo volcanic rocks are identified in the Shire and lower Zambezi valleys some 720 km to the south. The two patterns overlap in southern Malawi; younger igneous activity becomes more intense and the younger Rift fractures predominate to the north. In Kenya for example isotopically dated volcanic rocks, some of which are related to the Rifts, give ages of less than 10 m. y.

In conclusion it can be seen that a primitive system of Rift fractures (pre-Cretaceous, post-Karroo) is present and well exemplified in south-eastern Africa. This system has a fracture pattern (Fig. 2) markedly similar to the Great Rift Valley System of the central and north-eastern part of the continent. Morphologically they form troughs, in many respects like the East African Rifts, but which owe their present day topography primarily to the removal by erosion of the Karroo sediments and not to renewed movements along the Mesozoic fractures. It is not clear how far northwards the older fractures extend into East Africa but in part they are clearly overprinted by the younger (Tertiary to Recent) Rift faults. The lower Zambezi, Shire and Lake Malawi valleys form an intermediate area in which both fracturing and igneous activity on the two sets overlap. This also seems to be the transitional area between the tectonic pattern to the south which outlines the post-Mesozoic continental limits and the north where the fracture system occurs inland and cuts across the continental shield.

Acknowledgements

This work was carried out under the auspices of the Research Institute of African Geology at the University of Leeds during the tenure of an Oppenheimer Geological Fellowship. I would like to thank colleagues in the Institute for critical discussions.

References cited

ARCHAMBAULT, J.: Hydrogeologie du Sud du Save. — Bol. Serv. Geol. Minas Mozambique, **30**, 1962.

BAILEY, D. K.: Carbonatites of the Rufunsa valley, Feira District. — Bull. Geol. Surv. N. Rhodesia, **5**, 1960.

BLOOMFIELD, K.: The age of the Chilwa Alkaline Province. — Records Geol. Surv. Nyasaland, **1**, 95—100, 1961.

— : A major ENE dislocation zone in central Malawi. — Nature, Lond., **211**, 612—614, 1966.

COX, K. G., JOHNSON, R. L., MONKMAN, L. J., STILLMAN, C. J., VAIL, J. R. and WOOD, D. N.: The geology of the Nuanetsi Igneous Province. — Phil. Trans. Royal Soc., Lond., Ser. A, **257**, 71—218, 1965.

DE SWARDT, A. M. J.: Rift faulting in Zambia. — Report of UMC/UNESCO Seminar on the East African Rift System, Nairobi, 105—114, 1965.

DE SWARDT, A. M. J., GARRARD, P., and SIMPSON, J. G.: Major zones of transcurrent dislocation and superposition of orogenic belts in part of central Africa. — Bull. Geol. Soc. Amer., **76**, 89—102, 1965.

DIXEY, F.: Erosion and tectonics in the East African Rift System. — Quart. J. geol. Soc. Lond., **cii**, 3, 339—388, 1946.

— : The East African Rift System. — Colon. Geol. Min. Res., Supp. Bull. **1**, 1956.

DU TOIT, A. L.: The volcanic belt of the Lebombo — a region of tension. — Trans. Royal Soc. S. Afr., **18**, 189—217, 1929.

FLORES, G.: On the age of the Lupata rocks, lower Zambezi River, Mozambique. — Trans. geol. Soc. S. Afr., **LXVII** for 1964, 111—118, 1967.

GAIR, H. S.: A summary of the structure and tectonic history of the mid-Zambezi Valley. — Commission for Technical Cooperation in Africa south of the Sahara, Dar-es-Salaam, Geol., 123—127, 1965.

GREEN, D.: The Karroo System in Bechuanaland. — Bull. Geol. Surv. Bechuanaland, **2**, 1966.

GREGORY, J. W.: TheRift Valleys and geology of East Africa. — Seeley, Service and Co., Lond., 1921.

GOUGH, D. I., OPDYKE, N. D. and MCELHINNY, M. W.: The significance of palaeomagnetic results from Africa. — J. Geophys. Res., **69,** 12, 2509—2519, 1964. — (1964 a.)

GOUGH, D. I., BROCK, A., JONES, D. L. and OPDYKE, N. D.: The palaeomagnetism of the ring complexes at Marangudzi and the Mateke Hills. — J. Geophys. Res., **69,** 12, 2499—2507, 1964. — (1964 b).

HALES, A. L.: Research at the Bernard Price Institute. — Proc. Royal Soc. Lond., Ser. A, **258,** 1—26, 1960.

MANTON, W. I.: A rubidium-strontium study of the Lebombo-Nuanetsi Igneous Province, southern Africa. — Rept. on Scientific and Educational Programs 1964—1965, Graduate Research Centre of the Southwest, U. S. A., 1965.

McDOUGALL, I.: Potassium-argon age measurements on dolerites from Antarctica and South Africa. — J. Geophys. Res., **68,** 5, 1535—1545, 1963.

MILLER, J. A. and BROWN, P. E.: The age of some carbonatite igneous activity in south-west Tanganyika. — Geol. Mag., **100,** 3, 276—279, 1963.

NICHOLAYSEN, L. O., BURGER, A. J. and JOHNSON, R. L.: The age of the Shawa carbonatite complex. — Trans. geol. Soc. S. Afr., **LXV,** 293—294, 1963.

SCHURMANN, H. M. E., ATEN, A. H. and BOERBOOM, A. J.: Fourth preliminary note on age determinations of magnetic rocks by means of radioactivity. — Geol. en Mijnbouw, **22,** 4, 93—104, 1960.

SNELLING, N. J.: Age determinations on three African carbonatites. — Nature, Lond., **205,** 4970, 491, 1965.

—: Age determination Unit. — Ann. Rept., Overseas Geol. Survs., 1965, 44—57, 1966.

THIELE, E. O. and WILSON, R. C.: Portuguese East Africa between the Zambezi River and the Sabi River; a consideration of the relation of its tectonic and physiographic features. — Geogr. J., **LXVI,** 16—40, 1915.

VAIL, J. R.: Esboco geral da geologia da regiao entre os rios Lucite e Revue, Distrito de Manica e Sofala, Mocambique. — Bol. Serv. Geol. e Minas Mozambique, **32,** 5—17, 1964.

—: An outline of the geochronology of the late Precambrian formations of eastern central Africa. — Proc. Royal Soc., Lond. Ser. A, **284,** 354—369, 1965.

VAIL, J. R. and MALLICK, D. I. J. M.: The Mongolowe Hills nepheline syenite ring-complex, southern Malawi. — Records Geol. Surv. Malawi, **III,** 1961, 49—60, 1965.

VAN EEDEN, O. R., VISSER, H. N., VAN ZYL, S. J., COERTZE, F. J. and WESSELS, J. T.: The geology of the eastern Soutpansberg and the lowveld to the north. — Geol. Surv. S. Afr., Explanation of Sheet 42, 1955.

9

Reprinted from pp. 105–114 of *East African Rift System,* Upper Mantle Comm., UNESCO Seminar, Nairobi, 1965, 265p.

RIFT FAULTING IN ZAMBIA

A. M. J. de Swardt

Structural troughs of two different ages are present in Zambia. The older depressions were probably initiated during the Late Carboniferous, and movements along them continued up to the Jurassic and probably well into the Cretaceous. There is some evidence of recent warping and rejuvenation on some of the boundary faults. The old troughs are floored by Karroo sediments ranging in age from Permo-Carboniferous to Upper Triassic and possibly Jurassic. The younger troughs occur in the extreme north of the country, and the main movements along them took place in very recent times, though there are some indications of earlier, probably Pleistocene, warping and faulting. The alignment of the troughs is to some extent controlled by Precambrian structure, but there is no genetic relationship between them.

STRUCTURAL SETTING

The greater part of Zambia is underlain by Precambrian rocks; some of these have been orogenically deformed several times. The oldest structural trend varies from north-north-west to north-north-east and is preserved on the Copperbelt and in basement domes to the west, and in the Mkushi and Serenje areas, south and east of the Katanga pedicle. It is defined by a strong foliation and schistosity in granitic gneisses and metasediments. Traces of this old trend, which is referred to a Tumbide orogeny by Ackermann and Forster (1960), are also found in basement gneisses in the Southern Province of Zambia.

The Irumide trend is dominant over a large part of the older Precambrian and varies from east-north-east to north-east. Though it represents the primary deformation in some metasediments, it is in the main superimposed on earlier basement trends, which have been reorientated and largely obliterated. The Irumide orogeny has produced widespread refoliation in granitic gneisses and has in some areas transformed them into schists.

The youngest Precambrian orogeny, the Lufilian, affects mainly the Katanga System. In the north-west, near the Angola border, the trend of foliation and schistosity is east-north-east.

It veers through east-west to south-east on the Copperbelt. Southwards towards Lusaka, the trend is southerly, though it has been much distorted by later movements. In the Lusaka area and the Southern Province the trend varies from east-west to south-east. During the Lufilian orogeny the basement gneisses have been intensely refoliated and schisted, but this has occurred only near the Katanga-basement boundary and not on such an extensive scale as during the Irumide orogeny.

The Lufilian trends have been greatly distorted along two major zones of transcurrent dislocation (de Swardt, Garrard and Simpson, 1965). One of these runs north-east and coincides with the northern margin of the Zambezi trough, while the other (the Mwembeshi zone) runs east-north-east north of Lusaka and brings into juxtaposition Katanga sediments of different sedimentary and metamorphic facies. Between the Mwembeshi zone and the Copperbelt there are several other zones of dislocation with relatively small transcurrent displacement along which Katanga sediments have been refolded; these zones coincide with the Irumide trend in the basement.

Age determinations are still scanty but suggest ages of +1600 m.y. for the Tumbide orogeny, +1000 m.y. for the Irumide orogeny and +750 m.y. for the Lufilian orogeny.

OLDER TROUGHS

Previous Work:

The Karroo sediments of Zambia have been examined by many workers, and a large number of reports have been prepared, of which a high proportion, particularly those referring to work by exploration companies, are unpublished. Amongst the most important publications by earlier workers are those of Molyneux (1909), Dixey (1935, 1937, 1946) and Guernsey (1951).

Since World War II Tavener-Smith (1960) and Gair (1959) have mapped the Karroo rocks of the mid-Zambezi valley, and Gair (1960) has also investigated the western end of the Luano valley. Drysdall and Kitching (1963) remapped an area at the north-eastern end of the Luangwa valley first examined by Dixey (1937). Drysdall and Weller mapped a large area in the central Luangwa valley and prepared a paper on Karroo sedimentation in Zambia which takes into account the work of earlier investigators (Drysdall and Weller, in press). The summary that follows is based largely on their work.

Characteristics and Distribution:

In Zambia Karroo sediments are largely confined to troughs which are for the most part the down-faulted central zones of the basins in which they were deposited. These troughs show similarities with those of the East African rift system, but they are much older and are filled with Palaeozoic and Mesozoic strata rather than more recent deposits. The Karroo sediments were deposited in shallow water and there is no evidence for the existence at any stage of deep lakes like those which now occupy the Tanganyika and Nyasa rifts.

The mid-Zambezi valley was the first of the troughs to

138

mapped in detail and can be regarded as the type area, both
stratigraphically and structurally, for both Zambia and Rhodesia.
It is approximately D-shaped and runs north-east from Livingstone
to Chirundu and then east-north-east and east to join the lower
Zambezi trough. In section the mid-Zambezi basin is markedly asymmetrical, with steep dips and much marginal faulting on the north-
west margin, whereas south of the Zambezi the dips are far shallower,
and the southern margin is defined by the unconformable junction
between the Karroo and Precambrian. Even on the north-west margin,
however, the boundary is by no means always faulted, and the
formation of the trough can be seen to be largely due to downwarping,
which has resulted in quite steep dips in the basal sediments.
Complex faulting has occurred inside the basin where the trough
intersects the Lufilian trend almost at right angles, and there are
several horsts of Precambrian rocks.

A narrow depression links the mid-Zambezi basin with the
Rufunsa trough, which runs east-west and then swings south towards
Feira to join the Karroo basin of the lower Zambezi valley. Except
for a short break the trough is floored by Karroo sediments, which
appear to be faulted against the Precambrian on both sides.

The Luangwa valley runs north-east for 400 miles and is
colinear with the mid-Zambezi depression; it is an approximately
parallel-sided trough up to 60 miles wide but much narrower in the
south-west. In the north it is fault-bounded on both sides, but
further south it is asymmetrical in the same sense as the mid-Zambezi
valley, with little major faulting on the sout-east. Complex
faulting of the Karroo sediments has occurred in places within the
trough, especially in the north.

The Lukusashi valley is a much smaller trough than the
Luangwa valley and lies parallel to the north-west of its southern
extremity; the two are separated by a horst of Precambrian rocks
4 - 20 miles wide. In the south-west the Karroo outcrop and
boundary faults join up with those of the Luano valley, which according
to Gair (1960) is a complex faul-angle depression with much greater
downthrows on the northern boundary faults than on those partially
defining the southern margin.

Though the lower Kafue valley (Kafue Flats) is almost
completely covered by alluvium, marginal outcrops and borehole data
suggest that it is largely underlain by Karroo sediments which
extend westwards to join up with those of the Barotseland basin.
The trough is partly bounded by major faults, which have not been
rejuvenated in recent times.

North of Mumbwa, and about 55 miles from the northern
edge of the Kafue valley, there is a small area of Karroo sediments
apparently bounded by major faults trending east-north-east.

Relationship between Trough Formation and Sedimentation:

The relationship between sedimentation and warping and
faulting is summarised as follows by Drysdall and Weller (in press):

"Karroo sedimentation in Zambia provides an example of
tectonically controlled basin filling of downwarps on the
continental surface....
"Following the retreat of the Dwyka ice, shallow linear

downwarps began to form; the Barotseland basin is exceptional
in that it is not noticeably elongated. The pre-Karroo surface,
except in the Gwembe District of the mid-Zambezi valley and
the western end of the Luano Valley, was hilly, and the more
outstanding features were probably not completely buried until
late in Ecca times. The shallow troughs were soon filled with
debris, partly glacial in origin, stripped from the pre-Karroo
surface as a result of rejuvenation of the flanking areas
consequent upon the initial downwarping. Carbonaceous
sediments began to accumulate in widespread swamps but, with
the exception of the Kandabwe area, the tectonic environment
was too unstable to allow thick coal seams to form; thin
seams with a high ash content grading laterally and vertically
into carbonaceous mudstones are characteristic of the Ecca
successions. Late in Ecca times the swamps were flooded by
shallow fresh-water lakes which filled the depressions and
gradually expanded to encroach on adjacent areas that were
being progressively peneplaned. The lacustrine sediments
generally consist of calcareous mudstones with rare thin
limestones, alternating with sandstones and arkosic grits.
This alternation is a result of periodic rejuvenation of the
source areas following spasmodic subsidence of the basin or
possibly the earliest movements on the boundary faults.
Towards the close of Lower Beaufort times the lakes were
finally filled by sediments which were by no predominantly
argillaceous, suggesting that peneplanation of the flanking
areas was complete. Swamp conditions prevailed and some of
the marshes were inhabited by a varied reptilian fauna.

"After a break in sedimentation of unknown duration, the
first major movements on the boundary faults caused rejuvenation
of the positive source areas and the troughs were inundated with
coarse, often conglomeratic, detritus. Faulting resulted
from continued downwarping and eventual failure of the crust
along the hinge zones. Because the troughs were asymmetrical,
faulting was probably confined to the steeper limbs of the
warps. The Upper Karroo sediments are therefore thickest near
the faulted margins, where they may even have been banked
against the actual fault scarps, and become thinner across the
troughs. The basal conglomeratic grits are usually succeeded
by alternating grit and siltstone horizons, each grit-siltstone
unit representing a minor cycle of sedimentation initiated by
periodic movements on the faults. Deposition of another major
arenaceous horizon in the Zambezi and Luangwa Valleys followed
a second phase of major faulting.

"Towards the close of Karroo times, when the basins were
again filled, further warping resulted in the rejuvenation of
the positive areas flanking the troughs, and any remaining
sediments deposited on them during the expansion of the basins
in Lower Beaufort and again in late Stormberg times were
stripped off and redeposited in the basins as alluvial and
wind-blown sands. Similar sands are interbedded with the
first lava flows, but the intervals between subsequent flows
were so short that no sediments or even fossil soils separate
them."

Drysdall and Weller's work indicates that, though there

are important differences, the Karroo sequences in the various basins can be correlated; this suggests that major movements in the troughs took place more or less contemporaneously. Though most, if not all, of the Karroo succession is present in Zambia, no strict lithological correlation of the middle and upper parts of the succession can be made with those established in South Africa.

Age of the Faulting:

Warping and faulting probably started early in Karroo times, but most of the major faults clearly displace the Karroo sediments. However, there are indications that intense faulting took place before the outpouring of the Batoka (Stormberg) basalts, which are at the top of the Karroo succession. In the Livingstone area, in the south of Zambia, the lavas transgress over the sediments and over a large area rest directly on the Precambrian. South-east of Lusaka T. Cairney (personal communication) has recently found that, in addition to the basalts resting on the sediments in the Zambezi trough, an occurrence of Batoka basalt rests on the Precambrian directly above the main boundary fault. These relationships suggest that, at least on the margins of the troughs, the sediments were downfaulted and largely eroded on the flanks before the basalts were extruded.

Morphologically the Karroo troughs resemble the rift valleys of East Africa, but it seems that they owe their present form to the scooping out of the soft Karroo sediments subsequent to the capture of an inland drainage to the Indian Ocean. Thus the Zambezi is impaled on a high horst of Precambrian rocks at Kariba and runs in a deep gorge through dissected ancient rocks near the Mocambique border. This also applies to the Luano-Lukusashi and Luangwa valleys, and in the latter case there is certainly no evidence of recent rift faulting along the lower course of the Luangwa River where it cuts through high dissected Precambrian terrain. The lower Kafue valley, in which the river has only very recently been reversed and captured by the Zambezi, shows much gentler marginal relief than the mid-Zambezi and Luangwa valleys, and the marginal faults of the graben north of Mumbwa have little topographic expression.

There are references to recent rejuvenation on the boundary faults in both the mid-Zambezi and Luangwa valleys, but the extent and effects of such movements have not yet been fully elucidated.

LATER RIFT FAULTING

Rift faulting of very recent age is present in Zambia but is confined to the extreme north of the country. As far as the writer is aware no systematic study has been made of the faulting, and the following account is based on a study of air photographs and his own observations in the field. The most spectacular faulting occurs at the southern extremity of Lake Tanganyika. Another set of recent faults trends west-south-westwards from the Sumbu area in Cameron Bay of Lake Tanganyika to Lake Chishi, south of Lake Mweru Wantipa. On its eastern margin Lake

Mweru is bounded by a very recent fault of uncertain throw, which
runs south-south-west for a distance of nearly 70 miles. Faultline
scarps related to an earlier period of folding are also present in
the area.

Lake Tanganyika:

Lake Tanganyika is bounded on its south-west side by a
major recent fault which forms a high scarp. A few associated
faults to the west have small throws and do not continue far.
South and east of the lake there are faults with downthrows both
towards and away from it, but the plateau surface has been brought
below the level of the lake mainly by warping. An unusual aspect
of the southern end of Lake Tanganyika is the small number of
rift faults as compared with other parts of the East African
rift system.

Sumbu-Chishi Faults:

The faults occur over a stretch of country 10-20 miles
wide, and the throw is variable. Judging from the aerial photographs the maximum vertical displacement approaches 1,000 feet
and the dip of the fault planes ranges from 70° to vertical.
An important feature is the abundance of narrow horsts and graben,
sometimes less than a quarter of a mile wide. The direction of
throw is variable, but faulting with downthrow to the north
predominates in the south, while on the northern margin of the
faulted belt the country is downwarped to the south. The Sumbu-
Chishi faults are important in that they are very recent and not
obscured by lake deposits, and may illustrate the initial stages
in the formation of a rift valley.

Lake Mweru:

The eastern margin of the lake is a low but well-
defined fault scarp which is uninterrupted and in places remarkably
straight, and continues beyond the present northern limit of the
lake, where it is paralleled by another low fault scarp facing
east. Near the southern end of the lake another late rift fault
striking south branches off from the main fault.

Earlier Faulting and Warping:

South-east of Lake Mweru a much dissected scarp facing
north-west overlooks a narrow coastal plain. This may mark a
line of warping or represent a fault-line scarp related to an
earlier phase of movement along the Mweru boundary fault. Dip-
slope scarps, cut in Plateau Series (Katanga System ?) sediments
dipping south-east, run north-east on both sides of Lake Mweru
Wantipa. They are remarkably straight and may be fault-line scarps
related to a period of faulting much earlier than the Lake
Tanganyika and other recent faults in the area. Dissected fault

scarps running east-north-east and north-east are also present on the Congo border in the area north of Lake Mweru Wantipa.

Age of the Faulting:

The main faulting in the north is of very recent age, and in many cases the scarps have been little affected by erosion. Even the intense erosion with Lake Tanganyika as base-level has only caused a small retreat of the scarps. At least three stages of erosion can be recognised, indicating that movement on the faults was intermittent.

The country mainly affected by the rift faulting is separated by a strong erosion scarp from the main plateau surface of northern Zambia, which has relics on it of an even earlier erosion surface in the Abercorn area. The Sumbu-Chishi faults have dammed a north-west flowing drainage on the youngest regional erosion surface in the area and has diverted it to Lake Tanganyika. Warping to the south-east associated with the faulting has apparently also been responsible for the formation of Lake Mweru Wantipa. Even the earlier warping and faulting in the area affect the youngest regional surface.

North-west Warping:

Many years ago du Toit (1954) recognised the importance of warping on north-west axes and its effects on the drainage pattern in southern Africa. In Zambia this warping was responsible for the formation of the Bwela Flats in the north-east, near Chinsali; for the capture of the Chambeshi River, which was formerly the headwater of the Kafue, and the formation of Lake Bangweulu; for the formation of the Lukanga and Busango swamps in the central part of the country; and possibly also for the Kafue Flats and the reversal and capture of the Kafue to the Zambezi. One of the most important warps runs through the Copperbelt and coincides with the Kafue anticline, an important late-Lufilian structure which up-arched rocks which had already been strongly folded during the main phase of the orogeny.

Relationship between Trough Faulting and Precambrian Structure:

The mid-Zambezi trough appears to follow in general the Irumide trend and a major zone of transcurrent dislocation. However, south-east of Mazabuka it cuts at right angles across the Lufilian trend and it may be significant that in this area antithetic faulting is far more common than elsewhere in the basin. The lower Kafue trough seems to follow the Mwembeshi zone of dislocation, though in the east it veers south-east into the Lufilian direction. The small graben north of Mumbwa also strikes east-north-east and appears to correspond to one of the northern dislocation zones. The Rufunsa depression swings round in rough conformity to Precambrian structures, the Luano trough follows the Mwembeshi zone, and the Lukusashi valley is in line with the Irumide trend. In the south the Luangwa trough appears to terminate

against the Mwembeshi zone. Further north it is at a small angle
to the Irumide grain of the flanking Precambrian rocks; the
boundary faults follow the grain whenever possible, but in places
cut across it at right angles.

It seems that the alignment of the Karroo troughs is
largely controlled by Precambrian structures which are, however, of
different age and tectonic significance. There is no evidence for
a direct genetic relationship between the warping and faulting
and the older structures.

In the north the area affected by the late rift
faulting is largely underlain by gently dipping sediments of the
Plateau Series, and it is impossible to establish a connection between
the fault alignments and basement structures.

REFERENCES

ACKERMANN, E., and   1960   Grundzuge der Stratigraphie
FORSTER, A.   und Struktur des Irumiden-Orogens.
                                   21st Int. Geol. Congr. 18, 182-92.

DIXEY, F.   1935   The transgression of the Upper
                                   Karroo, and its counterpart in
                                   Gondwanaland. Trans. Geol. Soc.
                                   S. Afr. 38, 73-89.

_____   1937   The geology of part of the upper
                                   Luangwa valley, north-eastern
                                   Rhodesia. Quart. J. Geol. Soc.
                                   Lond. 93, 52-76.

_____   1946   Erosion and tectonics in the East
                                   African rift system. Quart. J. Geol.
                                   Soc. Lond. 102, 339-88.

DRYSDALL, A.R., and 1963   A re-examination of the Karroo
KITCHING,   succession and fossil localities of
                                   part of the upper Luangwa valley.
                                   Mem. Dep. Geol. Surv. N.Rhod. 1.

DRYSDALL, A.R., and   -   Karroo sedimentation in Northern
WELLER, R.E.   Rhodesia. (Occ. Pap. Dep. Geol.
                                   Surv. Zambia, 39) Trans. Geol.
                                   Soc. S. Afr. (in press).

GAIR, H.S.   1959   The Karroo System and coal
                                   resources of the Gwembe District,
                                   north-east section. Bull. Dep.
                                   Geol. Surv. N. Rhod. 1.

_____   1960   The Karroo System of the western end
                                   of the Luano Valley. Rep. Dep. Geol.
                                   Surv. N. Rhod. 6.

GUERNSEY, T.D. 1951 A summary of the provisional geological features of Northern
Rhodesia. Colon. Geol. Min. Resour.
1 (1950), 121-51.

MOLYNEUX, A.J.C. 1909 On the Karroo System in Northern
Rhodesia and its relations to the
general geology. Quart. J. Geol.
Soc. Lond. 65, 408-39.

SWARDT, A.M.J. DE, 1965 Major zones of transcurrent dis-
GARRARD, P., and location and superposition of
SIMPSON, J.G. orogenic belts in parts of Central
Africa. (Occ. Pap. Dep. Geol. Surv.
Zambia, 37) Bull. Geol. Soc. Amer. 76,
89-102.

TAVENER-SMITH, R. 1960 The Karroo System and coal resources
of the Gwembe District, south-west
section. Bull. Dep. Geol. Surv.
N. Rhod. 4.

TOIT, A.L. DU 1954 The Geology of South Africa,
3rd ed. edited and prepared by
S.H. Haughton. Edinburgh.

Part V

EASTERN AFRICA—WESTERN RIFT SYSTEM

Editor's Comments
on Papers 10 and 11

10 DIXEY
Excerpt from *The Nyasaland Section of the Great Rift Valley*

11 HARKIN
Excerpts from *The Rungwe Volcanics at the Northern End of Lake Nyasa*

NYASA-RUKWA ZONE

Paper 10 is an excerpt from an early article by Dixey on the Nyasa Rift Valley, the most southerly zone of the system. It provides a reasonably objective description of the geology of Lake Malawi (formerly Nyasa). The remainder of the paper deals with the structural and geomorphic history as interpreted by Dixey (to which reference has already been made, see Paper 4).

A controversial feature of this paper is Dixey's structural linking of the Lower Shire trough with the Nyasa rift valley via the Upper Shire valley. The Lower Shire valley is in continuity with the Zambezi rather than with the Nyasa trough (Vail, Paper 8). The initiation of the present form of the Nyasa rift valley must be related to the interruption of the late Cretaceous to early Miocene cycle and the displacement of the African surface, the end product of that cycle. The correlation of remnants of this surface on opposite flanks is discussed by James (1956). The early paper by Andrew and Bailey (1910), unusually comprehensive for its time, has formed the basis for all subsequent field work.

However, neither of these papers includes an adequate geological map. For this refer to the geological and structural map (scale 1:250,000) by Bloomfield (1968), which is chiefly concerned with the Precambrian. It portrays the relief of the lake floor but not that of the land. The earlier hydrographic chart by Rhoades and Phillips (1902) also includes cross-sections. The account of the geomorphology of Malawi by Lister (1965), which expands descriptions by King (1963), can be accepted as definitive and is more systematic than the accounts by Dixey (1937a, 1937b) and others. A report on a gravity survey (Andrew, 1974) contains good maps giving spot heights. Unfortunately none of these accounts specifically includes the eastern flank north of latitude 13°30' S, but the published maps of the Tanzania Geological Survey with explanatory notes, supplemented by the earlier work of Bornhardt and others, can be consulted.

The northern end of Lake Malawi from latitude 10°30' S is aligned with the Rukwa trough. The structural form is that of a gigantic fault-angle depression as far north as latitude 9°S, the southern end of Lake Rukwa (Holmes, 1965. Fig. 765). The main faults on the northeast flank follow the cataclastic zones of Proterozoic age (see Paper 5). The asymmetric and complex trough is partly occupied by the Rungwe volcanics (Harkin, Paper 11) and by largely infaulted Mesozoic and younger terrestrial sediments (Harkin, 1955; Spence, 1954). This is briefly described in Paper 11. Geomorphic expression of structure and tectonics, especially in relation to volcanic activity, are well described. This is also the definitive account of the vulcanism and petrology of the Rungwe volcanic field.

From latitude 9°S, the major faulting is on the southwestern flank, although the slightly sinuous fault bordering Lake Rukwa on the northeast has probably a considerable throw at least 400 m, but the total amount depends on the thickness of sedimentary infill (Teale and Harvey, 1933). The actual face of the fault was preserved beneath the lake waters, lowered by 200 m in recent times.

The southwestern flank fault, complex and stepped, is described in Paper 12, excerpts taken from a bulletin on the field geology by McConnell, published in 1950. The same bulletin describes two remnants of undoubted Karroo coal measures—on the Ufipa Plateau at 5,700 feet and in the floor of the Rukwa rift at 2,700 feet, the fault thus having 3,000 feet actual throw at this point.

Lake Rukwa is a shallow lake, frequently dessicated, and is floored by young sediments. Northward, as described by McConnell, the Rukwa rift valley dies out and does not connect with the Tanganyika rift valley via the Karema gap as indicated in some later publications.

REFERENCES

Andrew, A. R., and T. E. G. Bailey, 1910, The geology of Nyasaland, *Geol. Soc. London Quart. Jour.* **66**:189–252.

Andrew, E. M., 1974, Gravity survey of Malawi: field work and processing, *Inst. Geol. Sciences, Rep. No. 74/15*, 36p.

Bloomfield, K., 1968, The pre-Karroo geology of Malawi, *Malawi Geol. Survey Mem. 5*, 166p.

Dixey, F., 1937a, The geology of part of the upper Luangwa Valley, north-eastern Rhodesia, *Geol. Soc. London Quart. Jour.* **93:**52–74.

Dixey, F., 1937b, The pre-Karroo landscape of the Lake Nyasa Region, and a comparison of the Karroo structural directions with those of the Rift Valley. *Geol. Soc. London Quart. Jour.* **93:**77–93.

Harkin, D. A., 1955, The geology of the Songwe-Kiwira Coalfield, Rungwe District, *Tanganyika Geol. Survey Bull.* **27:**12–13.

Holmes, A., 1965, *Principles of Physical Geology*, rev. ed., Thomas Nelson, London, 1288p.

James, T. C., 1956, The nature of rift faulting in Tanganyika, *CCTA Comités Regionaux Géologie*, Dar-es-Salaam, 1956, pp. 81–94.

King, L. C., 1963, *South African Scenery*, 3rd ed., Oliver & Boyd, Edinburgh, 308p.

Lister, L. A., 1965, Erosion surfaces in Malawi, *Records Geol. Surv. Malawi* **7:**15–28.

Rhoades, E. L., and W. B. Phillips, 1902, Lieut. Rhoades' Survey of Lake Nyasa, *Geog. Jour.* **20:**68–70.

Spence, J., 1954, The geology of the Galula Coalfield, Mbeya District, *Tanzania Geol. Survey Bull.* **25:**19–21.

Teale, E. O., and E. Harvey, 1933, A physiographical map of Tanganyika—Notes on Physiography by Teale, E. O., *Geog. Rev.* **23:**402–413.

10

THE NYASALAND SECTION OF THE GREAT RIFT VALLEY

F. Dixey

[*Editor's Note:* In the original, material precedes and follows this excerpt.]

In the following account of the Nyasaland section of the Great Rift Valley it will be convenient to regard this section as comprising an upper rift occupied by lake Nyasa and a lower rift traversed by the Shire River, which drains the lake at the southern end.

The country in the immediate neighbourhood of the Rift Valley is made up largely of various gneisses and schists, with intrusions of granite and related rocks, which are all probably of pre-Cambrian age ; these ancient rocks are overlain by a few relatively small patches of sedimentary rocks belonging to several different geological periods. The earliest sediments are those of Karroo age, which are exposed both at the northern and at the southern end of the rift, where they are in general faulted down into the crystalline rocks. These sediments constitute the local coal-bearing strata, and they are thus of considerable importance to Nyasa-land (IX.). Moreover, at the foot of Mount Waller, on the north-western shores of lake Nyasa, they have recently yielded a valuable collection of reptilian fossils (V.) ; remains similar to these are well known from the Beaufort Beds of South Africa, but they have not previously been recorded from any part of Central Africa. It should be noted that the Rift Valley faults have cut right across the foliation of the crystalline rocks as well as the boundary faults of the Karroo Series.

The second group of sediments consists of the Dinosaur Beds of late Jurassic or early Cretaceous age (V.), which rest unconformably upon all older rocks ; the fossil Dinosaurs recently discovered in them are related to those of Tendagaru in Tanganyika Territory, 350 miles further east, which are at present being examined by an expedition sent out by the British Museum. It appears probable that the Dinosaur Beds accumulated in a depression, on the site of the northern end of lake Nyasa, which developed as a result of movements that took place some time before the main Rift Valley faulting began.

The third group of sediments consists of the Lacustrine Series (V.), which rests unconformably upon the Dinosaur Beds and is itself split into sub-groups by several unconformities. One sub-group has yielded fresh-water molluscs and, more recently, a few mammalian bones, and a preliminary examination of these fossils shows that the enclosing beds date back into Tertiary times ; the examination of the Lacustrine Series may be expected therefore to yield valuable data in connection with the development and structure of the Rift Valley. It is interesting to observe that the Beaufort reptiles, the (?) Cretaceous Dinosaurs, and the Lacustrine mammals all occur on the north-western shores of lake Nyasa within but a few miles of one another, and also that it is the discovery of these fossils within the last twenty months that has made possible the recent important revision of the geology of this region.

During the deposition of the Lacustrine Series, probably towards the end of the time represented by the mammal-bearing beds, volcanic activity became manifest at the northern extremity of the Nyasan Rift Valley, and consequently a great quantity of alkaline lavas and tuffs accumulated there upon the floor of the rift. A subsequent fall of several hundred feet in the level of the lake gave rise to a wide but variable fringe of lacustrine deposits of Recent geological age.

2. The Nyasan Rift Valley.

The Nyasan Rift Valley, which is about 50 miles in mean width, takes the form of a great trough that runs almost due north and south over a total distance of nearly 400 miles. It intersects various plateaux of high average altitude; to the north these plateaux range from 7000 to 9000 feet above sea-level, whereas those to the south are from 3000 to 5000 feet. The floor of the rift, on the other hand, descends near the northern end to a depth of several hundred feet below sea-level, with a maximum descent of 700 feet below this datum, whereas near the southern end much of it stands at 800 to 1000 feet above sea-level; accordingly, the rift is very much deeper at the northern than at the southern end, and the maximum depth occurs in the north-eastern part, where the scarp of the Livingstone Mountains overlooks the bottom of the trough from the extraordinary height of more than 10,000 feet. Moreover, the Nyasa basin as a whole is asymmetric, as has been pointed out by Professor Gregory (II., p. 305); the eastern side is due to one main fault, whereas the western part of the downthrown block has been dropped in steps by parallel faults, with the steps sloping to the east.

The walls of the rift are generally steep, and locally even precipitous; frequently the descent from the plateau margins is broken by one or more fault-steps, and additional steps, relatively broad and low, extend across the floor of the rift. Usually these fault-steps are inclined towards the rift, but not infrequently they are inclined in directions parallel with it. Although the trend of the rift as a whole is north and south, the walls themselves have assumed a zig-zag course as a result of the intersection of two sets of faults that run respectively north-north-west and north-north-east; furthermore, these faults sometimes terminate against, and are therefore older than, additional faults that follow a northerly course. Running parallel with the main rift there are, particularly on the western side, minor rifts standing at various elevations; these rifts are separated by rectilinear crystalline ridges ranging up to 30 miles in length, and at the northern end of the lake three such rifts may be seen to run side by side over a distance of nearly 30 miles.

The greater part of the floor of the main rift is occupied by lake Nyasa, around which extends a fringe of lake plains of varying width. The slope of the plain lying along the north-western shores of the lake is continued for many miles below lake-level almost to the sub-lacustrine foot of the Livingstone Mountains, where the maximum depth of 2200 feet has been recorded.* Farther south, between the mouth of the South Rukuru River and Nkata Bay, the adjacent section of the lake floor is tilted in the opposite direction, towards the foot of the western plateau, which rises abruptly from the floor of the lake in a manner similar to that of the Livingstone Mountains. In a third and shallower section extending to the southern end of the lake, the floor is gently concave, and here a well-developed drowned topography may be observed along many parts of the coast.

The lake plains are sometimes built up of great thicknesses of sand and clay of late Recent age, as in the country west of Domira Bay, for example, but sometimes they consist of even crystalline platforms that dip gently, except where traversed by young faults, below the level of the lake, as in the area west of Kota Kota ; these crystalline platforms are usually overlain by thin patches of pebbly gravel. The north-western plain, however, is traversed by a number of crystalline fault-ridges running parallel with the lake-shore, and between these ridges lie narrow strips of the sediments that make up the Dinosaur Beds and the Lacustrine Series.

Towards the close of the period of deposition of the Lacustrine Series, the crystalline ridges were more or less completely buried under these sediments ; as the waters of the lake receded, the streams that ran towards the lake over the surface of that time gradually cut their way down to the buried ridges, and then in due course incised deep narrow gorges through them ; at the same time the secondary streams opened up wide transverse valleys in the soft sediments lying between the ridges. Consequently, in this region the rivers of to-day may be observed to run directly across the grain of the country in their progress from the foot of the western scarp to the lake-shore, so that instead of flowing along the broad shallow troughs running parallel with the coast they first enter the troughs on one side through a deep narrow gorge and then depart from them again by a similar gorge situated almost directly opposite. Some of the weaker streams were naturally unable to maintain their courses across the hard ridges, so that after a time they were obliged to abandon the gorges they had already cut, and these now remain only as wind-gaps ; the streams were then obliged to find an easier course to the lake through the softer rocks, or else they were captured by the headward erosion of the tributaries of neighbouring more vigorous streams.

In many places old storm beaches and lines of sand-dunes run parallel with the shore, and between them swamps and lagoons may often be observed ; ridges of similar form may continue below the water-line, and these sometimes make landing very difficult for the lake craft.

* In ' The Tanganyika Problem ' (London, 1903), p. 123, J. E. S. Moore gives 2580 feet as the maximum depth of lake Nyasa.

Lacustrine deposits and raised beaches show that the level of the lake has varied considerably from time to time. The older lacustrine deposits range from 200 to 1000 feet above the present level of the lake, while· gravels and raised beach deposits of very late age stand at the following successive levels : 400, 150, 25, and 15 feet. The lake appears to have been able to effect but very little lateral erosion into the crystalline rocks that abut upon its shores, although severe storms are of common occurrence at certain seasons of the year ; this is probably due to the circumstance that the rather wide seasonal and periodical variations in the level of the lake, combined with the instability of level indicated by the lake beaches, have rarely if ever enabled the waves to concentrate their efforts within suitable limits.

The fact that the water of the present lake is fresh, and that the only fossils found in the older lacustrine deposits are of fresh-water origin, indicates that the lake has had a more or less continuous outlet to the sea almost from the beginning, in spite of the vicissitudes it has undergone.

3. The Shire Rift Valley and Associated Fractures.

An examination of the uplands bordering the Shire Rift shows that there are several additional branches of the rift system almost as important as that of the Shire itself (VII.). Actually, the rift system in this region appears to comprise two main series of fractures that run approximately N.N.W. and N.N.E. (see Figs. 1, 3). For instance, the Shire Rift itself follows these two directions in turn ; it may thus be regarded as consisting of (1) an Upper Shire Rift, running N.N.E., that extends from Malombe to the Murchison Cataracts, and (2) a Lower Shire Rift, running N.N.W., that extends from the Murchison Cataracts to Port Herald. Running parallel with the Upper Shire Rift is the long depression that includes lake Chilwa and the adjacent low country, and also the elongated Palombe–Tuchila plain that extends from the Chilwa plain past the western foot of the Mlanje Mountains to the lower course of the Ruo River ; this depression not only passes, at its northern end, into lake Chiuta and the upper Lujenda valley,* but also it stands in practically the same line as the Urema Sunkland of E. O. Teale and R. C. Wilson,† which lies a little further south across the Zambezi.

Running parallel with the Lower Shire Rift, on the other hand, is the depression that extends towards the S.S.E. from the southern end of lake Nyasa and includes lake Malombe, that part of the Upper Shire Rift that it crosses obliquely, the even plain north of Chikala Hill, lake Chilwa, and the eastern foot of the Mlanje massif. Lakes Malombe and Chilwa are thus seen to occupy depressions situated at the intersection of a N.N.W. rift by two N.N.E. rifts, and the Murchison Cataracts to

* See also F. E. Studt, " The Geology of Katanga and Rhodesia," *Proc. Geol. Soc. S. Africa*, 1913, p. 44 ; and J. E. S. Moore, ' The Tanganyika Problem' (London, 1903), p. 48.

† " Portuguese East Africa between the Zambezi River and the Sabi River," *Geogr. Journ.*, 46, 1915, p. 286.

be situated in the broken country at the angle in the Shire Valley in which the Upper Shire Rift and the Lower Shire Rift intersect. Moreover, the Shire Highlands are seen to constitute a broadly diamond-shaped upland area bounded by four branches of the rift system. This upland area however is not of uniform height, but is tilted towards the east in a direction parallel with its shorter axis. Consequently, a traverse from west to east across the broadest part of the Shire Highlands and the

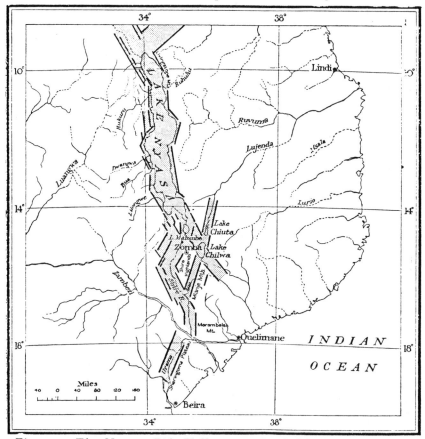

Figure 1. *The Nyasan Rift Valley: associated fractures and down-faulted areas.*

adjacent part of Nyasaland would show (Fig. 2) an abrupt descent from the Kirk Mountains into the wide floor of the Shire Rift, followed by an abrupt but stepped rise to the Shire Highlands, and a long gentle slope down to the foot of the magnificent step-faulted scarp forming the western face of the Mlanje massif.

It would appear that the rift system described has resulted from faulting that took place at several distinct periods separated by appreciable intervals. For instance, the Upper Shire Rift did not assume its present form until after the Malombe–Chilwa Rift was well established and its floor reduced to an even plain ; this plain was then fractured along the line of the low Makongwa scarp that runs for nearly 20 miles in a northerly direction from near Chikala Hill, and forms the eastern wall of that part of the Upper Shire Rift.   Doubtless connected with these movements is the " drowned " coast of lake Nyasa.   At an even later date additional fracturing left the Mpimbi–Namitembo valley plain standing 250 feet above the modern floor of the Upper Shire Valley.

The walls of the Shire Rift Valley are bold and well defined throughout a large part of their course ; at different points they are respectively high and precipitous, broken into ascending fault-steps, or merely of irregular form.   South-east of lake Malombe, however, the eastern wall is practically absent over a distance of nearly 20 miles, or is at least represented only by a low scarp of relatively late age, the Makongwa Scarp, that separates this part of the Rift Valley from a branch of slightly older date. The next gap of importance lies on the western side of the valley between the southern end of the Kirk Mountains and the ridge of crystalline rocks

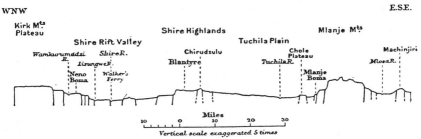

Figure 2. *Section across Southern Nyasaland from Kirk Mountains to Mlanje Mountains.*

that runs southwards from Namalambo Hill near Chiromo to Port Herald.   Since this gap is occupied exclusively by rocks of Karroo age, it has doubtless originated in consequence of the low resistance to erosion of these rocks as compared with that of the Crystalline Series.   Opposite the Chiromo–Port Herald ridge the Rift Valley widens considerably, and below this point the course of the rift is indicated mainly by the bold western scarp of the isolated Marambala Mountain.

11

Reprinted from pp. 3–5, 28–31, and 169–170 of *Tanganyika Geol. Survey Mem. II*, 1960, 172p.

THE RUNGWE VOLCANICS AT THE NORTHERN END OF LAKE NYASA

D. A. Harkin

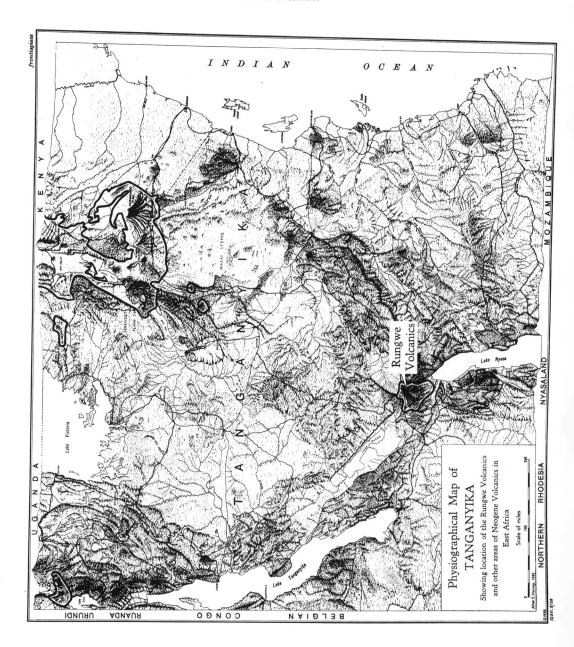

Physiographical Map of TANGANYIKA

Showing location of the Rungwe Volcanics and other areas of Neogene Volcanics in East Africa

[Editor's Note: In the original, material precedes this excerpt.]

(c) PHYSICAL FEATURES

The physical features of the region closely follow the geology, with the rift scarps and volcanoes as controlling factors. The Y-shaped form of the Nyasa, Rukwa and Ruaha troughs is bounded on its three major sides by high country of pre-Karroo formations.

The eastern boundary is possibly the most impressive feature of its kind in Africa, the rift scarp at the north-eastern end of Lake Nyasa. From lake level at 1,568 feet, what appears to be a sheer wall when viewed from the west, rises to altitudes of well over 7,000 feet in the Livingstone Mountains. Northwards from the Livingstones to the Elton Plateau, where volcanics cover the pre-Karroo formations, there is a gradual rise in the land surface to a maximum of just over 9,000 feet at the western edge of the Plateau, on the edge of the fault scarp. The highest point is CHALUHANGI (s26, 9,623 ft.), probably an old basaltic cone, at the south-western corner of the Plateau. The Elton Plateau has a slight tilt towards the north-east, that is, away from the rift scarp, and is bounded in this direction by the northern part of the Kipengere Range, made up chiefly of Bukoban (Buanji) quartzites, with altitudes around and over 9,000 feet south of the Ndumbi Gap (y24). ISHINGA (p14, 8,821 ft.) marks the northern extremity of the Range. Except at two places, the Matamba Gap (u18) and the Ndumbi Gap (y24), this ridge has acted as a barrier to the lavas spreading eastwards over the Elton Plateau. At the northern end of the Elton Plateau, the land surface falls away rapidly along an east to west line to the Buhoro Flats (Ruaha trough), at about 4,000 feet.

Unlike the north-eastern side of Lake Nyasa, the north-western boundary of the Nyasa trough is ill-defined in Tanganyika. While the ancient gneisses lie near the edge of the lake in Nyasaland, their boundary is further to the west in Tanganyika. This is reflected in a widening of the trough north of the Songwe River. The Igembe Hills, south of the Songwe River in Nyasaland, and the Bundali Hills, rising to over 6,000 feet in the west, are the major features. Like the Livingstone Mountains in the east, the Bundali Hills have a gradual increase in altitude northwards to the high, rolling, plateau-like country of Umalila with levels generally between 7,000 feet and 8,000 feet (MBOGO, 8,064 ft. is just west of the mapped area). The northern part of Umalila has a covering of volcanics and slopes gently westwards. In the north, it gives way along the low, dissected, south-western Rukwa Escarpment to the Rukwa trough, at just over 4,000 feet.

In the north the Mbeya Block, the wedge-shaped mass of ancient rocks, is highest at its southern apex where MBEYA MOUNTAIN (J8, 9,272 ft.) stands above the rest of the Block, which is here about 8,000 feet (NYANUWA (U7, 8,012 ft.)) and has a covering of volcanics, giving a small area of plateau-like country at the top of the Usangu Scarp. The Mbeya Block is bounded on its western and eastern sides by clear-cut scarp features falling to the Rukwa and Ruaha troughs along north-westerly and north-north-easterly lines respectively (Plate II).

Other major topographic features are formed by crystalline rocks occurring as two inliers in the middle of the Nyasa trough. The more southerly inlier forms a rift scarp south of Kiejo (m35). Near to Massoko Lake (k42), where it attains its maximum relief approaching 1,000 feet, this youthful feature looks like a miniature of the main Nyasa rift scarp which runs parallel to it about 10 miles east. It diminishes rapidly both to north and south. Volcanics blanket its extremities and its topography has been modified in

159

places by lavas of the Kiejo suite, which have flowed down the scarp face. The second inlier forms KALENGALENGA (V29, about 5,400 ft.), a hill of knife-edged ridges and spurs to the west of Rungwe. There seems little doubt that this inlier is tectonic in origin, although there is no direct evidence. Other outcrops of gneiss are so small as to have no effect on the topography of the region.

Along the western boundary of the Nyasa trough, prominent features have resulted from sedimentary rocks of Karroo and Cretaceous age, and the same is true to a lesser extent in the Rukwa trough. The major area of Karroo rocks is that of the Songwe-Kiwira Coalfield (Harkin, 1955a), where the north to south strike of the beds is reflected in the Kabulo-Ivogo Ridge (about 4,000 ft.) and in lesser features such as the low (200 ft. high) Ngana Ridge (e60).

Cretaceous sediments have a more widespread distribution along the western side of the Nyasa trough and in the Rukwa area. In the extensive outcrop south of the Igale Pass (M24) they give rise to hilly country reflecting the strike of the rocks. Elsewhere their topography is of lesser importance and subordinate to that of the succeeding volcanics.

Within the troughs the most prominent features, which are due to the volcanics, may be divided as follows:

(a) The arc-shaped Poroto ridge which lies transversely across the northern end of the Nyasa trough, from Umalila in the west to the Elton Plateau in the east. Most of the ridge lies around 7,000 feet. It is slightly concave towards the south and is made up of a number of volcanoes, of which the major and highest point is NGOZI (V18, 8,602 ft.), occupying a central position and holding a large deeply-set crater lake.

The Poroto ridge forms a barrier completely separating the Nyasa trough from the Rukwa and Ruaha troughs, so that a belt of high ground, enclosing the northern end of the Nyasa trough, sweeps round from the Livingstone Mountains through the Elton Plateau, the Porotos, Umalila and the Bundali Hills.

Immediately south of the Poroto arc are situated the major central volcanoes:

(b) RUNGWE (d28, 9,713 ft.), occupies a central position in the trough, just south of the Porotos.

(c) KATETE (j27, about 8,500 ft.), an eroded volcano on the eastern side of the trough, lies just east of Rungwe.

(d) KIEJO (m35, 7,135 ft.), lies a little to the east of the centre of the trough and south-east of Rungwe.

(e) TUKUYU (b37, 5,101 ft.), is an extensive, low, dome-shaped volcano south-west of Rungwe.

Lavas flowing northwards from the Poroto ridge have banked up against the southern apex of the Mbeya Block and flowed around it into the Rukwa and Ruaha troughs. To the south of the Poroto ridge, the flow of lavas is generally southwards, towards the alluvial plain bordering Lake Nyasa. They abut against the steep Nyasa scarp in the east, but in the west have encroached in places on to the gneiss of the slightly gentler slopes rising to the Bundali Hills.

The Poroto ridge divides the Rukwa and Ruaha drainage systems in the north from the Nyasa system in the south. The Rukwa-Songwe drainage into the Rukwa dischargeless lake includes streams flowing off the northern part of Umalila and the western side of the Porotos. That part of the Porotos to the east of Mbeya and much of the Elton Plateau are within the Ruaha system, which flows to the Indian Ocean.

South of the Porotos all drainage is to Lake Nyasa and ultimately to the Indian Ocean *via* the Shire and Zambezi rivers. Within the Nyasa trough the main rivers are:—

(a) Lufirio, rising on the north-western edge of the Elton Plateau and following the eastern edge of the trough to the Lake.

Plate II

Looking to the **Mbeya Range** and **Rukwa Trough** from the north flanks of the **Porotos** near Igale. Two young cones of melanephelinite in middle ground

161

(b) Kiwira, which rises close to the source of the Lufirio, bu tflows westwards between Rungwe and the Porotos before pursuing a course, which follows closely the western side of the trough, to the Lake.

(c) Mbaka, draining the south of Rungwe and the central part of the trough.

The Livingstone Mountains and the southern part of the Elton Plateau are within the Lake Nyasa drainage, while the Nyasa-Songwe, which forms the Tanganyika-Nyasaland boundary over most of its course, includes the streams of southern Umalila and the Bundali Hills.

The Rukwa dischargeless lake was probably caused by reversal of drainage following the formation of the Poroto ridge. Previously, Rukwa trough drainage may have been mainly to Lake Nyasa, the level of which is about 1,000 feet lower than Lake Rukwa. This was recognized by early German workers (Scholz, 1914, p.83) and is also referred to by Spence (1954, p.19).

(d) Geological Setting

The following table shows the geological formations, other than the Rungwe Volcanics, which are represented in the region.

NEOGENE		Lacustrine deposits.
—————————— Unconformity ——————————		
CRETACEOUS		Red sandstones, siltstones, grits, etc.
—————————— Unconformity ——————————		
	K5	Green mudstones with magnesian limestones and calcareous sandstones.
KARROO[7]	K3	Red mudstones and siltstones with inter-calated feldspathic sandstones.
	K2	Coal measures.
	K1	Basal conglomerate and sandstones.
—————————— Unconformity ——————————		
BUANJI SERIES (BUKOBAN)		Quartzites, red shales, etc.
—————————— Unconformity ——————————		
UKINGA SERIES (KARAGWE-ANKO-LEAN?)		Phyllites, etc., with development of granitic rocks in places.
—————————— Unconformity ——————————		
UBENDIAN		Gneisses, schists, etc.

1. Ubendian, Ukinga Series, Buanji Series

Only an outline is necessary of the rocks of the pre-Karroo "complex," which are outside the scope of the present study and have not been examined in detail. Both the Ubendian and the Ukinga Series are of Precambrian age, while the Buanji Series may be late Pre-cambrian or early Cambrian.

(i) *Ubendian.*—The general picture of the pre-Karroo geology gained by the most recent work (Harpum, 1955) in Ukinga and Uwanji, which lie to the east of the Rungwe Volcanics, is one of a large anorthositic igneous complex of batholithic dimensions, centred in Ukinga, which has intruded a series of geosynclinal sediments. Subsequent events in Ubendian times included high-grade regional metamorphism and migmatization, with emplacement of extensive migmatitic granite. The metasediments and the anorthosite complex were metamorphosed, although the anorthosite acted as an effective granite resister.

———————————

[7]The division of the Karroo beds into K1, K2, K3 and K5 follows the classification used by Stockley (1931) in describing the rocks of the Ruhuhu Coalfields, which is the type area in Tanganyika.

162

[*Editor's Note:* Material has been omitted at this point.]

III.—TECTONIC RELATIONSHIPS OF THE VOLCANICS

The frontispiece illustrates the association of the Rungwe Volcanics, and also of most of the Neogene volcanics in East Africa, with the rift system. It is furthermore significant that the Rungwe pile has arisen at the junction of the so-called "Western" (Rukwa), "Eastern" (Ruaha), and Nyasa rifts, whereas for several hundreds of miles along these rifts, there is no comparable volcanicity.

The structural pattern shown by the rocks underlying the volcanics has already been discussed (*vide* Geological Setting, p. 5). It is clear from the faulting, distribution and disposition of the rocks, that the structures established in Precambrian times have controlled the subsequent tectonic history of the region.

In considering the major fault movements in Neogene times which have given rise to the East African rift system as it is known today, it becomes evident that the "graben" concept of a rift valley is not typical in the region being discussed.

The German workers recognized that although the eastern edge of the Bundali Hills is a tectonically controlled feature, it no longer represents the main fracture on the western side of the Nyasan rift system.   Large-scale faulting appears to follow the more westerly course of the Nyasa-Songwe River (along the Tanganyika-Nyasaland boundary) for some distance before swinging north-westerly again, in the region to the west of the Bundali Hills and Umalila (Lehmann, 1924, p. 3).   The frontispiece illustrates the concept of the Ubendian rocks of the Bundali Hills and Umalila thus forming a large horst within the Nyasa-Rukwa rift system.

In the Nyasa trough (which in Tanganyika is considered to be the depression between the Nyasa Scarp and the Bundali horst) the main fracture of the rift system in the west roughly corresponds to the western boundary of the trough as far north as the Kasimulo Hill area (1 59).   In Nyasaland, faulting on both sides of the trough throws down towards the depression, although that in the west does not approach in magnitude the large Nyasa rift scarp in the east.   Westwards from the Nyasa Scarp to the Bundali Hills, however, the fault pattern is one of step-faulting across the trough, with downthrows consistently to the west, even near to the edge of the Bundali Hills horst.   The Nyasa Scarp is itself made up of a stepped series of normal faults.   Bailey Willis (1936) in discussing the Nyasan rift, describes it thus:—

> ". . . . my own examination was confined to the Livingstone Mountains and north-eastern coast.   There, however, the facts are obvious and admit of no misinterpretation.   Climbing from the lake at Lumbira (Alt. Langenburg) or any point in the vicinity, one ascends the spurs of a mountain range, as it seems; but the front, on which the spurs are but slight projections between ravines, is composed of a succession of step-faults.   The downthrow is in every case toward the lake basin, the upthrow increased the mountain height, ..."

The Ubendian scarp in the middle of the Nyasa trough is a replica in miniature of the Nyasa Scarp and throws down to the west.   Further west, in the Songwe-Kiwira Coalfield, the largest fault (d51), also in a north-westerly direction, shows a large downthrow to the west, probably about 2,000 feet (Harkin, 1955a).   Faulting noted in the Cretaceous outcrop south of the Igale Pass (M24), though of smaller dimension, also shows the step-fault pattern, with downthrow to the west.

A similar situation exists in the Rukwa trough, where the Mbeya Scarp, on the eastern side, is a fresh, clear-cut, young feature similar to the Nyasa Scarp, but much smaller.   The western side of the trough, however, bordering the Umalila part of the horst referred to previously, is marked by a mature, dissected, low escarpment where there has probably been little movement since post-Karroo, pre-Cretaceous times (Spurr, 1953, pp. 7-9).   In verbal communications, geologists of the Mbeya Exploration Company, who are engaged in detailed exploration of this area, have informed the writer that they have not located any faulting along this escarpment, so that the tectonic setting here is, not unexpectedly, a continuation of that obtaining further south, on the edge of the Bundali Hills.

The Usangu Scarp bounding the Buhoro Flats (Ruaha depression) in the west is a youthful feature and evidently owes its origin to movements during the Neogene rift episode. It is not, however, so youthful-looking as the Nyasa Scarp.   It is significant that the considerable rise in the land surface to the south of the Buhoro Flats is along an easterly line although there is here no regular scarp feature.   Harpum (1955, p. 27), considers this boundary to be made up of a zig-zag pattern of faults, formed by rift movements (later than those connected with the Nyasa rift) along a pre-existing fracture system normal to the main Nyasa rift, with new fractures developed as connecting links.

No evidence was found in the region suggestive of compressive rather than tensional forces giving rise to the rift faults.

Practically all of the faulting shown on Map G.S. 1178 (in pocket) follows the Nyasan trend (north-westerly) or the Songwe trend (east to west) which have been referred to previously (*vide* Geological Setting, p. 5).   Not shown on this map, however, is a considerable amount of faulting in the crystalline rocks, postulated by workers in the Mbeya Block and in the Ukinga (Livingstone Mountains) area, on the fringes of the region being discussed.

Quennell (1951, p. 2) has shown, mainly from a study of air photographs, that in the Mbeya Block, there is a general swing in the direction of faulting from north-westerly (Nyasa trend) in the west to easterly (Songwe trend) in the east. This appears to be a parallel of the tectonic pattern in the Bundali Hills, which have easterly strikes and faulting in the east, but show a return to the north-westerly strike direction in the west.

An important direction of faulting found by Harpum (1955) in Ukinga, east of the mapped area, is at right-angles to the Nyasan trend, that is, approximately north-easterly. Harpum considers that movement along these faults in "rift times" was almost completely radial and caused very little displacement of the main Nyasa rift scarp (op. cit., p. 397). From a study of the air photographs and from field evidence further east, Harpum has outlined a series of these faults cutting the crystalline rocks of the Nyasa Scarp.

Map G.S. 1178 shows that the distribution of volcanic centres has been controlled to a large extent by the under-lying tectonic pattern and especially by the Nyasan direction. It is also significant that the volcanics have arisen mostly in the eastern part of the Nyasa trough where there was the greatest fault displacement.

The most prominent tectonic line suggested by the structural and volcanological features is that linking Ngozi, Rungwe and Kiejo and there are other obvious parallel lines, especially south of Kiejo, in what was evidently a structurally weak area.

Field evidence has shown that Katete may have arisen on one of the rift faults (vide p. 10), possibly on an extension of the one at the base of the Nyasa Scarp in the south. The Scarp ankaramite may be situated on one of the step-faults of the scarp.

The distribution of volcanic centres along the northern flank of the Porotos and on the edge of the Buhoro Flats would seem to reflect the swing to the Songwe trend from the Nyasan trend in the pre-volcanic formations, and is therefore a continuation of the Mbeya Block tectonic pattern.

Although not so well illustrated as the Nyasan trend, it is likely that tectonic lines more or less normal to the Nyasan trend, have exerted control in the distribution of the volcanoes. The Poroto ridge appears to be aligned along an east-north-easterly line, and Harpum (1955) has traced as far as the Ndumbi Gap (y24) a large fault along this direction, the extension of which further to the west could pass through Katete and Rungwe.

Should this be the case, the three largest volcanic foci in the province would be located at the intersections of tectonic lines along the Nyasan trend and that normal to it.

Another tectonic line is suggested by the group of explosion craters lying across Rungwe, which seems to connect with the northerly line of the Usangu Scarp. Just north of the mapped area, the Usangu Scarp runs in a north-north-easterly direction,

With the possible exceptions of Katete, the Scarp ankaramite and some small cones at the foot of the Ubendian scarp south of Kiejo, none of the volcanic centres is situated on rift faults exposed in the volcanic area. Rather it appears that they may be built up in the middle of the faulted block, as is possibly the case with Kiejo and Rungwe. It is possible that in large-scale normal or block-faulting, such as seems to have occurred here, while the major faults formed closed structures after movement, tension cracks opened in the fault block itself, which would permit the easy upwelling of lava. These cracks are also located on structural lines, however, as is indicated in Map G.S. 1178, where the continuation of the line joining the Kiejo, Rungwe and Ngozi volcanoes coincides with faulting in the ancient rocks of the Mbeya Block.

The major movements of the Rift episode in post-Miocene times appear to antedate the bulk of the volcanics and were probably instrumental in initiating the activity. In some cases, however, there is evidence of further movement on the rift faults. In discussing Katete (p. 10) it was considered that movement on the main Nyasa rift scarp is probably the most likely explanation for the relatively high level of the lavas of the Elton Plateau, which are thought to have come from Katete and perhaps other centres within the Nyasa trough.

The Ubendian scarp south of Kiejo may have taken its present form at about the same time as the latest major movement of the Nyasa Scarp. Its relationship to the Tukuyu lavas and the Kiejo older basaltic (Kb) group, could not be ascertained. Although it has suffered no major movement since before the formation of Rungwe, the air photographs show one of the small scoria cones situated on it (n43) to have suffered very minor movement in post-Kiejo younger basaltic (Kb1) times.

[*Editor's Note:* Material has been omitted at this point.]

VIII.—THE EAST AFRICAN RIFT VOLCANICS

While a review is not within the scope of the present work, the application of the petrological concepts discussed for the Rungwe Province to the other East African rift volcanics can be considered on the basis of some general characteristics.

A comprehensive overall picture of the large mass of volcanics associated with the eastern rift of Tanganyika and Kenya is not readily gained although a considerable volume of literature is available, notably the early work of Gregory (1900, 1921), that of Campbell Smith (1931, 1938) and the numerous contributions by members of the Kenya Geological Survey. In Tanganyika, a short preliminary account on the work of the first Sheffield University Kilimanjaro Expedition (1952) has appeared (Wilcockson, 1956), and a memoir on the results of this and the subsequent expedition (1957) is in preparation. James (1956, 1958) has studied the carbonatite centres in the Northern Province but the work of N. J. Guest (1953) over a large area of the Northern Province volcanics is not yet published.

The literature indicates basaltic lavas to be abundantly represented in Kenya and Tanganyika; Guest has noted an extensive series of "basic andesites" (basaltic lavas according to the mineralogical assemblage) as generally older than some of the large central volcanoes associated with carbonatite activity, and the Sheffield University expeditions have observed that trachybasalts made up the bulk of Kilimanjaro, but normal olivine–basalts and phonolites (including rhomb-porphyries) are also present. In addition to rock types that can be matched in Rungwe, the eastern rift field in Kenya contains rhyolites and other saturated lavas more typical of calc-alkaline assemblages, and a greater range and variation in types is evident, which probably results, at least in part, from the relatively greater size of the field.

A notable feature is the large volume of phonolitic and trachytic lavas in the Kenya sequence; as well as being prominent in the large volcanoes like Mount Kenya, sheets of these rocks, especially the phonolites, are of wide lateral extent.

Unlike the Rungwe Province, there is no marked interval of time separating the volcanics showing basaltic affinities from those associated with carbonatite activity in the eastern rift areas; both types have been erupted during Tertiary and Recent times, so that there is neither isolation in space nor time.

Greater variation in the basaltic layer (whatever its ultimate derivation) due to the larger size of the field, and the more or less contemporaneous activity of basaltic magma and that producing the carbonatites, with resulting greater interaction, may account for the heterogeneity of the eastern rift rock assemblages as compared with the Rungwe lavas.

The western rift volcanics in Uganda and the Belgian Congo and known chiefly from the work of A. D. Coombe, Holmes and Harwood (1937) and later contributions by Holmes (1950), in the Bufumbiro and Toro/Ankole fields, which are distinctive on account of the potash-rich character of the association.   In discrediting the conception that the potash-rich lavas of the western rift were somehow the result of compression tectonics whereas the eastern rift soda-rich volcanics were related to a tensional environment, Bowen (1938) noted that in the North and South Kivu volcanics there was no consistent predominance of potash over soda.   Considerable work in these fields by Belgian Congo geologists (e.g., Meyer, 1954; Tazieff, 1951), much of the data and results of which are not yet published, has confirmed the "mixed" nature of the western rift province.   The South Kivu volcanoes have erupted a more normal alkaline olivine–basalt association, although there are minor amounts of potash-rich rocks, and some basalts with tholeiitic affinities are also present.   To explain certain distinctive ultrabasic potash-rich lavas in the Toro-Ankole field near Ruwenzori, Holmes (1950, p. 786) proposed the reaction of carbonatite magma with granitic rocks, and the occurrence of carbonatite has since been reported in northern Kivu near the south-west termination of the Lake Edward rift (Béthune and Meyer, 1957).   The presence of a basaltic magma and one which produced carbonatite is therefore also demonstrable in the western rift.   While the occurrence of potash-rich rocks indicates some special local conditions or environment, the broad outline of the petrogenesis of this part of the East African rift province would appear also to demand the consideration of both magma types.

The last major group to be considered comprises the late Tertiary group of carbonatite-type volcanoes; Elgon, Kadam, Napak, Moroto and Toror, which are aligned along the Kenya border in eastern Uganda.   Two of these centres, Napak and Elgon, are the subjects of memoirs by King (1949) and Davies (1952) respectively.   Although James (1956) considers that carbonatites in East Africa are generally associated with block faulting, notably that of the rift episode, Davies (1952, p. 50) saw little evidence to suggest that major faulting is related to this group of volcanoes.   Neither are lavas showing basaltic affinities present in the group, the general position of which is separated from the eastern rift alkaline and calc-alkaline volcanics, lying to the east in Kenya.   It may therefore be that the absence of large-scale rift faulting in this area has precluded the generation and/or the eruption of basaltic magma; alternatively, the supposed basaltic layer may be absent here.

It is concluded that the solution of petrogenetical problems of the East African rift volcanics offers least difficulty if the two apparently related magma types—the basaltic and that producing the carbonatites—are taken as the fundamental basis for consideration.

[*Editor's Note:* Material has been omitted at this point. Only references cited in these excerpts are reprinted here.]

REFERENCES

Béthune, P. de, and A. Meyer, 1957, Carbonatites in Kivu, *Nature, Lond.,* 179, pp. 270-271.

Bowen, N. L., 1938, Lavas of the African rift valley and their tectonic setting, *Amer. J. Sci.,* Ser. 5, 35, pp. 19-33.

Davies, K. A., 1952, The building of Mount Elgon (East Africa), *Mem. geol. Surv. Uganda,* 7.

Gregory, J. W., 1900, Geology of Mount Kenya, *Quart J. geol. Soc. Lond.,* 56, pp. 205-223.

Gregory, J. W., 1921, The rift valleys and geology of East Africa, Seeley, Service and Co. Ltd., London.

Guest, N. J., 1953, The geology and petrology of the Engaruka-Oldonyo L'Engai-Lake Natron area of Northern Tanganyika. Unpubl. Ph.D. Thesis, Sheffield University.

Harkin, D. A., 1955a, The geology of the Songwe-Kiwira coalfield, Rungwe District, *Bull. geol. Surv. Tanganyika, 27.*

Harpum, J. R., 1955, Ukinga series in south-west Tanganyika, *Unpubl. Rep. geol. Surv. Tanganyika.*

Holmes, A., 1950, Petrogenesis of katungite and its associates, *Amer. Min.,* 35, pp. 772–792.

Holmes, A., and H. F. Harwood, 1937, The petrology of the volcanic area of Bufumbira, *Mem. geol. Surv. Uganda,* 3, part 2.

James, T. C., 1956, Carbonatites and rift valleys in East Africa, *Rep. 20th Int. Geol. Congr. Mexico* (Abstr., also unpublished account.).

James, T. C., 1958, Summary of silicate rocks associated with carbonatite bodies in Tanganyika, *C.C.T.A. East-central, West-central and Southern reg. Comm. Geol.,* Leopoldville, 1958, pp. 307–308.

King, B. C., 1949, The Napak area of southern Karamoja, Uganda, *Mem. geol. Surv. Uganda,* 5.

Lehmann, E., 1924, Das Vulkangebiet am Nordende des Nyassa als magmatische Provinz, Reimer, Berlin (original in *Z. Vulkanol.,* 4, p. 175).

Meyer, A., 1954, Notes vulcanologiques, *Mém. Serv. géol. Congo Belge et Ruanda Urundi,* 2.

Quennell, A. M., 1951, The Lupa Goldfield, Tanganyika Territory, *Min. Mag. Lond.,* 85, pp. 341–347.

Scholz, E., 1914, Beiträge zur Geologie des südlichen Teiles Deutsch-Ostafrikas, *Mitt. dtsch. Schutzgeb,* XXVII, Bd. 1, pp. 49–67.

Smith, W. Campbell, 1931, Classification of rhyolites etc. from part of Kenya Colony, *Quart. J. geol. Soc. Lond.,* 87, pp. 212–258.

Smith, W. Campbell, 1938, Petrographic description of volcanic rocks from Turkana, Kenya, *Quart. J. geol. Soc. Lond.,* 94, pp. 507–553.

Spence, J., 1954, The geology of the Galula Coalfield, Mbeya District, *Bull. geol. Surv. Tanganyika,* 25.

Spurr, A. M. M., 1953, The geology of the Songwe River, Mbeya District, *Unpubl. Rep. geol. Surv. Tanganyika.*

Stockley, G. M., 1948, Geology of north, west and central Njombe District, Southern Highlands Province, *Bull. geol. Surv. Tanganyika,* 18.

Tazieff, H., 1951, L'éruption du volcan Gituro, *Mém. Serv. géol. Congo Belge et Ruanda Urundi,* 1.

Wilcockson, W. H., 1956, Preliminary notes on the geology of Kilimanjaro, *Geol. Mag.,* 93, pp. 218–228.

Willis, Bailey, 1936, East African plateaus and rift valleys, *Publ. Carneg. Inst.,* 470.

Editor's Comments
on Paper 12

12   McCONNELL
Excerpts from *Outline of the Geology of Ufipa and Ubende*

TANGANYIKA ZONE

The geology of the Tanganyika Rift Zone is not described comprehensively in any one publication. The eastern margin of the southern half as far as 6°S latitude is briefly described by McConnell in Paper 12. Willis (1936) gives an incomplete but useful description of the lake margins and applies a datum plane technique, the plane being the mid-Tertiary surface. Capart (1949) describes the geomorphology and the submarine configuration and sedimentology of the lake floor from echo sounding. Veitch (1935) describes the Lukuga Gorge through which flows the Lukuga River, the outlet of the lake, and the relationship of the African or mid-Tertiary surface in former Katanga and the Congo to the present lake level.

Cahen (1953, 1954) briefly describes the western system but with greater emphasis on the Kivu to Albert zone. The Tanganyika Zone is described and discussed by Dixey (1946). Regional geological mapping is uneven in its coverage. The southern end is covered by Zambia Geological Survey mapping. The Tanzania Survey maps cover the eastern margin to the Burundi border. The mapping of Burundi and of the western margin in Zaire is on smaller scale.

Many features of the Tanganyika Rift Valley remain unstudied, but their explanation is necessary for an understanding of the zone. These include: (1) the southern termination in faulted Precambrian basement; (2) the termination in the north and absence of structural connection with Kivu (see, however, Paper 13); (3) presence on both flanks of the African (mid-Tertiary) "datum" surface (Willis, 1936) at different levels (200 m higher on the west at latitude 6°S) signifying continuity before the later rift faulting; (4) differences in altitude of remnants of this datum surface along the lengths of the flanks; (5) the prerift westward flow of the principal rivers (for example, Malagarasi and Rungwa via the Lukuga) on the mid-Tertiary surface (Wayland, 1952; Teale, 1932; Cooke, 1957) with subsequent truncation, capture, and reversal (this also applies to the Kivu-Albert zone); (6) lowering and raising of lake level from 1200 m above sea level (Veitch, 1935) down to minus 80 m (Capart, 1949), and thence up to 771 m (present level); (7) the presence of two deeps in the rift lake floors exceeded only by Lake Baykal, minus 539 m and minus 699 m, respectively (Cahen, 1954); (8) the differences in the nature of the coasts—emergent, submergent, fault—along the lengths of both flanks, generally without transverse matching; (9) the absence of any sediments older than Miocene that could have been deposited within the trough; (10) the complete absence of vulcanism associated with the rift faulting.

The Upemba trough is in Katanga and is described by Cahen (1954, p. 467) and Cornet (1905).

REFERENCES

Cahen, L., 1953, Esquisse tectonique du Congo belge à l'echelle du 1/3,000,000, *Comm. de Géologie du Ministre de Colonies,* Brussels.

Cahen, L., 1954,*Géologie du Congo Belge,* Vaillant-Carmanne, Liège, 577p.

Capart, A., 1949, *Exploration Hydrobiologique de Lac Tanganika (1946-1947). Resultats Scientifiques,* vol. 2, Inst. Royale Sciences Naturelles de Belgique, Brussels, pp. 1-16.

Cooke, H. B. S., 1957, Observations relating to Quaternary environments in East and Southern Africa, *Geol. Soc. S. Africa Bull.* **60**(annex): 1-73.

Cornet, J., 1905, Les Dislocations du Bassin du Congo. I: Le Graben de l'Upemba, *Soc. Géol. Belgique Annales* **32**:205-234.

Dixey, F., 1946, Erosion and tectonics in the East African Rift System, *Geol. Soc. London Quart Jour.* **102**:339-388.

Teale, E. O., 1932, The eastern shore of Lake Tanganyika and its vicinity, *Tanganyika Geol. Surv. Ann. Rep., 1931,* pp. 22-24.

Veitch, A. C., 1935, Evolution of the Congo Basin, *Geol. Soc. America Mem. 3,* pp. 91-98.

Wayland, E. J., 1952, The study of past climates in tropical Africa, *1st Pan African Congr. Prehist. Proc.,* Oxford, 1947, pp. 56-66.

Willis, B., 1936, Studies in comparative seismology: East African plateaus and rift valleys, *Carnegie Inst. Washington Pub. 470,* 343p.

12

OUTLINE OF THE GEOLOGY OF UFIPA AND UBENDE

R. B. McConnell

PART II PHYSIOGRAPHY

(a) *General*

Ufipa and Ubende lie in Western Tanganyika between the 6th and 9th south parallels ; they form a strip of country 200 miles long by about 60 miles wide, oriented north-west and bordered by Lake Tanganyika on the west and by the Rukwa Rift Valley and its north-west prolongations on the east. The Ubende country extends to the north of the strip described and also to the east of the Rukwa Valley to include the Mpanda mineral area. The area is astride the Western Rift Valley System and includes the impressive troughs of Lake Tanganyika and Lake Rukwa. In contrast to these are the highlands of Ufipa and Ubende with open grass downland and clear streams. Between the two are extensive bush-covered lowlands dotted with inselbergs and occasional open grassy plains, called *mbuga*,* marking the course of the seasonal rivers.

Five main units can be distinguished as follows : (1) the eastern slope of the Tanganyika Rift Valley ; (2) the high grasslands of the Ufipa Plateau in the south-west ; (3) the highlands of Western Ubende and the Mahali Mountains in the north-west ; (4) the low-lying closed basin of the Rukwa Valley stretching from south-east to north-west ; and (5) Northern and Eastern Ubende and the Mpanda country lying to the north and north-east of the Rukwa Valley.

The whole area lies just outside the equatorial climatic belt and rain falls sporadically from November to April : the rainfall is generally low, the average ranging from about 20 inches to 45 inches and more in the highlands. The dry weather lasts without a break from May to October, although showers may fall in the highlands from September onwards. During the rains the Rukwa Valley becomes flooded over its whole length, but it dries out in June except for a few areas of permanent swamp.

Most of the country is covered with orchard bush and, except for the Ufipa Plateau and the Mahali Mountains, is infested with tsetse-fly. Sleeping sickness is largely overcome by moving inhabitants to concentration areas.

(b) *Ufipa Plateau*

The physiography of the Ufipa Plateau is largely controlled by its geological structure. The whole of Ufipa is probably an uplifted mass or horst pinched between the rift valleys of Rukwa and Lake Tanganyika by a mechanism as yet imperfectly understood. The uplifted mass is a rectangle some 150 miles long and 50 miles across, whose general altitude is more than 5,000 feet above sea level with high plateaux at a level of 6,200 and 7,000 feet, from which summits rise to 8,000 feet. The level of Lake Tanganyika is 2,534 feet and that of Lake Rukwa 2,602 feet, although the solid bottom of each rift is at a very much lower level. The long axis of the Ufipa rectangle is directed N.35°W. and the altitude falls gradually both to the north-west and the south-east from a height of land in the region of

***Footnote :** Seasonal swamps characteristic of the highlands of Central Africa. Called *dambo* in Northern Rhodesia.

Sumbawanga, Mkundi and Mtantwa Mountain. The drainage of the area is mainly longitudinal, although, of course, each scarp has its own lateral consequent drainage which is rapidly capturing the longitudinal rivers. Longitudinal faults, directed north-west or north-north-west, originated in at least three main periods, the pre-Cambrian, the Mesozoic and the late Tertiary: they have been responsible for dividing the country up into a number of physiographical units succeeding each other from south-west to north-east as follows :—

Tanganyika Slope
South-West Highlands
Namanyere-Mkoe Plains
Luiche-Mtembwa Trench
North-East Highlands
Rukwa Scarp

Tanganyika Slope. The slope, descending from the plateau to Lake Tanganyika, is broken by a number of steps which owe their origin to a series of parallel faults. The slope is formed by parallel strips of granite, granite-porphyry and sandstone. These formations give rise to different scenery, vegetation and water-supply conditions.

The South-West Highlands crown the summit of the Tanganyika Slope and consist of the Kate-Mtantwa Highlands in the north-west and the Mbaa Highlands to the south-east. They are composed of granite-porphyry and granite and are of a rugged nature. Their summits are grass-covered, but their rocky slopes afford shelter to *miombo* woodland* of various types. Some patches of rain forest survive in the Mbaa Mountains. Streams from them appear to be permanent. The Kate-Mtantwa Highlands are slashed longitudinally by the curious Ipeta-Mfuizi Trench, which is probably a small rift valley.

The Namanyere-Mkoe Plains are a wide grassland area a dozen miles across between the highlands, but they broaden out to a plateau at Chala and Namanyere. A mantle of laterite covers the substratum of granite and gneiss. The impervious bedrock is indicated by the clay-bottomed mbugas and swamps which form in the shallow drainage lines. The country is rich and supports many cattle. With a considered water supply policy it would probably be capable of increased production.

The Mwazye Highlands are composed of gneiss and are about 30 miles long by 10 miles across : they appear to be the result of the erosion of a raised plateau 6,500 to 7,000 feet high. These highlands give rise to a number of permanent streams which serve to irrigate farms on the periphery. Patches of rain forest remain in sheltered valleys below but grassland is widespread at high altitudes and *miombo* bush prevails lower down. To the north the large inselbergs of Kalulunga and Chala, besides other small ones, continue the line of the Mwazye Highlands.

The Luiche-Mtembwa Trench is composed of two valleys separated by a height of land, one valley trends north-north-west past Sumbawanga and the other south-east to Mpui and the lowlands. That to the north-west, the Luiche Valley, is connected by a slight slope with the Mkundi Plains, a

*Footnote : A widely spread type of orchard bush composed mainly of the genera *Brachystegia* and *Isoberlinia*.

grassland area merging through the inselbergs of the Chala Highlands with the Namanyere Plains. The valleys are fairly well watered, but the impervious substratum favours the formation of *mbuga* and papyrus swamps. The valleys tend to be dry owing to the high run-off and the long dry season.

The North-East Highlands form a rim along the top of the Rukwa Scarp. In the Mbui Mountains there is a high plateau at nearly 7,000 feet. The precipitation in these mountains must be considerable and there is often mist. The result of these conditions is the development of wide areas of rain forest which have been declared a forest reserve. Large permanent streams descend from these summits, but little use is made of them. Other summits, such as Malonje Mountain, rise to over 7,000 feet.

The Rukwa Scarp is an impressive one, rising as it does 2,500 or 3.000 feet from the Rukwa Plains. It is being strenuously attacked by erosion, but the long straight facets between the principal attacking streams show that it is of comparatively recent origin. The face of the scarp is rocky and steep but supports a flourishing *miombo* woodland.

(c) *The Karema Depression*

Separating the Ufipa Plateau from the Ubende Highlands to the north lies the low, bush-covered area of the Karema Depression. Since Scholz (1914) this depression has been regarded as a down-faulted rift, but it is due to other causes. It is bounded to the south by the northern slope of the Ufipa Plateau, but to the north it merges into the Luega-Msenguse Flats. Both these low-lying areas are underlain by rocks belonging mainly to the Ubende series.

There is reason to believe that before the formation of the Rukwa and Tanganyika rift valleys in late Tertiary times, the Rungwa River flowed into the Congo by way of the Lukuga Valley, and the Karema Depression is probably an abandoned segment of this former great Congo river valley. Temperley has shown that Lake Rukwa emptied into Lake Tanganyika by way of this trough, at the time of the high water level caused by a pluvial period.

(d) *Ubende Highlands*

The Ubende Highlands are bounded on the south by the Karema Depression, to the north they are divided into parallel physiographical features trending north-west which may be enumerated as follows, starting from Lake Tanganyika :—

> *The Mahali Mountains*
> *The Central Lowlands*
> *The Mwesi Hills*
> *The Luegele-Katuma Trench*
> *The Sifuta Hills*

The Mahali Mountains form the north-western corner of the area. Structurally they are a continuation of the Ufipa Plateau, the intervening sector was carried down by the rift faults. They form a narrow ridge

30 miles long and 6,000 to 8,000 feet in altitude culminating in **Kungwe**
Mountain (8,450 feet). The mountains are grass-covered and carry
remnants of rain forest: there are considerable areas where cultivation
is possible and where stock-raising could be encouraged. The Nyende
Plateau lies between Lake Tanganyika and the Mahali Mountains, to
which it forms a comparatively low-lying extension.

The Central Lowlands are formed by the featureless *miombo* flats of
the Luega, in Msenguse, Lufubu and Lugonezi rivers, standing at about
3,300 feet altitude with occasional low hills which rarely rise above 4,000
feet: between these flats and Lake Tanganyika lies a range of low coastal
hills. To the north, towards the headwaters of the Lufubu and Lugonezi,
the country becomes more hilly and culminates in the dissected plateau
of the Mugansa Hills, about 5,000 feet in altitude, which descend rapidly
to the shore of Lake Tanganyika at Kungwe Bay. All this area is underlain by rocks of the Ubende Series.

The Mwesi Hills rise in a steep scarp from the lowlands to the west;
they are largely grass-covered with summits rising above 5,000 feet. They
form a narrow belt composed of the Wakole Series of metamorphic rocks
and extend north-westwards from the Karema Trough to the summit of
Kakungu overlooking Kungwe Bay.

The Luegele-Katuma Trench forms a narrow continuation of the
Rukwa Rift Valley, to which it is related structurally. Between the
Luegele and Katuma Rivers it rises to a height of 4,500 feet.

The Sifuta Hills are grass-covered and rise to heights of over 6,000
feet. They are composed of the sandstones and shales of the Bukoban
System.

To the north of the area described, rocks of the Bukoban System outcrop continuously.

(e) *Rukwa Rift Valley*

The Rukwa Rift Valley is a closed basin due to rift faulting of late
Tertiary date; it is 200 miles long, 20 to 30 miles wide and is elongated
north-west: the north-western half lies in the area covered by this paper.
The south-western fault scarp of the valley, where it limits the Ufipa
Plateau, is one of the most impressive in East Africa, rising in places to
a height of 3,000 feet. Elsewhere the scarps are less pronounced and
not so high.

The Rukwa Valley terminates to the north-west in the Ubende Highlands between Mpanda and Sibwesa: it appears to bifurcate here, but the
geological evidence for this is not complete. It is continued to the north
by the depression occupied by the head-waters of the Nyamanzi River
and to the north-west by the Luegele-Katuma Trench. This trench is the
geological continuation of the Rukwa rift, but was not much faulted in
Tertiary times and so did not develop into a rift valley.

The floor of the Rukwa valley is filled to a depth of some 500 feet
(borehole evidence) with coarse alluvium and lake beds covered with
mbuga clay: the seasonal swamps are marked by wide expanses of grass
interspersed with patches of bush on the drier areas. The Katuma,
Rungwa and other smaller rivers flow into it, and the floor is generally
waterlogged or flooded from November to May. The water is carried
south-eastwards to Lake Rukwa by the Kafufu River which is dry from
June to October.

(f) *The Liande Ridge*

The Liande Ridge is a very remarkable feature of the Rukwa Valley. It is a curving sand spit 21 miles long, ¾-mile wide and over 100 feet high stretching north-east right across the valley, and lying about level with the Karema depression. The Katuma River is dammed by this ridge and forms Lake Katavi: the outlet from the lake is diverted north-east-wards parallel to the ridge, until it meets the flank of the valley, where it has cut a gorge through the end of the sand spit, and continues its course to Lake Rukwa. This so-called "sandy ridge" was found and mapped by the writer in January 1945, when a route for a permanent road across the Rukwa Valley was being sought. It now forms an all-weather link in the through road from Uvinza and Mpanda to the Rhodesias.

The true nature of the Liande Ridge as a sand spit, formed by wave action, was discovered during a subsequent survey by B. N. Temperley of the Tanganyika Geological Survey and described in reports. Temperley has not yet published his account of the origin of the ridge, and so the writer gives his own interpretation, which differs somewhat. The sand composing the ridge is of aeolian character and is thought to have been accumulated in the Katavi area by the south-east trade wind during a dry period. A great pluvial period followed and the level of Lake Rukwa rose until it overflowed into Lake Tanganyika through the Karema depression, as Temperley has shown The formation of this outlet led to the establishment at Katavi of a locus of equilibrium between the movement of Lake Rukwa, and the south-easterly flow of the rivers draining the well-watered Ubende Highlands. At this point of equilibrium, the waves whipped up by the south-east trade wind during its 180 mile passage up Lake Rukwa, piled the accumulated aeolian sands into a typical river-mouth spit, which is now the Liande Ridge. The sand would thus owe its original deposition to wind action, and its accumulation into the form of a spit, to the waves of a high-level Lake Rukwa: the site of the spit was determined by the position of the overflow channel to Karema.

[*Editor's Note:* Material has been omitted at this point.]

R. B. McConnell

PART IV STRUCTURE

(a) *General*

The structure of Ufipa and Ubende is dominated by the adjacent rift valleys, as may be well seen on the new geological map of Tanganyika Territory (Scale 1:4,000,000, 1946) which incorporates the results of the writer's field work. The rift-valley faults trend north-west to north-north-west and the geological formations are arranged in strips parallel to these two directions. The highland areas between the rifts are usually clearly bounded by faults, and may be regarded as uplifted or *horst* blocks. The following structural units, comprising down-faulted areas and highlands, may be distinguished from south-west to north-east.

> *Lake Tanganyika Rift Valley.*
> *Ufipa Plateau* with its structural continuation to the north-west, and the Mahali Mountains.
> *Central Lowlands of Ubende* and *Karema Depressions.*
> These are regarded as the same structural unit and as a north-western continuation of an original Rukwa Rift.
> *Mwesi Hills.*
> *Rukwa Rift Valley,* bifurcating once on the latitude cf Karema and again on that of Mpanda.
> *Western Border of Tanganyika Shield,* including *Mpanda.*

(b) *The Western Rift Valley System*

The Great Rift Valleys were defined by Gregory in his great work (republished in 1921) on the geology of East Africa and divided into a central and a western system. These are now called the Gregory and the Western Rift Valley Systems, respectively. The former includes the Abyssinian Rift, the Lake Rudolf Rift, the Rift Valley of Kenya and dies out in eastward-facing scarps in Central Tanganyika. The latter is arcuate in shape and includes Lakes Albert, Edward, Kivu and Tanganyika, and at the southern end of the arc, Lake Rukwa in a parallel rift.

(i) *Lake Tanganyika Rift Valley.* The magnificent deeply-sunk valley of Lake Tanganyika forms part of the Western Rift Valley, it is 380 miles long, 20 to 40 miles wide and is one of the deepest lakes in the world, reaching to 1,600 feet below sea level. On either shore are steep mountains rising in many places to 8,000 feet and, on the western shore to 10,000 feet. The lake is divided into a northern and a southern basin by a sharp bend opposite the town of Albertville: a sublacustrine ridge connects the Mahali Mountains with the Ugoma Mountains on the western shore and separates the two deep basins.

The origin of this division into two basins has caused much discussion. Krenkel has dealt with it (1925) and Bailey Willis (1936) in his memoir on the rift valleys has given a good description of the phenomenon, but the final explanation will only be given by detailed geological mapping on the ground. The writer became convinced during the course of his own field work, that the plan of the rift valleys had been laid down in pre-Cambrian times, and the present fault scarps were merely revivals of

177

ancient faults (McConnell, *loc. cit,*). This is not the place for a discussion on the theory of rift valley structure, and the subject will be dealt with more fully in another publication. The writer believes, however, that the southern portion of the Tanganyika Rift Valley was continued north-westwards by the Lukuga Graben in the Belgian Congo, and that the northern portion was originally in continuity with the Rukwa Rift Valley by way of the Central Lowlands of Ubende. The Tertiary movements have formed a cross-rift, between Albertville and Kungwe Mountain, from one ancient rift to another, leaving above water the unrevived portions, in this case the Lukuga Graben and the Central Lowlands of Ubende.

The rift faults can be rarely seen or directly inferred from geological evidence. Their presence is usually indicated by rectilinear scarps with hanging valleys and broad facets between the rapidly eroding ravines. One of the rift faults can, however, be very well seen on the shore of Lake Tanganyika near Kirambo, a port 50 miles west of Sumbawanga. Here the fault separates granite-porphyry from Bukoban sandstone thrown down on the lake side. The attitude of the fault is vertical and it is filled with 20 feet of fault mylonite.

The eastern shore of the southern basin of Lake Tanganyika is chiefly formed by the uplifted blocks of the Ufipa Plateau and the Mahali Mountains, but between these a block has sunk, causing the lake to widen near Karema and impinge on the Ubende Lowlands.

(ii) *The Rukwa Rift Valley.* The Rukwa Rift Valley is 200 miles long, and from 20 to 30 miles wide; it forms a closed basin striking north-west and south-east. To the south-east it is partly filled by the Rungwe volcanic mountains near Mbeya, and to the north-west it dissipates itself in the Ubende Highlands. The eastern wall of the rift is very pronounced near Mbeya but to the north-west becomes less clearly defined, whereas, except for the vicinity of the Karema depression, the western wall is always a high escarpment.

The structure of the northern termination of the Rukwa Rift Valley is complicated and the writer has not worked it out in detail. One branch of the valley fades out in the wide plain around the headwaters of the Nyamanzi River. This plain is surrounded by a scarp of Bukoban sandstone which is due entirely to erosion, so it is probable that the faults die out very quickly. The gneiss hills bordering the valley on the east at Mpanda can, however, be seen to continue to the north-north-west for some miles so that it is probable that a pre-Cambrian fault may have existed here which has not been revived further north.

The second branch of the Rukwa Rift continues to the north-west as a narrow trench occupied by the head-water of the Katuma River, flowing south-east to the Rukwa, and by the Luegele River, flowing north-west to Lake Tanganyika. This trench has not been much faulted by the Tertiary movements but faulting and shearing have been severe at earlier periods. This branch also shows bifurcation, as one line of the faulting turns to the north-north-west. The pre-Cambrian faulting is indicated by (a) the production of wide and continuous zones of migmatite, containing at Wantindi a lens of porphyritic granite, and at Mahumwe a narrow belt of similar granite. and (b) the down-faulting of a number of wedges of the Bukoban rocks (See page 16).

The writer believes that the Karema Depression is a branch of the Rukwa Rift Valley which existed in former times, but has only been mildly revived by the Tertiary and Recent faulting; this conception is discussed in the next paragraph.

(iii) *Central Lowlands of Ubende and Karema Depression.* The Karema Depression was regarded by Scholz (1914) as a transverse rift valley running east and west between the Rukwa and Tanganyika Rifts. Detailed mapping shows, however, that while there is in fact a well-marked scarp, running north-west and south-east, bounding the depression to the south, there is no fault scarp to the north but continuity of slope with the Central Lowlands of Ubende.

The geological maps published with this Bulletin show the structural continuity between this zone and the Rukwa Rift Valley. The Mahali Mountains are the structural equivalent of the Ufipa Plateau, and the high block of the Mwesi Hills wedges out south of Wakole Mountain. The Central Lowlands of Ubende therefore represented the floor of the Rukwa Rift Valley during the pre-Cambrian period of faulting, but they escaped to a large extent the late intensive faulting which has produced the present striking feature.

The structure of the Central Lowlands has not been worked out in detail. The strikes are generally parallel to the elongation of the zone, but the dips are very variable. The zone is formed almost entirely of rocks of the Ubende Series and there is an indication that the Ikola limestone may be the centre of a syncline and thus one of the youngest formations. If, as seems most probable, the Ikola limestone is the equivalent of the crystalline limestone of Malambo, then the gondite and metamorphic rocks of Manyoro and Ipungulu form an anticlinal core, or an upthrust block, separating the two limbs of an anticline.

The gondites of Manyoro form a vertical bed striking north-east, that is, at right angles to the quartzites and limestones of Malambo, and they are thus separated from the latter by a fault. In Ipungulu Hill the metamorphic rocks of this series have a normal north-west strike, and there is thus a definite indication of a fold pitching to the south-east but the dips in Ipungulu are to the north-east and not to the south-west as they would be in a normal anticlinal structure. The connection between Ipungulu and the Wakole area is obscured by much laterite, but a detailed survey of this region would be of great interest.

A remarkable feature of the Central Lowlands are the many belts and lenses of migmatite elongated north-west. There is reason to believe that these have formed in pre-Cambrian times along zones of shearing (McConnell *loc. cit.*) and that this peculiarity is due to the fact that the whole zone was once the floor of a rift valley.

The eastern boundary of the Central Lowlands is formed by the scarp of the Mwesi Hills. All along this scarp great crushing and granulitization have taken place, and in pre-Cambrian time, zones of migmatite have formed. From the shearing and local faulting it is believed that this boundary is a thrust fault with a steep hade.

The Central Lowland belt thus appears to be bounded to the south-west and north-east by great thrust faults which have served as ramps for the uplifted blocks of the Mahali Mountains and the Mwesi Hills, and, as explained above (page 24) it continued in the northern segment of Lake Tanganyika which were revived by the latest period of rift faulting.

(c) *The High Blocks*

Between the branches of the southern termination of the Western Rift Valley, which have just been described, are a series of high blocks formed chiefly by the older and more metamorphosed rocks of the Basement System.

(i) *The Ufipa Plateau.* The main lines of the structure of the Ufipa Plateau were defined in the physical description (pp. 3 & 22). The plateau is divided into strips striking north-west and bending west-north-west towards the northern termination. To the south-west the plateau descends in a series of faulted steps to the lake. This slope is made up of strips of granite, granite-porphyry and Bukoban sandstone but has not yet been examined in detail. The Ipeta Valley is a minor rift valley developed between the Kate granite and the granite-porphyry of Mtantwa Mountain. The most striking feature is the long narrow shape of the Kate granite which is described on page 14.

The plateau itself and the eastern scarp are formed by the Ufipa Gneiss Complex (page 13), the pre-Cambrian structure of which has not been properly worked out: the principal folding and faulting were on north-west lines, parallel to the present rift valleys, and cross-strikes may be due to pitching folds.

Zones of vertical and steeply dipping strata, striking north-west are thought to be old shear zones formed prior to the older granite, as they have been in part migmatized. One of these zones runs north-west up the Mtembwa valley, past Sumbawanga and Namwele, to follow the Mtozi valley into the Rukwa Rift at Milumba. At Namwele this shear has been revived, probably in Cretaceous times, and forms the boundary fault (McConnell, 1947, p. 12) of the Karroo basin, with a computed throw of more than 3,000 feet. It is vertical and contains lenses of fault mylonite. The same fault has again been revived in late Tertiary or Quaternary times, and a scarp varying in height from 20 to 200 feet has been formed which can be followed for nearly 100 miles from north of Sumbawanga south-eastwards into Northern Rhodesia. The natives call this scarp the Kanda. The Abercorn-Kipili road crosses it just west of the Mtembwa River at Mpui, and again 3½ miles north of Sumbawanga.

The eastern scarp of the Ufipa Plateau is very impressive, as it rises steeply for 3,000 feet and more, from the Rukwa Valley. It has embayments and segments with different strike directions which indicate that the fault is complex. In numerous places along the scarp, and parallel to it in direction, a well-developed jointing and shearing system exists. This system is usually vertical but may dip steeply to the north-east or east-north-east. From the attitude of this jointing it is assumed that the rift fault is either vertical or dips steeply towards the Rukwa Valley. The western boundary of the Muze coalfield is formed by the rift fault, and some planes appear, dipping at 45° to 60°, but trenching has proved that these planes are not due to faulting.

At its northern end the Ufipa Plateau descends gradually towards the Karema Depression where it is finally terminated by a low scarp running west-north-west: this scarp also separates the Ufipa gneiss from the less metamorphosed rocks of the Ubende Series, and, therefore, is the true limit of the Ufipa Plateau. The junction between this fault and the boundary faults of the Rukwa Rift Valley takes place to the east of Milumba and is of a confused nature. The various faults have been marked as exactly as possible upon the geological map, but a more detailed survey of this area is required.

(ii) *The Mahali Mountains.* This high block, which is uniformly composed of gneiss (See page 14), is the structural equivalent of the Ufipa Plateau, the intervening portion between Karema and Kibwesa was down-faulted and forms a bay of Lake Tanganyika. It constitutes a unit dipping regularly to the south-west at 45° to 60° which appears to be thrust up over the Ubende rocks of the Central Lowland belt; along this thrust plane much shearing has taken place. Zones of granitization and two small lenses of porphyritic granite (see geological map) are formed parallel to this fault, and indicate that great movements took place along it in pre-Cambrian times. The block is broken longitudinally by a steep normal fault which has lowered the Nyende Plateau towards the lake on the south-western side. At the north-western termination of the Mahali Mountains the strike changes locally and the gneiss dips steeply to the west. There is thus an indication that the cross rift described on page 23, which joins the northern and southern basins of Lake Tanganyika, originated also in pre-Cambrian times. For the reasons already given, the writer believes that the Mahali Mountains are continued to the north-west, on the opposite side of Lake Tanganyika, by the Ugoma Mountains in Belgian Territory.

(iii) *The Mwesi Hills.* This range of hills is formed by the metamorphic rocks of the Wakole Series. The belts of schist are separated by ribs of quartzite, and the strikes are nearly always north-west parallel to the elongation of the zone. The dips are usually vertical or very steep. The south-eastern termination of the range, towards the Karema Depression, has not been closely examined. The north-western end at Kakungu Hill, is very crushed and granulitized, it is finally pinched out by a powerful zone of faulting directed north-north-west, which is continued in the eastern shore of the northern basin of Lake Tanganyika. To the north-east the schists of the Mwezi Hills are bounded by a wide zone of migmatite belonging to the Rukwa Rift Valley system.

The Mwesi Hills are thus regarded as an uplifted block wedging out at both ends, and included in that portion of the Rukwa Rift between the Rukwa Valley proper and the northern segment of Lake Tanganyika which has not been revived by the latest phase of rift faulting.

[*Editor's Note:* In the original, material follows this excerpt.]

Part VIII List of References

DANTZ, Dr. 1902-3. *Mitt. Deut. Schutzgebiet.* 16, Bd. II, pp. 145-6 Berlin.

GREGORY, J. W. 1921. The Rift Valleys and Geology of East Africa, London.

KRENKEL, 1925. Geologie Afrikas.

McCONNELL, R. B. 1947. The Geology of the Namwele-Mkomolo Coalfield, *Tanganyika Geol. Surv., Short Paper 27.*

Rift and Shield Structure in East Africa. *Paper read to Int. Geol. Congress, Session 18, LONDON, 1948.* In press.

POUSSIN, de la V., 1936. La Stratigraphie des terrains anciens dans la region des Grands Lacs africains. *Bull. Acad. roy. Belgique, 10,* p. 1,100.

and R. B. McCONNELL, 1948. The Mpanda Mineral Field of Western Tanganyika. Symposium on the Geology, Paragenesis, and Reserves of the Ores of Lead and Zinc. pp. 135-143. *Int Geol. Congress, Session 18.*

SCHOLZ, E. 1914. Beitrage zur Geologie der sudwestlichen Graben-gebiete Deutsch-Ostafrikas. *Der pflanzer, 10,* Nr. 2.

STOCKLEY, G. M., 1931. Report on the Geology of the Ruhuhu Coalfields. *Tang. Geol. Surv., Bull. 2.*

1938. The Geology of Parts of the Tabora, Kigoma and Ufipa Districts, North-West Lake Rukwa. *Tang. Geol. Surv., Short Paper 20.*

1939. Outline of the Geology of the Uruwira Mineral Field. *Tang. Geol. Surv., Short Paper 22.*

TEALE, E.O., 1927-30-32, *Ann. Reports, Tang. Geol. Surv.* 1936. Provisional Geological Map of Tanganyika with Explanatory Notes. *Tang. Geol. Surv. Bull. 6.*

TEMPERLEY, B. N. 1938. Geology of the Country around Mpwapwa. *Tang. Geol. Surv., Short Paper No. 19.*

TEMPERLEY, B. N., 1947. Kabulwanyele Nickel, *M. R. P. No. 43.*

WILLIS, B., 1936. East African Plateaux and Rift Valleys. *Carnegie Inst., Publ. 470, Washington.*

Editor's Comments
on Papers 13 and 14

13   PEETERS
Excerpt from *General Trends in the Geomorphology and Origin of the Basin of Lake Kivu*

14   SWARDT and TRENDALL
Excerpt from *The Physiographic Development of Uganda*

KIVU-ALBERT ZONE

Lake Kivu, the subject of Paper 13, is the southernmost of the rift valleys of the Kivu-Albert zone. It is not immediately recognizable as a rift valley, although faulting was initially responsible, because its geomorphic history and the influence of volcanic activity tend to dominate. The paper is an abridgement of a translation of Peeters (1959). The original paper contains more geographical detail than necessary for our purpose. Peeters refers to the proximity of the lake to the intersection of the two main strike directions: the Albert strike (NNE) and the Tanganyika strike (NNW). This is in line with the controversial hypothesis discussed above in general terms (Furon, 1963; Boutakoff, 1933).

In this paper, Peeters, by implication, rules out a structural relationship between the Kivu rift valley and the Tanganyika graben, there having been from early times a watershed ridge between Kivu I and Tanganyika, breached by the Ruzizi River in recent times. His second conclusion is that after Kivu II resulted from the truncation of the Kivu I by capture, and, before the building of the dam by the Virunga volcanics, it did not flow into the Rutschuru valley and Lake Edward; instead it flowed to the west into the Moesa River, a tributary of the River Oso. The latter conclusion is in opposition to the views of others who regard Lake Kivu as the former headwaters of the Rutschuru River and the floor of the lakes as continuous with the valley floor beneath the volcanics (Sahama and Meyer, 1958).

Although there is structural continuity, albeit by an en-echelon fault pattern, along this rift zone between Kivu and West Nile (north of

Albert), Belgian authors and others exclude Lake Kivu and write of the Lake Edward-Semliki-Lake Albert Rift. This is because of topographical dissimilarity, the differences in sedimentary deposits, and the view of the former continuity of Kivu with Tanganyika, now discarded.

Degens et al. (1973) is an account of the geomorphology and structure (based on geophysical survey) and also contains the geochemistry and biology of Lake Kivu.

As in the case of both the Tanganyika and Malawi rift zones, the Kivu-West Nile zone is flanked by different countries on either side. No one article furnishes a comprehensive account. The bibliography, pp. 53–82 in Anonymous (1965), is comprehensive.

Paper 14, which is an excerpt from a definitive account of the physiography of Uganda, conveniently summarizes the geology, geomorphology, and structure of the eastern flank and of the West Nile region. A somewhat older but enduring account with emphasis on the sediments of the western flank is that by Lepersonne (1949), which is in French. Further contributions on the stratigraphy include Hopwood and Lepersonne (1953), Heinzelin (1955), Gautier (1967), Lepersonne (1970) and Bishop (1965). Cahen (1954, pp. 340–348) describes the structural history in relation to sedimentation within the troughs and the geographic cycles (p. 419). The Ruwenzori Mountain range is described by McConnell (1959). It is a tilted horst block, one of three in the Western Rift System, the others being Mbeya Mountain, Paper 11, and Mahali Mountains, Paper 12.

This rift zone has been studied geomorphologically to a greater degree than has any other. The approach is best made via the works of a general nature: Cooke (1957); Dixey (1946) and King (1963). For greater detail concerning the whole length of the western flank refer to Cahen (1954), and for the eastern flank papers of the Uganda Survey are available. These include Wayland's researches during 1920 to 1934 (1931, 1952), Pallister (1954), and Bishop and Trendall (1967). The list of references with Paper 14 is comprehensive.

The West Nile Province, a key region, has been related to the history of the Albert rift by Ruhe (1954); Lepersonne (1956); Gautier (1965); Hepworth (1962, 1964) and de Swardt and Trendall in Paper 14. The questions at issue are the dating of the planation surface designated "P. III"; whether it belongs to the end-Tertiary subcycle (Wayland and Lepersonne) or to the mid-Tertiary (Bishop, Ruhe, Hepworth); the nature of the West Nile slope—whether it is step-faulted, warped with benches, or polycyclic. Comparison of the age of the oldest sediments in the rift zone with that of the floor on which they rest furnishes data on the relationship, stratigraphy, geomorphology, and structure. Whiteman (1971, pp. 168–169) supports Hepworth's explanation.

The westward flow of rivers across the former site of the Tanganyika and Kivu-Albert zones has already been mentioned. This subject is also discussed in the unexcerpted part of Paper 14.

With regard to structure of the zone, in general the fault scarps—on the west, the Kisali, the Ruindi, and the west Albert and, on the east, the west Ruwenzori—are the dominating features while those facing them are generally lower and subdued and even nonexistent. Holmes (1916) remarked on this pattern. Thicknesses of sediments in the Semliki and Albert troughs have been estimated on the basis of gravity surveys (unpublished reports by J. M. Brown) and by drilling (Harris et al., 1956; Davies, 1951) and in Paper 14. McConnell (1959) and Pallister and Hepworth (1956) give some detailed information on the rift fault planes.

The volcanics of the Kivu-Albert zone comprise: (1) the basalts south of Kivu (Meyer, 1954); (2) the Virunga (Sahama and Meyer, 1958; Sahama, 1962); (3) the Bufumbiro (Holmes and Harwood, 1937); (4) southeast Ruwenzori (Holmes and Harwood, 1932); and (5) Fort Portal (Nixon and Hornung, 1973).

The Western Rift System terminates against the Aswa mylonite belt, which is a Precambrian structure trending NW–SE and unrelated to the Cenozoic rift faulting (see Paper 7, p. 778).

REFERENCES

Anonymous, 1965, Report on the geology and geophysics of the East African Rift System, *I.U.M.P. Sci. Rep. 6*, Nairobi University Press, Nairobi, 265p.

Bishop, W. W., 1965, Quaternary geology and geomorphology of the Albertine Rift Valley, Uganda, *Geol. Soc. America Spec. Paper 84*, pp.293-321.

Bishop, W. W., and A. F. Trendall, 1967, Erosion surfaces, tectonics and volcanic activity in Uganda, *Geol. Soc. London, Quart, Jour.* **122:**385-413.

Boutakoff, N., 1933, Le coude du systeme de fractures du graben central Africain au lac Kivu et sa ramification dans la cuvette congolaise, *Soc. Belge Géologie, Paléontologie et Hydrologie* **43:**80-85.

Cahen, L., 1954, *Géologie du Congo Belge*, Vaillant-Carmanne, Liège, 577p.

Cooke, H. B. S., 1957, Observations relating to Quaternary environments in East and Southern Africa, *Geol. Soc. S. Africa Bull.* **60**(annex):1-73.

Davies, K. A., 1951, The Uganda section of the Western Rift, *Geol. Mag.* **88:**377-385.

Degens, E. T., R. P. von Herzen, H.-K. Wong, W. G. Deuser, and H. W. Jannasch, 1973, Lake Kivu: Structure, chemistry, and biology of an East African lake, *Geol. Rundschau* **66:**245-277.

Dixey, F., 1946, Erosion and tectonics in the East African Rift System, *Geol. Soc. London Quart. Jour.* **102:**339-388.

Furon, R., 1963, *Geology of Africa*, Oliver & Boyd, Edinburgh, 377p.

Gautier, A., 1965, Relative dating of peneplains and sediments in the Lake Albert Rift area, *Am. Jour. Sci.* **262:**537–547.

Gautier, A., 1967, New observations on the later Tertiary and early Quaternary in the Western Rift, in *Background to Evolution in Africa,* W. W. Bishop and D. Clark, eds., University of Chicago Press, Chicago, pp. 73–87.

Harris, N., J. Pallister, and J. M. Brown, 1956, Oil in Uganda, *Uganda Geol. Survey Mem. 9,* 33p.

Heinzelin, J. de, 1955, Le fossé tectonique sous le parallele d'Ishango, in *Mission J. de Heinzelin de Braucourt,* Congo Belge Inst. Parcs Nat., Brussels, pp. 1–150.

Hepworth, J. V., 1962, The relative ages of plateau and plain in West Nile District as indicated by Quaternary erosion surfaces, *Uganda Geol. Survey Rec. 1957–58,* pp. 37–45.

Hepworth, J. V., 1964, Exploration of the geology of sheets 19, 20, 28, and 29 (southern West Nile), *Uganda Geol. Survey Rep. 10,* Entebbe, 123p.

Holmes, A., 1916, Notes on the structure of the Tanganyika-Nile Rift Valley, *Geog. Jour.* **48:**149–159.

Holmes, A., and H. F. Harwood, 1932, Petrology of the volcanic field east and south-east of Ruwenzori, Uganda, *Geol. Soc. London Quart. Jour.* **88:**370–442.

Holmes, A., and H. F. Harwood, 1937, The petrology of the volcanic area of Bufumbiro, S. W. Uganda and of other parts of the Birunga Field, *Uganda Geol. Survey Mem. 3,* pt. 2.

Hopwood, A. T., and J. Lepersonne, 1953, Presence de formations d'age Miocene inferieur dans le fosse tectonic du Lac Albert et de la Basse Semliki (Congo Belge), *Soc. Geol. Belgique Annales* **77:**83–116.

King, L. C., 1963, *South African Scenery,* 3rd ed., Oliver & Boyd, Edinburgh, 308p.

Lepersonne, J., 1949, Le fosse tectonique Lac Albert-Semliki-Lac Edouard: Resume des observations geologiques effectuees in 1938, 1939, 1940, *Soc. Geol. Belge Mem.* **72:**M3–M88.

Lepersonne, J., 1956, Les aplanissements d'erosion du nord-est du Congo Belge et des regions voisines, *Colonial Royal Acad. Sci.* **4:**1–110.

Lepersonne, J., 1970, Revision of the fauna and the stratigraphy of the fossiliferous localities of the Lake Albert-Lake Edward Rift (Congo), *Mus. Royale Afrique Cent. Annales,* Ser. 8, **67:**169–207.

McConnell, R. B., 1959, Outline of the geology of the Ruwenzori Mountains *Overseas Geol. Mineral Resour.* **7:**245–268.

Meyer, A., 1954, Notes vulcanologiques: Les basalts du Kivu meridional, *Congo belge et Ruanda-Urundi Serie Géol. Mem.* **2:**23–52.

Nixon, P. H., and G. Hornung, 1973, The carbonatite lavas and tuffs near Fort Protal, Western Uganda, *Overseas Geol. Mineral Resour.* **41:**168–179.

Pallister, J. W., 1954, Erosion levels and laterite in Buganda Province, Uganda, *19th Intern. Geol. Congr. Proc.* **21:**193–199.

Pallister, J. W., and J. V. Hepworth, 1956, Notes on mylonite and rift faulting in Uganda, *CCTA Comités Regionaux Géologie,* Dar-es-Salaam, pp. 95–97.

Peeters, L., 1959, Traits generaux de la géomorphologie et de la genese du bassin du lac Kivu, *Soc. Royale Belge Géographie Bull.* **83:**66–75.

Ruhe, R. V., 1954, Erosion surfaces of central African interior high plateaus, *Congo Belge Inst. Nat. Etudes Agronom. Pub.,* Sci. Ser. **59:**1–40.

Sahama, Th. G., 1962, Petrology of Mt. Nyiragongo, a review, *Edinburgh Geol. Soc. Trans.* **19:**1–28.

Sahama, Th. G., and A. Meyer, 1958, The volcano Nyiragongo—a progress report 2, *Congo Belge Inst. Parc Nat.,* 85p.

Wayland, E. J., 1931, *Summary of Progress of the Geological Survey of Uganda for the Years 1919–1929,* Gov. Printer, Entebbe, 44p.

Wayland, E. J., 1952, The study of past climates in tropical Africa, *1st Pan-African Congr. Prehist. Proc.,* Oxford, 1947, pp. 56–66.

Whiteman, A. J., 1971, *The Geology of the Sudan Republic,* Clarendon Press, Oxford, 290p.

13

GENERAL TRENDS IN THE GEOMORPHOLOGY AND ORIGIN OF THE BASIN OF LAKE KIVU

L. Peeters

*This excerpt is from the translation by R. Halligan, Como, W. Australia, of Traits generaux de la geomorphologie et de la genese du bassin du lac Kivu in Soc. Royale Belge Géographie Bull. **83**:66–75 (1959)*

Lake Kivu lies to the north of the Tanganyika and south of the Edward-Albert rift valleys. Its level (1,463 m) is much higher than that of the other lakes and it has an indented coastline and many islands. It has a different origin being a drainage basin dammed by the Virunga volcanics. The lake now discharges southward into Lake Tanganyika.

The floor of the basin is largely of Precambrian crystalline metamorphics—schist, quartzite, and granitic gneiss. Ancient lavas occupy the central area and young lavas the northern area.

Geomorphologically there are four units: (1) fault-controlled; (2) youthful relief on basement; (3) areas of ancient lavas; and (4) areas of young volcanic landscape.

1. In the areas of the first unit, faulting was normal, the result of tensile stresses. It still continues. The original drainage basin, later drowned, occupied a wide graben, modified by erosion and sedimentation, but with original scarps, now retreated and eroded, still in evidence. This landscape is best developed on the western side of the lake. Four stages are recognized in remnants of the succeeding graben floors, separated by fault scarps now dissected. Headlands in the northwest and Idjwi Island were defined by this faulting. On the eastern side of the lake, faulting is less evident, but three levels can be recognized. In the south is a prominent scarp that rises from the lake shore to the Congo-Nile divide. Faulting has influenced the forms northeast of the lake , while Virunga volcanics lie on faults belonging to an underlying mosaic.

There have been successive phases of faulting, some predating and some postdating the oldest lava flows. Faulting similarly affected the younger Virunga lava flows. In general, the faulting of the basement rocks on the east is younger than on the west where erosion is more advanced.

2. The areas of Precambrian basement rocks, outside those masked by volcanics or influenced by faulting, have youthful relief, strongly influenced by structure. Deep weathering prevails. Recent fall in the level of the lake is responsible for rejuvenation of streams, giving rise to waterfalls and other features.

3. The central part of the basin contains the ancient lavas, but they are also found in the east and on Idjwi Island. The basalt and trachyte lava flows infilled the valleys, and successive flows buried the drainage patterns, including laterized surfaces, that were developing on earlier

volcanics. The pre-existing relief on the basement rocks was buried and drainage history is not easily decipherable. The gentle initial slopes on the new formations, temporary lava dams and factors such as permeability and lessened surface flow have retarded dissection and erosion that, however, continued as before on the unaffected areas. Hot springs and travertine deposits as well as caverns and underground streams belong to these areas.

4. The volcanic landscape of the Virunga volcanic areas consists of two forms, the volcanoes and the lava plains. They filled three low-lying areas: that to the NNE between Kivu and Edward; the E-W Bufumbira embayment extending into Uganda; and that to the NW, which includes the Moeso Basin and the Mokoto lakes. The major volcanoes, with summits higher than 3000 m, are Nyamlagira and Nyaragongo (both active) and six others (extinct). Some cones are deeply dissected but others have young volcanic morphology. In the surrounding lava plains are many small craters, some formed under water. Virunga volcanic activity, still active, has had a long history, commencing before the formation of the lake, and continuing during its formation.

The formation of the lake was by the damming of the drainage system flowing from south to north, and this is shown by the general fall in elevation of shoreline features (bays and headlands), northward slopes, and the drowning of islands in the northern part of the lake.

The earliest hydrographic network is known as river Kivu-I. It was in existence before the earliest volcanic activity. It extends far to the south of the present lake basin, to Mount Muhi. The flow was northward along the upper waters of the Mugera River, a tributary of the modern Ruzizi. A watershed divide, recognizable today, at that time separated the Kivu-I basin from that of Lake Tanganyika, which had not then developed as a rift valley. The down-cutting of the deep gorge of the lower Ruzizi through the former divide is a very recent event and is related to lowering of the northern graben of Tanganyika. The age of the divide and its erosional, rather than tectonic, origin is established. It extended eastward to the Ruanda slope, east of Lake Kivu. Comparative biological studies of lakes Tanganyika and Kivu demonstrate that there was no connection between the lakes at that time.

Basaltic volcanism (see 3 above) began before the diversion or capture of Kivu-I, and temporarily dammed the upper reaches to form a lake that overflowed as the Ruzizi River to the developing Tanganyika graben. Following this beheading of river Kivu-I, the second phase, Kivu-II, developed as two branches separated by what was to become the island Idjwi. The eastern tributary is seen as having its source on the Ruanda slopes.

It was formerly accepted that the flow of Kivu-II was to Lake Edward along the tectonic depression. However, based on faunal evidence it appears that the river did not in fact connect with Lake Edward. From the levels of the reconstructed thalweg of Kivu-II and from knowledge of subsequent earth movements, it appears probable that the drainage was to the northwest along the depression now occupied by the Mokoto lakes and the upper Moeso River basin. The floor of the gap was later not only uplifted but also raised by lava flows from Nyamlagira (see above), the latter causing the damming of Kivu-II and the formation of the lake. Its highest level was between 1,550 and 1,600 meters and there is evidence that this was maintained for a long period.

The waters of Lake Kivu were captured by, or overflowed into, a tributary of the Ruzizi, which then proceeded vigorously to lower its bed. The resulting lowering of the lake waters rejuvenated the streams flowing into the lake and caused accelerated erosion (see above). The gorge of the lower Ruzizi continued to be cut down by the increased flow while the upper Ruzizi was unable to keep pace.

That the formation of the lake was very recent is demonstrated by faunal evidence and the accumulation of gas in the lake waters.

14

Reprinted from pp. 273-284 of *Overseas Geol. Mineral Resour.* **10**:207-288 (1969)

THE PHYSIOGRAPHIC DEVELOPMENT OF UGANDA

A. M. J. de Swardt and A. F. Trendall

[*Editor's Note:* In the original, material precedes this excerpt. Maps (other than Map 1) and plates cited in this excerpt have not been reproduced.]

Late Erosion Cycles

Erosion Cycles Associated with the Western Rift

THE LAKE EDWARD AND LAKE GEORGE AREA

Along a fairly well-defined line running north-east from a point near the southernmost point in Uganda through Lake Karenge to the divide between the Katonga and Rusangwe rivers, the flat-bottomed valleys end abruptly and the streams continue in steep-sided valleys or are reversed to the east (Map 5). This line roughly coincides with an axis of flexure or warping; to the south-east of this the land falls gradually towards Lake Victoria, while to the north-west it slopes towards the rift valley. There is evidence, especially in Kigezi, that the edge of the area affected by the first rift erosion cycle has advanced beyond the line of warping.

In the belt of country immediately to the north-west of the warp erosion has been intense and has thus produced a closely spaced array of steep-sided valleys which are encroaching on the gently undulating surface to the south-east with its flat-bottomed valleys. Though the relief is much greater in the more intensely eroded area, the valleys of the new drainage system also have flat bottoms for part of their courses, with narrow incised stretches in between. The hills in this area rise to the same general level, which slopes visibly towards the rift. The new drainage has largely destroyed the old, though some stream-courses follow scooped-out and deepened flat-bottomed valleys. The Rusangwe and its tributaries were rejuvenated. The point of rejuvenation moved up one of the old tributaries of the Katonga, now the north-flowing part of the Rusangwe. In this area the new drainage is in the same direction as the old, and the flat-bottomed valleys are gradually more incised downstream until they become narrow and youthful. It is therefore not easy to draw the boundary between the two erosion surfaces with any precision. In southern Toro District tilting at the northern end of the Lake Edward trough has rejuvenated the Nsonge drainage.

A second late erosion cycle subsequently worked back from the rift, and thus produced an even more broken topography with deeply incised streams, but with the interfluves still rising to more or less the same height. An appreciable break is sometimes present between the two recent surfaces; such a break was noted and described by Combe (1934b) in an area south of the Buhwezu plateau. The boundary between the

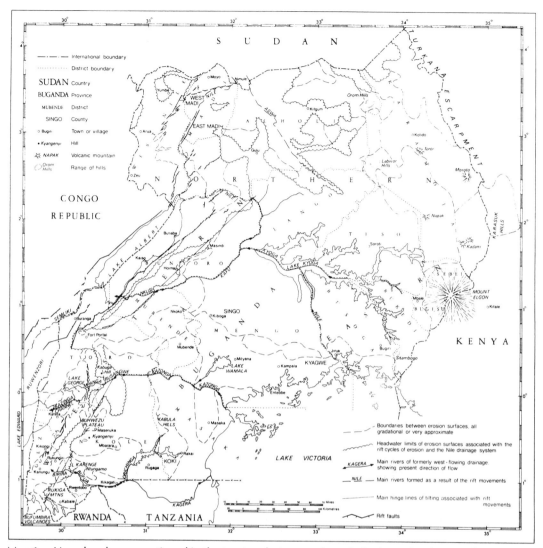

Map 1.   Uganda: places mentioned in the text in relation to selected physiographic features from Map 7.

surfaces is irregular, but can be traced in southern Toro, western Ankole and northern Kigezi. In southern Kigezi, however, erosion during the rift cycles has produced a very steep and rugged topography, and the effects of the two cycles of erosion cannot be distinguished (Plate V, fig. 2).

The visible rift faulting initiated a third erosion cycle, which produced deep gorges cut to the base of the rift scarps, or, where the faults have been rejuvenated, to the base of the upper part of the scarps. The effects of this cycle can be seen for a few miles at most from the rift; the cycle is evidently of very recent origin compared with the older cycles of erosion.

In the extreme south-west of Uganda downwarps on north-west axes have lowered the topography from a maximum elevation of over 8,000 ft. to less than 6,500 ft. This depression is undoubtedly connected with the extrusion of the Bufumbira volcanic rocks (Maps 1 and 7) and is comparable to the depression between Elgon and Kadam (Trendall, 1961). The warping was accompanied by tilting to the south-east and, together with the vulcanicity, was responsible for the formation of lakes Mutanda and Mulehe in Kigezi and Lake Bulera in Rwanda. Several rivers which formerly flowed north-west, and still have flat bottoms in their upper reaches, were diverted by stream capture to Lake Mutanda as a result of the warping (Plate II, fig. 1).

The wide and deep Kazinga Channel, with a negligible drop between lakes George and Edward, must have been cut by a large and active stream. It is in a direct line with the Rusangwe and lower Mpanga. It is not certain whether the rejuvenated section of the old Katonga had a sufficient flow to erode such a large channel, or whether the Katonga, at least in part, still flowed to the west when the channel was formed. Since it is cut in the Semliki Series, which were probably deposited after the main warping had rejuvenated the drainage to the rift, this second suggestion would entail a very late date for the final reversal of the Katonga.

In parts of Ankole and Kigezi, tilting on the main north-easterly axis and warping on axes trending north-west have led to the ponding back of the rivers and the deposition in places of upwards of 100 ft. of sediments. Sediments have also been deposited in the Katonga valley, though probably long before the actual reversal of the drainage. Late rejuvenation has caused the rivers to incise the valley deposits. In some valleys rivers have incised the valley deposits along some stretches, but elsewhere the deposits are unaffected and the valleys are swampy. This may be the result of late warping which is still going on.

THE LAKE ALBERT AREAS AND WESTERN ACHOLI

As in the Lake Edward section of the Western Rift, three late erosion cycles affected the country to the south-east of Lake Albert. Many of the valleys of the first cycle still follow the west-south-westerly course of the wide-valley drainage and some have forested flat bottoms, which are progressively more incised downstream. The second cycle has produced steep-sided grassy valleys which drain directly towards the rift. Erosion of the third cycle, affecting the upper dissected part of the rift scarp, cut back for only a mile or so away from the scarp, and most of the rivers are hanging above the rejuvenated lower part of the scarp.

The Nkusi and some of its tributaries occupy wide U-shaped valleys cut well below the general level of the surrounding country. These were probably eroded with the lake as base level and represent a general lowering of the wide flat-bottomed valleys of the old drainage.

Many of the smaller streams draining the rift scarp do not persist on the sedimentary flats below, and the larger ones have cut gullies in the lake flats which become gradually shallower towards the present lake. The smoothness of the sloping surface of the lake beds indicates that it is the original top of the sediments in the Lake Albert basin. Near the lake shore there is evidence of former shore-lines, and along the present beach a cliff has been cut in the sediments.

The boundary between the wide-valley drainage and the later drainage related to the rift runs north-eastwards and then swings north to cross the Nile at the Karuma Falls. It continues northwards as a low but well-defined scarp, which was noted by Wayland (1934a), who referred to the country to the east as the Gulu peneplain. East of the scarp the valleys are typically flat bottomed, but to the west they are narrow and being actively eroded. The south-west-flowing rivers in western Acholi may in part be the incised equivalents of the old drainage, but the pattern is denser than that of the wide-valley drainage east of the scarp, and the south-west-flowing rivers are clearly encroaching on the area of wide-valley drainage by headward erosion; all along its course the Juma, which joins the Nile above the falls, is very close to being captured. The south-westerly drainage becomes gradually more youthful downstream and it is impossible to trace the boundary separating that part of the drainage affected by two rift erosion cycles from the two main rift surfaces recognisable south-east of Lake Albert. Hepworth (personal communication) has found evidence of several erosion surfaces post-dating the lower laterite in the area between Gulu and the Nile. Thin laterites are present on some of them. The late rift faulting in this area had a small throw and did not produce a significant rejuvenation of the drainage.

Very recent intense gullying, which is clearly later than the normal drainage on the sediments, has occurred and is still occurring on both sides of the Nile below the Murchison Falls, but ceases downstream. This may be due to the comparatively recent warping which has led to a deepening of the Nile below the falls.

West Nile and Madi Districts

An important relic of the wide-valley topography is preserved in East Madi, at nearly the same elevation as the topography of the younger drainage systems to the east and south. To the north, however, the country is much more incised and, especially in granite country, a close jointing pattern has been etched out to produce a bare and very broken topography.

In the southern half of West Nile, Hepworth (1962) has described a complex of recent erosion surfaces due to the intermittent warping of the end-Tertiary peneplain. The whole of the country to the east of the Congo–Nile watershed was tilted to the east and south-east, and a strong warp occurred on a line (Hepworth's inselberg line) immediately to the west of the escarpment that separates the plain and plateau portions of the warped peneplain. At least one of the inselbergs shows a smooth surface sloping at 10° towards the plain. The height of the escarpment and the height difference between the recent erosion surfaces decrease northwards, and north of Arua it is difficult to differentiate between minor erosion surfaces.

The gap between the lower laterite surface on the plateau along the Uganda–Congo border and the same surface around Yumbe was caused by recent erosion along a belt of country where the tilt to the east was steeper; this is a continuation of the warp described by Hepworth and the evidence in this area confirms his conclusion that the end-Tertiary (lower laterite) surface was warped to form the plateau and plain and was subsequently modified by the resulting rejuvenation of the drainage.

THE ASWA CYCLE OF EROSION

The Nile below Nimule is a very narrow stream which has deeply incised the mylonites and flaser gneisses of the Aswa fault zone (Maps 5 and 6). All its tributaries in this area are narrow and youthful. This youthful stretch of the Nile is continued upstream in a south-easterly direction by the Aswa River, which, up to a few miles west of Paranga, is a narrow, straight, swift stream with a rocky bed, which follows the Aswa mylonite zone. Above the point where it is joined by the Agago, all the tributaries on the north-east bank occupy flat-bottomed valleys and clearly existed during the wide-valley erosion cycle. The tributaries on the south-west bank, on the other hand, are narrow and active and developed during a younger cycle which is encroaching on the south and south-westerly wide-valley drainage of the Gulu area. On the air photographs it can be clearly seen that the Aswa has cut back across an old south-westerly flowing drainage system. On the north-east, where the streams which are now tributaries to the Aswa have maintained their original direction of flow, the wide-valley topography has been preserved, whereas on the other side, where the tributaries run in the opposite direction to the old direction of drainage, an entirely new drainage system has developed. The west-flowing stretch of the Agago, down to the point where it turns north-west before joining the Aswa, is a meandering stream which follows an old flat-bottomed valley. The meanders are only slightly incised and have incorporated the lower parts of tributaries to the old valley. The impression gained is not that of a river which is cutting its way back by headward erosion, but rather of a stream with a low gradient which has developed meanders as a result of an increase in the amount of water flowing in it. The same is true of the meandering upper course of the Aswa, which occupies an old flat-bottomed valley as far as a point about ten miles south-east of Paranga.

The preservation of an old drainage system on the north-east bank of the Aswa, and the meandering upper courses of the Aswa and the Agago, suggest that these rivers have been deflected to the north-west by a warp or rise running parallel to but about 15 miles to the south-west of the straight section of the Aswa. This warp would also explain the rejuvenation of the south-flowing wide-valley drainage in the Gulu area and the disruption of the old Pager–Ome River. The westerly to west-north-westerly deflection of both the old Agago and the old Aswa, which formerly joined the Tochi and Koli Rivers respectively, may be related to early warping on this line when the wide-valley cycle was still in progress. The junction between the straight and meandering sections of the Aswa is in line with the warp axis of the Albert section of the rift, and the low gradient on the upper Aswa and on the Agago is in part due to tilting to the south-east.

THE PEDIPLAIN OF EASTERN KARAMOJA

The presence of separate relics of a belt of high country along the eastern border of Uganda has already been noted. Much of this belt has been reduced to a plain which shows no break with the plains of central Uganda (Map 7; plate IV, fig. 2).

The drainage on it, however, is quite different from that further to the west. The valleys are only slightly lower than the interfluves and the streams show a braided pattern. Closely spaced tributaries may run parallel for long distances before joining at an acute angle. The soil cover is thin. Hilly areas are separated from the plain by steep scarps and the topography is typical of an area that has attained an advanced stage of pediplanation (King, 1953). Pediplanation has also occurred in and around the Labwor Hills.

THE ZONE OF AGGRADATION IN WESTERN KARAMOJA AND NORTH-EASTERN TESO

Over an extensive tract of country, stretching from north of Mount Elgon to near Kotido, the wide-valley topography west of the pediplain has been covered by recent sediments. Scattered islands of the old topography are still visible and flat-bottomed valleys can be recognised in a few places. Near the margins of this recent sedimentary belt, the flat-bottomed valleys become progressively wider until the flats of neighbouring valleys coalesce.

A change of climate probably initiated the pediplanation of the former highlands to the east, and the sluggish streams flowing west were unable to cope with the resulting greatly increased sedimentary load, which was thus shed to the west of the former highlands and reworked by the wide braided rivers which still flow on the sedimentary flats.

Across central and northern Uganda from the Turkana Escarpment westward to the Nile, five different erosion surfaces can be recognised though there is little or no break in slope or elevation between them. These, from east to west, are (1) the pediplain; (2) the sedimentary flats; (3) the wide-valley topography, interrupted by (4) the Aswa surface; and (5) the surface related to the West Nile depression (Map 7). These surfaces, with the exception of the wide-valley topography, may be in part contemporaneous, though they differ in their mode of origin.

The Development of the Western Rift and its Relation to the Erosion Surfaces of Uganda

EARLY TILTING

Tilting to the south-east appears to have started at an early date. In south-central Uganda the difference in elevation between the highest hill on the older laterite surface and the floor of flat-bottomed valleys is 400 to 600 ft. In eastern Mubende District the difference in an area with rocks of comparable lithology is 800 to 900 ft. In Bunyoro the difference is over 1,000 ft. A similar relationship exists in the south-west in rocks of comparable lithology. In the Koki highlands the maximum relief is about 1,000 ft. This increases to 1,500 ft. or more in the Buhwezu plateau and in the hills east of Ntungamo, and locally to over 2,000 ft. in Kigezi. Though part of this difference in relief can be ascribed to deeper dissection downstream along the old westerly drainage, it is certain that the main cause was gradual tilting to the south-east as the lower laterite surface was formed. This tilting may even have started during the cycle of erosion which produced the upper laterite surface, but much of the differential tilting occurred during the lower laterite cycle, since the incision in valley floors during the wide-valley cycle has been small in comparison, generally about 50 to 100 ft., and this may have been due to the ponding back of the streams during the final stages of tilting.

THE RIFT VALLEY SEDIMENTS AND THE SURFACE ON WHICH THEY REST

The Western Rift Valley is divisible into three parts arranged in echelon: from north to south, the West Nile, Lake Albert, and Lake Edward/Lake George sections. Each is filled by a variable thickness, locally over 8,000 ft., of (so far as can be assessed from surface exposures in Uganda) Plio-Pleistocene sediments: the Kaiso Series and Semliki Series (Maps 6 and 7). At the western edge of the valley, in the Congo Republic, some basal beds are known to be Miocene (Hopwood and Lepersonne, 1953) and it is possible that the lowest parts of the succession are even older.

There is strong evidence that the rift sediments were deposited on the lower laterite surface (Plate VI, fig. 1). In the Lake Edward trough, in the area to the north and east of Kateta, thick cellular laterite occurs on pre-Cambrian rocks close to the margin of the sediments in an area where there is no boundary fault on the south-east side of the rift. Laterite is absent in the country to the east, which has been affected by the two rift cycles of erosion. To the south-west, near the Congo border, the sediments occupy embayments between hills formed by Karagwe-Ankolean rocks, and isolated hills form islands in the sediments. The tops of the hills in this area have been eroded well below the level of the upper laterite surface, remnants of which occur further south at an elevation of about 8,000 ft. It seems reasonable to infer that the laterite in the Lake Edward trough has only recently been exposed from under a cover of sediments, and it obviously cannot be correlated with the upper laterite.

In the Lake Albert section of the rift, Lepersonne (1949) recorded a coarse feldspathic conglomerate cemented by laterite at the base of the Kaiso Series, and a ferruginous zone was intersected at the junction between sediments and Basement Complex in a borehole drilled in the Butiaba area by the African and European Investment Company Limited (Harris, Pallister and Brown, 1956). To the north-east, near the termination of the Albert trough, Brown recorded scattered flat-topped hills capped by laterite which project through the sediments. He regarded these as portions of the down-warped Bunyoro–Acholi erosion surface that now form islands in the sediments.

In West Nile District, Macdonald (*in litt.*) recorded laterite at the base of the sediments and considered that the 'end-Tertiary' surface in the area at least in part underlies the sediments. His maps show isolated areas of sedimentary rocks surrounded by rocks of the Basement Complex with many laterite relics, whereas no laterite is present on the sediments themselves. The laterite is best preserved near the sediments and almost certainly represents the exhumed floor on which they were deposited.

THE DEVELOPMENT OF THE RIFT VALLEY

THE LAKE EDWARD, LAKE GEORGE AND LAKE ALBERT BASINS

The lithology of the Kaiso Series and the lateral impersistence of beds within it suggest that it was deposited in a shallow lake, and the fineness of the sediments indicates that the rivers which brought the material were mature. The great mass of sediments in the Western Rift, with a maximum thickness of 6,000-8,000 ft., an estimate based on geophysical measurements (J. M. Brown, personal communication), and with a proved thickness near the south-eastern margin of the Albert trough of over 4,000 ft. (Harris, Pallister and Brown, 1956), could not have been derived by erosion of the narrow belt between the rift warps and the rift itself; the erosion along this belt has not lowered the surface more than a few hundred feet at most. The bulk of the sediments must thus have been deposited when the drainage of Uganda still flowed west. If the sediments are later than the lower laterite surface, they were derived from the erosion which modified this surface and produced the flat-bottomed valleys.

That deposition kept pace with subsidence is shown by the fact that there are no signs that the wide-valley drainage was rejuvenated while it was still flowing west, except perhaps to the south-east of Ruwenzori. The lithology of the sediments suggests that the deposition may at times have occurred in a series of small lakes, and it is quite possible that the accumulated water was still at times released to the west into what is now the Democratic Republic of the Congo across a subsiding trough corresponding closely to the present Rift Valley.

In western Kigezi there is no bounding fault on the south-east side of the rift valley, and the sediments were deposited in a deepening basin gradually let down by faulting on the Congo side, with a hinge on the same line as the main faults of the Edward trough further to the north-east. The tilting, which had already started during the lower laterite of cycle erosion, became more pronounced and a warp or flexure developed about 20 miles to the south-east of the hinge line. The country between the warp and Lake Victoria was tilted to the south-east, and the rivers were ponded back and eventually reversed, while the sedimentary basin in western Kigezi was sharply depressed (Map 7). The extent of the warping is indicated by the fact that the base of the sediments to the east of Kateta stands at over 4,800 ft., while the top of the sediments is now below the level of Lake George at 2,995 ft.

Further north in this southern section it is not certain whether the boundary faults existed at the time of the main sedimentation or whether they developed later in response to the intense movements that occurred when the late flexures were formed. It is evident, however, that the present fault scarps are young, as the rejuvenation of drainage associated with them has not extended far.

Though an early steep downwarp may have existed along the south-eastern margin of the Albert trough, the great thickness of sediments right up to the edge of the basin indicates that a fault or series of faults existed from an early stage. This is confirmed to the north-east, in the Murchison Falls area, where the recent scarp on the main rift fault-line dies out and the surface on sediments and Basement Complex stands at the same elevation, though the boundary between them is a straight line which clearly represents an old fault. As in the Edward basin, marked flexures were formed at a late stage on an axis about 20 miles south-east of the rift. The tilting towards the trough west of the axis of flexure can be seen on the rift shoulders and on the exposed top of the sediments, both of which have an appreciable slope towards Lake Albert. The tilting in the opposite direction east of the axis of flexure caused the reversal of the Kafu and the formation of Lake Kyoga. At a late stage faulting occurred in two stages; the lower parts of the scarps were formed so recently that only the most powerful streams have cut to the base of the scarps.

The tilting towards Lake Edward and Lake Albert initiated the two rift cycles of erosion already described. The valleys of the first of the cycles are now in part swampy, and even the deeply incised valley of the Birira River is wide and swampy with meanders, above a stretch where the river is deep and actively eroding. Combe (1929, 1932) has described gravel beds up to several hundred feet above the present streams in deeply incised valleys in south-west Kigezi and near the rift shoulder in southern Toro; these may have been deposited during the first rift cycle It is possible, therefore, that the axes of flexure gradually moved north-west, with a concomitant increase in the tilting towards the rift. This led to a ponding back of the streams of the first erosion cycle, and the development of a later and more active cycle near the rift shoulders.

The wide U-shaped valley of the Nkusi River was cut during the rift cycles of erosion; its shape and width may be due to the fact that the Kafu River was still flowing west during the early stages of the formation of flexures. If the fine sandstones in the Nkusi and Musizi valleys belong to the lake succession (Harris, Pallister and Brown, 1956), the lake at one time must have extended along them and so stood at a much higher level in relation to the surrounding topography than when the valleys were cut with the lake as base-level. This suggests either large variations in the level of the lake or else considerable downwarping of the belt between the axis of flexure and the rift before the late faulting occurred. The late faulting may have been accompanied by an appreciable rebound of the downwarped section.

The depressions occupied by Lake Edward and Lake Albert, with their impressive scarps rising to over 1,800 ft. near the south-west end of Lake Albert, are very recent topographic features. The evidence of erosion surfaces in Uganda and the extreme youth of the scarps show that for the greater part of their existence the depressions were shallow. If the lake sediments rest on the lower laterite surface, the topography over a large part of Uganda had attained practically its present form before the rift valleys were initiated, since the country has been only slightly lowered by subsequent erosion of this surface.

The West Nile Basin

The West Nile trough is the smallest of the three sections of the Western Rift valley which are represented in Uganda. Normal sedimentary contacts between sediments in the trough and rocks of the Basement Complex are more common on both sides of the basin than elsewhere in the rift valley in Uganda, and drilling has shown that the sediments thicken rapidly away from the margins (Pallister, 1955; Hepworth and Biggs, 1961). Much of the rift faulting in this area is very late, but the fault running east-north-east parallel to the Nile in the Moyo area is probably the oldest rift fault with physiographic expression in Uganda; wide embayments have been cut in the scarp, extending back for many miles. The fault has been rejuvenated at a late stage.

The West Nile depression was formed mainly by warping and tilting to the east and south-east. On the east side of the basin tilting has not been so marked. The bounding fault in West Madi may have existed from an early stage, in which case the basin was let down by faulting on the south-east rather than on the north-west in contrast to part of the Lake Edward trough, where the main bounding fault is on the north-west side.

Ruwenzori

The Ruwenzori Mountains (McConnell, 1959b) are an elevated block, which slopes steeply and disappears below the sediments of the rift both at its northern and southern ends. On the west side, in Uganda, it is bounded by two overlapping faults with a maximum throw of probably well over 20,000 ft. The throw decreases steadily to the north and the fault, after swinging east for a short distance, is continued in the lake sediments beyond the northern nose of the mountains. The lower part of the fault scarp shows faceted spurs and, as on the eastern margins of the rift, the faults have been rejuvenated (Plate VI, fig. 2).

The northern half of the mountain block is bounded on the east by a single major fault. In the central part, the mountain spurs slope below the sediments of the Lake George trough, but in the south again the hills rise sharply from the plain and their edges may mark a much indented fault-line scarp.

Some writers (*e.g.*, Dixey, 1946b; Busk, 1945) have regarded Ruwenzori as an ancient residual mass. All the evidence now available, however, shows that it is of recent origin and attained its present elevation largely during the final phases of movement along the rift valley (McConnell, 1959b).

The ridges on the southern part of the mountain have a uniform easterly slope. The deep dissection of the mountain block has not destroyed all traces of old surfaces: some relics still remain. The best example of such a relic occurs above Kilembe. This sloping relic, about $2\frac{1}{2}$ miles long and $1\frac{1}{2}$ miles wide, has a drainage pattern which, though the valleys are more deeply incised, is similar to the pattern formed during the wide-valley cycle of erosion in the Fort Portal area. It is very likely, therefore, that the sloping relics on Ruwenzori represent the uplifted and tilted lower laterite surface, and that the flat-bottomed valleys were already in course of development before the main uplift occurred. The formations of the Buganda–Toro System continue without change in lithology from eastern Toro across the mountains (Map 6), and there is no reason to believe that a large transverse erosional residual would have been preserved between two mature west-flowing rivers with no major highlands upstream in their basins; at least, none as far east as the present eastern shore of Lake Victoria. Nor is it likely that the Ruwenzori Mountains represent the tilted Buganda (= upper laterite) surface, as this would hardly have been preserved on the mountains but not on the plains immediately to the east.

It is important to appreciate the position of the Ruwenzori block in relation to the rift valley troughs. It is situated between the north-eastern end of the Lake George basin and the south-western end of the Albert trough. The tilting of the block may therefore represent an accentuation of the general rise to the rift caused by depression of the northern end of the Lake George basin. This would not, however, account for the total uplift of the block between its boundary faults, which probably has to be explained by some isostatic mechanism (McConnell, 1959b, p. 265).

The easterly course of the Mpanga River in the Fort Portal area suggests that during the development of the flat-bottomed valleys the area now occupied by Ruwenzori was already rising to the north-west faster than the rest of the country. The morphology of the boundary scarps, however, shows that they are in the main of very recent origin, probably partly earlier than, but largely contemporaneous with, the first stage of the faulting which produced the rift scarps elsewhere. However, a scarp may have existed along the line of the western boundary fault from an early stage in the development of the Albert trough.

Minor Faulting in the Western Rift

In the West Nile trough and in the Albert basin there are many faults of small throw which are so recent that they still have topographic expression in the weakly consolidated lake sediments. These minor faults have no consistent direction of throw and trend in all directions, though most of them run more or less parallel to the main rift faults. The throw of some is very small; they are little more than sinuous cracks in the sediments.

The area around the Kazinga Channel is criss-crossed with many minor parallel and intersecting faults, which are again, for the greater part, roughly parallel to the main rift faults. Miniature horsts and graben are common. In western Kigezi, where there is no bounding fault on the south-east, a large number of minor faults trend north-east with movement towards the south-east; they may reflect faulting of the rift floor as at least one fault has brought gneisses against down-faulted lake sediments. The direction of throw is opposite to what one would expect near the edge of a rift trough and may be due to reverse faulting or underthrusting. This suggests that subsidence in the central part of the basin did not keep pace with the warping and flexure of the rift shoulder during the latest stages of movement.

THE PATTERN OF RIFT FAULTING IN RELATION TO PRE-CAMBRIAN STRUCTURES

Dixey (1956) and McConnell (1951, 1959b) suggested that the rift valleys follow ancient structures in the pre-Cambrian that have been rejuvenated at intervals. Evidence now available from Uganda does not support this view. Mapping and reconnaissance surveys in recent years have shown that the Buganda–Toro System, with its basal quartzites and amphibolites, can be traced continuously from the Jinja area across the Ruwenzori Mountains (Map 6). The thrust described by McConnell (1959b) marks the boundary between the Basement Complex and the Buganda–Toro System; the rocks of the Buganda–Toro System have been closely folded and thrust during at least two phases of movement. The dextral displacement of the beds which McConnell supposed to be related to an ancient fault on the line of the existing Wasa fault represents in fact a gentle cross-fold on a steep south-east-plunging axis, the continuation of which can be traced across the rift into southern Toro and northern Ankole. The evidence available at present indicates that the lithology and structure of the Buganda–Toro System are the same on Ruwenzori as in the country to the east of the northern end of the Edward trough, and that the mountains occupy an area with no special structural characteristics.

The fold belt of the Buganda–Toro System thus crosses Ruwenzori from east to west. In south-west Uganda a folded belt of the Karagwe–Ankolean System runs north-west almost at right angles to the rift. The fold belts of two pre-Cambrian orogenies thus cross the rift and show no signs of having been affected by movement along a proto-rift.

At the northern end of Ruwenzori, the Wasa fault follows an ancient north-trending belt of refoliated gneisses with mylonite bands, some of which have been invaded by pegmatites. The original foliation can be seen in places and strikes east-north-east. On the south-east side of Lake Albert, however, the foliation is nearly everywhere at a small angle to the rift; near Butiaba, flaser gneisses and mylonitic rocks strike at an angle of approximately 30° to the rift faults. In West Nile District the foliation intersects the line of the rift at all angles, and ancient mylonite zones are in places cut at a narrow angle by the recent faults.

There is no indication that the area of the rift valley is occupied by rocks any different from those found elsewhere in Uganda or that the rifts are bounded by ancient structural lines. Charnockites are not confined to the vicinity of the Western Rift: they are widespread elsewhere in the Basement Complex of Uganda and certainly do not have any significance in relation to pre-Cambrian phases of movement and deep-seated metamorphism along the belts now occupied by the Western Rift; nor are crushing and mylonitisation more common near the rift than elsewhere; nor do they provide evidence that the Western Rift Valley originated by compression as suggested by Groves (1932): belts of mylonite and zones of refoliation are common in the Basement Complex throughout Uganda. The most important mylonite belt, which is erroneously shown on some maps as belonging to the rift system of faults, follows the north-west course of the Nile below Nimule, and is continued along the straight course of the Aswa.

There is no evidence in Uganda to support the theory of compression; morphologically the Western Rift Valley is comparatively recent and all the major faulting associated with it is normal. To argue about conditions at depth in our present state of knowledge would be fruitless. There is also no clear evidence of Mesozoic faulting along the Western Rift in Uganda, and a correlation of faults by the appearance of their associated mylonites (McConnell, 1959b, p. 264) would appear to be a decidedly risky procedure.

[*Editor's Note:* In the original, material follows this excerpt. Only references cited in this excerpt are reprinted here.]

REFERENCES

Busk, H. G., 1945, On the normal faulting of rift valley structures, *Geol. Mag.*, Vol. 82, No. 1, pp. 37–44, fig., refs.

Combe, A. D., 1929, Summary of the field work carried out in 1928, *A. Rep. geol. Surv. Uganda, 1928*, pp. 9–22, refs.

Combe, A. D., 1932, The geology of south-west Ankole and adjacent territories with special reference to the tin deposits, *Mem. No. 2, geol. Surv. Uganda*, 236 pp., figs., photos., refs., col. geol. maps.

Combe, A. D., 1934b, Notes on the geology of the northern half of Igara, western Ankole, *A. Rep. geol. Surv. Uganda, 1933*, pp. 16–19, refs.

Dixey, F., 1946b. Reference omitted in original publication.

Dixey, F., 1956, The East African Rift System, *Bull. Suppl. No. 1, Colon. Geol. Miner. Resour.*, 71 pp., figs., photos., refs., maps.

Groves, A. W., 1932, Relation of petrological evidence to tectonic theories, *A. Rep. geol. Surv. Uganda, 1931*, pp. 17–18.

Harris, N., Pallister, J. W., and Brown, J. M., 1956, Oil in Uganda, *Mem. No. 9, geol. Surv. Uganda*, 33 pp., fig., refs., maps.

Hepworth, J. V., 1962, The relative ages of plateau and plain in West Nile District as indicated by Quaternary erosion surfaces, *Rec. geol. Surv. Uganda, 1957–58,* pp. 37–45, figs., refs., col. geol. map.

Hepworth, J. V., and Biggs, E. R., 1961, Investigation into the dry boreholes in the Rift Valley area of West Nile, *Rep. No.* JVH/16, and EB/7, *geol. Surv. Uganda,* 11 pp., ref. geol. map.

Hopwood, A. T., and Lepersonne, J., 1953, Présence de formations d'âge miocène inférieur dans le fossé tectonique du Lac Albert et de la Basse Semliki (Congo belge), *Annls Soc. géol. Belg.,* Vol. 77, *Bull. Nos.* 1, 2 and 3, pp. 83–113, figs., photos, refs. [In French.]

King, L. C., 1953, Canons of landscape evolution, *Bull. geol. Soc. Am.,* Vol. 64, No. 7, pp. 721–751, figs., photos., refs.

Lepersonne, J., 1949, Le fossé tectonique Lac Albert-Semliki-Lac Édouard, Résumé des observations géologiques effectée en 1938-1939-1940, *Annls Soc. géol. Belg.,* Vol. 72, pp. M1–M88, figs., photos., refs., maps. [In French.]

McConnell, R. B., 1951, Rift and shield structure in East Africa, *Rep. 18th int. geol. Congr., London, 1948,* Pt. 14, pp. 199–207, fig., refs., map.

McConnell, R. B., 1959b, Outline of the geology of the Ruwenzori Mountains, *Colon. Geol. Miner. Resour.,* Vol. 7, No. 3, pp. 245–268, fig., photos., refs., geol. map.

Macdonald, R., *in litt,* Explanation of the geology of sheets 11 and 12, *Rep. geol. Surv. Uganda.*

Pallister, J. W., 1955, Notes on the northern termination of the Western Rift, *Rev. geol. Surv. Uganda, 1953,* pp. 49–52, refs., map.

Trendall, A. F., 1961a, Report on the geological results of some recent work in Bugisu, *Rep. No.* AFT/5, *geol. Surv. Uganda.* (Unpublished.)

Trendall A. F., 1961b, Explanation of the geology of sheet 45 (Kadam), *Rep. No.* 6, *geol. Surv. Uganda,* 46 pp., figs., refs., maps, col. geol. map.

Wayland, E. J., 1934a, Peneplains and some other erosional platforms, *A. Rep. geol. Surv. Uganda, 1933,* pp. 77–79, refs.

Part VI

EASTERN AFRICA—EASTERN RIFT SYSTEMS

Editor's Comments
on Paper 15

15   KENT, HUNT, and JOHNSTONE
Excerpt from *The Geology and Geophysics of Coastal Tanzania*

EAST TANZANIA

Paper 15, an excerpt from a paper by British Petroleum geologists Kent, Hunt, and Johnstone, is concerned with a pattern of rift faulting named the Eastern Zone by Krenkel (1925). Krenkel used the term "Central Zone" for what is now the Eastern or Gregory system.

The faults and rift valleys that extend SSE from Kilimanjaro are the stronger of two elements of the zig-zag pattern of NNW and NNE fractures that delineate the coast. Results of geophysical surveys and drilling supplementing intensive surface geology of the coastal belt are recorded in Paper 15.

The NNE-trending fractures are of two ages, the older being Karroo or post-Karroo in age. They have been discussed in Part III. McKinlay (1963) and Spence (1957) can also be referred to.

According to Cliquet (1957), these two dominant fracture trends, NNE and NNW, are also found in Madagascar, the former being followed by the concealed Karroo troughs. Incidentally, this structural relationship gives support to du Toit's model (Smith and Hallam, 1970) of the restoration of Madagascar to a position east of Tanzania and Kenya rather than off Mozambique (Embleton and McElhinney, 1975). Brenon (1956) describes the graben of Lake Alaotra in central Madagascar.

REFERENCES

Brenon, P., 1956, Le graben du lac Alaotra Madagascar, *CCTA Comités Regionaut Géologie,* Dar-es-Salaam, 1956, pp. 107–116.

Cliquet, P. L., 1957, La tectonique profunde du sud du Bassin de Morondora, *CCTA Comités Regionaux Géologie,* Tananarive, 1957, pp. 199–219.

Embleton, B. J. J., and M. W. McElhinny, 1975, The Palaeoposition of Madagascar: Palaeomagnetic evidence from the Isalo Group, *Earth and Planetary Science Lett.* **27:**329–341.

Krenkel, E., 1925, *Geologie Afrikas,* vol. 1, Borntraeger, Berlin, 1918p.

McKinlay, A. C. M., 1963, The coalfields and the coal resources of Tanzania, *Tanganyika Geol. Survey Bull.* **32,** 82p.

Smith, A. G., and A. Hallam, 1970, The fit of the southern continents, *Nature* **225:**139–170.

Spence, J., 1957, The geology of part of the eastern Province of Tanganyika, *Tanganyika Geol. Survey Bull.* **28:**1-62.

15

THE GEOLOGY AND GEOPHYSICS OF
COASTAL TANZANIA

P. E. Kent, J. A. Hunt, and D. W. Johnstone

[*Editor's Note:* In the original, material precedes this excerpt.]

Post-Palaeozoic Tectonics
Regional Structure

Continuing work on the crystalline Basement rocks of East Africa
has confirmed Holmes's interpretation of the metamorphic complex
nearest the Indian Ocean as a distinctive belt, the Mozambique belt. It
is distinguished by metacalcareous sediments and has a strongly marked
north or north-easterly strike (McConnell, 1967) which has been altered
by an orogeny dated as late Pre-Cambrian or Cambrian. Kennedy (1965)
has emphasised that this is part of a peripheral marginal belt of Africa,
implying that in Cambrian times the continent was already the broad
structural entity known today.

The survey operations described here shed no light on the geology of
the Basement rocks, which were out of reach, buried beneath great
thicknesses of sediment in the coastal zone (Figs. 40 and 41). The nearest
area described in detail is that surveyed by Sampson and Wright (1964)
in the Uluguru Mountains west of the Mvuha Rift, which shows a series
of thrust-folded masses of migmatic granulites and gneisses, pyroxene
granulites and, at the top of the series, a metacalcareous suite, intruded
as a whole by pegmatites. Thrusting is from east to west, and indicates
orogenic conditions in a period long preceding the deposition of the
Karroo. The eastern boundary of the Basement is now a normal fault-
zone, probably of early Karroo age, against which the terrestrial Karroo
deposits in the Mvuha Rift dip generally westward into the fault.

In relation to the sedimentary belt the most important features of the
Basement rocks are the parallelism of the Mozambique belt to the
Indian Ocean margin, implying (as Kennedy stressed) the great age of
the African continental block, and the development of structural trends
identical with those found in the marginal Karroo rifts in the Mesozoic
belt not only as far east as the large islands but also in the submerged
banks still further offshore.

After the early Palaeozoic East Africa remained free from orogenic
movements. Tectonics appear to be entirely related to Basement faults
and epeirogenic tilting. The single example in Tanzania of a normal
surface anticline (Mandawa) proved to be related to a salt bulge, whilst
all the other folds can be ascribed to draping over concealed deep faults.

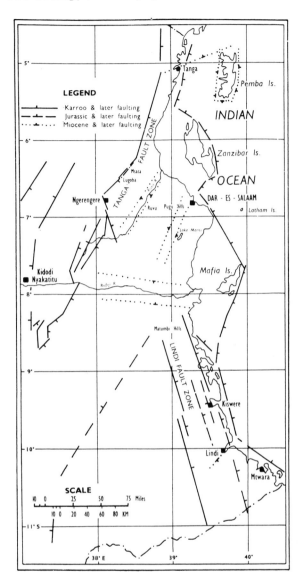

FIGURE 42. Major fault trends. The Mesozoic origin of the faults bounding the continental shelf is hypothetical.

The geometry of faults in the coastal zone indicates general steep dips, with no evidence of low-dipping fault planes which might be directly indicative of the regional crustal extension or stretching which should be associated with continental drifting.

The modern structure of coastal Tanzania was blocked out in Karroo times. The overall fault pattern (Fig. 42) fits the regional African pattern as interpreted by Furon (1963, fig. 3), with dominant NE–NNE and NW–NNW trends. The spread of this pattern across Africa far beyond the limits of the classic Rift Valley area is thus further emphasised. A major fault-zone of Karroo age trending NNE limits the main Basement outcrop and together with another trending NNW from Mozambique

209

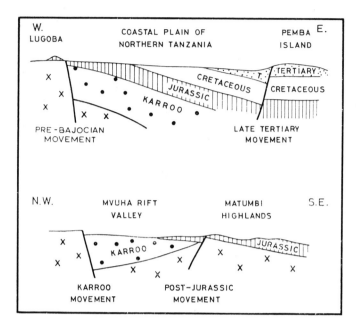

FIGURE 43. Two-phase development of graben structure, shown diagrammatically.

and down-throwing to the east, controls the structure of southern coastal Tanganyika. The second fault may possibly terminate the first fault zone, possibly with an ancient dextral shear which might conceivably displace the first fault from west of Dar-es-Salaam to north of Mombasa. For convenience the zone trending NNE is referred to as the 'Tanga trend' and the fault-zone trending NNW as the 'Lindi trend'.

The 'Lindi trend' is discontinuously seen, and possibly dies out in the Matumbi highlands. It may be cut off by the Utete east-west shear belt, mentioned below, or alternatively may be deeply buried by later sediments in the Rufiji downwarp.

In a discussion following the symposium on the World Rift System held in Ottawa in 1965, F. K. North (1966) made the suggestion that development of one master-fault commonly predated graben information where this two-sided second phase developed at all. In the coastal area difficulty arises in full analysis of movements because the older beds are deeply buried in most of the basins, but a two-phase history is indicated for the faults of the Mvuha Karroo rift, and on the coast itself the late movement on the western flanking fault of Pemba contrasts with the early (Karroo to Bajocian) movement of the facing Tanga fault (Fig. 43). In contrast to this the Mandawa graben appears to have developed symmetrically.

As noted above, there is evidence for an anomalous but strongly marked E–W fault belt, the Utete line, immediately south of the Rufiji (about latitude 08° 05'S), which could qualify for the hypothetical equatorial shear zone postulated by Krause (1966). This is the only known E–W lineament of any size between Mozambique and the equator.

Dixey (1956) and others have shown that rift-valley faulting in Nyasaland and the adjoining areas began in Karroo times and continued during the Mesozoic. In the coastal zone clear stratigraphical evidence of early movement is provided by the Triassic salt mass at Mandawa and Pindiro, developed to a thickness of 10 000 ft (3000 m) in a linear depression having the general characters of a graben, with abrupt flanks across which Middle Jurassic and later rocks transgressed onto structural highs (Kent, 1965). The faults associated with this graben trend N and NNE. Gravity data point to an analogous evaporite filled downwarp or graben west of the Msanga high, between the upper Ruvu and the Rufiji.

A comparable history of faulting, less closely defined as the beds of the downthrown side are only dated by inference, is shown by the Tanga fault zone near Msata and Lugoba. There Middle Jurassic rocks transgress from the basin area onto the gneisses of the upthrown side of the fault in a relationship which would be regarded as normal for a minor fault-line scarp, except for the aeromagnetic evidence of a sedimentary thickness measured in tens of thousands of feet on the immediate down thrown side. These sediments are interpreted as Karroo on the evidence of the thick beds of this age in the corresponding position 60 miles (100 km) to the north on the Pangani River.

It is not certain how far these Middle Jurassic faults had already developed during Karroo times (Fig. 44). For the Mandawa–Pindiro graben it seems most reasonable to envisage a progressive subsidence through most of the Trias to provide the trap mechanism in which the thick salt accumulated, but there is a notable rarity of coarse beds in the sequence which could have reflected corresponding uplift of the flanks. Pebbly sandstones in the deepest part of the penetrated sequence, and a conglomeratic phase within the Bathonian are the only symptoms of basement exposure in the flanking highs. The angular unconformity indicated within the salt series at 6600 ft (2000 m) may perhaps indicate a particularly pronounced subsidence. At Mandawa, at Msata and also, judging by sedimentary grade, at Ngerengere and in the Matumbi Hills it seems that the structurally high areas were close to sea level in Bathonian–Bajocian times, so that the marine Jurassic transgressed onto a landscape which was probably undulating but was certainly not a mountainous dissected fault-line scarp. Only at Ngerengere coarse clastics underlying the fossiliferous Middle Jurassic beds indicate in Liassic times erosion of a fault scarp adjoining the subsiding basin, but this appears to have been quickly lowered. This unusual circumstance of down-faulting without a corresponding uplift of the positive areas, in contrast to the balanced upward and downward movements inland, may reflect the special case of fault movement close to the developing Indian Ocean basin.

The Upper Jurassic represents a period of relative tranquillity, characterised in post-Callovian times by fine-grained sediments and an apparent absence of stratigraphic breaks away from the actual shoreline areas (Aitken, 1961). Crustal instability, not yet fully analysed, is however indicated in the Lower Cretaceous by the widespread arenaceous development throughout coastal East Africa and by the Aptian and Albian phase of regression and transgression known at intervals from Lindi to the Dar-es-Salaam embayment, and reflected further north still by the transgressive Cenomanian of the Wami River. This phase is likely to indicate uplift of inland areas contemporary with Lower Cretaceous fault and block movements which have been identified in the Rift zone by Dixey (1946, 1956) but there is as yet no firm knowledge of fault history at this time in the coastal belt.

During the Upper Cretaceous and early Tertiary a long period of exceptional stability intervened. There appears to have been a regression in Turonian times in the south, but there was no significant variation in the long sequence of essentially argillaceous sediments, deposited to great thickness. The rocks appear to represent the steady erosion of a relatively mature hinterland, with no coarse material symptomatic of active fault-scarp movement or major gradients.

Movements on the established Tertiary structures were, however, well developed during the Palaeogene, for the seismic surveys of the large islands show that the Lower Tertiary beds are much more strongly flexed and faulted than the late Oligocene and Miocene beds near the surface. It appears that the islands of Pemba, Zanzibar and Mafia were already taking shape at this time.

The analysis of rates of sedimentation recorded above shows an abnormally high rate for the Middle Eocene, implying accentuated erosion in the hinterland which in turn suggests that movements, probably faulting, were widespread.

Direct stratigraphic evidence of fault movement in the Lower Miocene is provided by the Ras Tapuri breccia on the Lindi fault, a melange of huge slipped blocks of Oligocene and Lower Miocene in a Lower Miocene matrix. On the coast there is thus clear evidence of fault movement (with consequent non-compressional flexing) at the end of the Palaeogene and the beginning of the Neogene periods. Evidence of a corresponding episode of earth movement inland, leading to a flood of arenaceous sediments which transgressed across the argillaceous Palaeocene and Eocene, occurs in the Dar-es-Salaam embayment. The evidence consists of 4000 ft (1200 m) or more of arkosic Pugu Sandstone which is Lower Miocene at the base and possibly Pliocene in the upper part, thick sandy Miocene sediments in many places further south along the coast, and the 8353 ft (2546 m) of arenaceous (deltaic) Miocene beds proved by the Zanzibar deep boring.

This heavy sedimentation in Neogene times is known from seismic data to be associated with a major phase of structural development. Limited ability to distinguish microfaunal subdivisions in beds with strongly varying facies hindered correlation at the time of the surveys, so that detailed analysis of the movements is a matter for the future. There is, however, evidence of an unconformity between the (Lower?) Pliocene and the Lower and Middle Miocene on Zanzibar and Pemba and widespread angular unconformity over the whole length of the coast between the Pleistocene–Recent raised coral reefs and the underlying Miocene and Pliocene.

The structural features of the Neogene are clearly related to broad tilting and folding which is in turn controlled by fault and block movements. The surface structures of the large islands are due to late movements of older features, and it is reasonable to regard the similar but less well documented late structures on the mainland, particularly in the Dar-es-Salaam embayment, as of corresponding posthumous type. On the evidence of the structural history of, say, Zanzibar, the surface N and NW striking faults and warps of the Dar-es-Salaam embayment are likely to have originated at least as early as Eocene.

In summary, very large N–S and NNE–SSW faults are known to have developed during later Karroo (pre-Middle Jurassic) times; other faults with this strike were active in the early Tertiary and with the smaller NNW trending Lindi fault underwent major movements in early Miocene. Many minor faults, predominantly NNW and NE in trend, together with

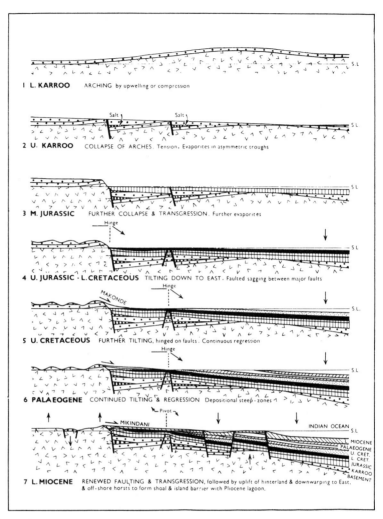

FIGURE 44. Developmental history of faulting, associated with hypothetical arch collapse.

gentle folds which are probably symptomatic of fault control, have been mapped in the Neogene rocks. Some, at least, are likely to have had extended histories like the faulting on the islands.

On the evidence available fault movement in the Tanzanian coastal belt appears to be episodic. BP-Shell workers have failed to find evidence that movement of any of the faults has been continuous over a long period in such a way as to be a dominant mechanism in the basinal subsidence. This contrasts with the current interpretation of development of coastal basins in the United States of America (Shelton, 1968).

Implications for Rift Valley
Interpretation

It is evident from the foregoing account that the structures of the Tanzanian coastal zone agree closely in strike and in date of movements with the structures of the classical rift valleys. It follows that they are likely to be related to allied causes, and it is therefore appropriate to summarise some of the salient features arising from the foregoing description.

1. No evidence of true compressional structures has been found in the coastal sediments.

2. Faults tend to be steep, and show no obvious evidence of major extension or stretching of the crust locally. The dominant feature appears to be vertical movement.

3. The irregular pattern of the major faults is not easy to reconcile with an origin related to deep continent-wide horizontal shear (transcurrent) structures in the Basement Rocks.

4. The existence of structures not significantly different in time or type from those of the Rift Valleys in this area as well as in West Africa (Furon, 1963) suggests that the whole width of the African continent was affected by similar stresses at the same time. This is not obviously compatible with the suggestion that the Rift Valleys originated directly as a continental continuation of a mid-ocean-ridge type of structure.

5. The Karroo, Miocene and later phases of movement previously well documented for the Rift Valley belt are clearly defined on the coast. There are additional indications of uplift giving rise to the early Cretaceous rudaceous phase and transgressive relationships, and of a major uplift beginning in the Middle Eocene, for which evidence should be sought inland.

6. Within the coastal area the major Karroo faults are not known to have been reactivated on a comparable scale subsequently. This feature is particularly marked in the case of the NNE Tanga trend, which had a pre-Middle Jurassic throw in excess of 10 000 ft (3000 m) but was subsequently only associated with minor warping. The less strong NNW Lindi fault trend shows local spectacular evidence of Lower Miocene movement at Lindi, but is mostly represented by a large concealed pre-Cretaceous feature shown up further north by seismic and gravity data. Topographically expressed faults and fault scarps seawards of these major features have undergone late-Tertiary movements, as have many faults further inland. There is thus reason to believe that the vertical throw on the old faults flanking the main coastal basin was several thousands of metres, and that these faults could not thereafter adjust to further stresses, which were taken up along different lines nearer the edge of the continental shelf as it was progressively built out.

[*Editor's Note:* "Appendix: List of boreholes" has been omitted here. Only references cited in this excerpt are reprinted here.]

REFERENCES

Aitken, W. G., 1961, Geology and palaeontology of the Jurassic and Cretaceous of southern Tanganyika, *Bull. No. 31, geol Surv. Tanganyika,* 144pp.

Dixey, F., 1946, Erosion and tectonics in the East African Rift System, *Jnl geol. Soc. Lond.,* Vol. 102, Pt. 3, pp. 339–379.

Dixey, F., 1956, The East African rift system, *Colon. Geol. Miner. Resour., Bull. Suppl. No.* 1, 71pp.

Furon, R., 1963, *The geology of Africa,* 2nd Ed., 377 pp. (Edinburgh and London: Oliver and Boyd.)

Kennedy, W. Q., 1965, The influence of Basement structure on the evolution of the coastal (Mesozoic and Tertiary) basins of Africa, pp. 7–16 in *Salt basins around Africa,* 122 pp. (London: Institute of Petroleum.)

Kent, P. E., 1965, An evaporite basin in southern Tanzania, pp. 41–54 in *Salt basins around Africa,* 122 pp. (London: Institute of Petroleum.)

Krause, D. C., 1966, Equatorial shear zone, *Pap. No. 66–14, geol. Surv. Can.,* pp. 400–443.

McConnell, R. B., 1967, The East African rift system, *Nature, Lond.,* Vol. 215, No. 5101, pp. 578–581.

North, F. K., 1966, *Discussion in* The World rift system, T. N. Irving (Ed.), *Pap. No. 66–14, geol. Surv. Can.,* p.447.

Sampson, D. N., and Wright, A. E., 1964, The geology of the Uluguru Mountains, *Bull. No. 37, geol. Surv. Tanganyika,* 69pp.

Shelton, J. W., 1968, Role of contemporaneous faulting during basinal subsidence, *Bull. No. 52, Am. Ass. Petrol. Geol.,* pp. 399–413.

Editor's Comments
on Papers 16, 17, and 18

16 DAWSON
Neogene Rift Tectonics and Volcanism in Northern Tanzania

17 WILLIAMS
Excerpts from *The Volcanics of the Gregory Rift Valley, East Africa*

18 BAKER, MOHR, and WILLIAMS
Excerpts from *Geology of the Eastern Rift System of Africa*

EASTERN RIFT SYSTEM

North-central Tanzania (Southern Gregory)

In Paper 16 Dawson, who spent some years working on the rift volcanics of northern Tanzania, briefly describes the rift faulting of this region. From the Kenya border southward into central Tanzania, Gregory's type rift valley—the graben form seen from the Kedong scarp in Kenya—is replaced by block-faulted forms—horsts, generally tilted, fault-angle depressions, and the related asymmetric graben. This pattern of faulting is characterized by splaying out from the Kenya rift valley, mainly as a westerly divergence, into central Tanzania, with volcanism ceasing. The volcanic activity that accompanies the faulting is described in Paper 16. This description complements the work by Williams (Paper 17) on the Kenya volcanics.

This area or zone of rift faulting and volcanism has been described in Paper 18 (p. 25) as the most southerly of the sections of the Eastern Africa Rift System and named the north Tanzania divergence. Also included is the Pangani graben, which belongs to the Eastern zone of Tanzania.

Southward in central Tanzania and thence southwestward towards Mbeya and the Rukwa trough, the faulting is discontinuous and generally not even en echelon. There are broad intervening unfaulted areas of Precambrian basement (Dundas, 1965, pp. 95–100; Kennerley, 1971; Teale and Harvey, 1933; Cooke, 1957).

It is frequently stated or implied in the literature that the Gregory or Eastern Rift System connects structurally with the Western Rift

System but this is not the case. The faulting of west-central Tanzania is of an entirely different style and order of magnitude from that of the western rift system. The faults are subdued or are replaced by monoclinal flexures of the Tertiary surfaces modified by scarp retreat. This has an important negative bearing on three concepts: (1) that there is a structural relationship between the Eastern Rift System and the Karro troughs of Zambia; (2) that there is continuity of the so-called global rift system along the eastern flank of Africa; and (3) that the Eastern Rift System plus the southern part of the Western Rift System could be regarded as an incipient or actual plate junction.

Kenya (Gregory), Kavirondo, Turkana Zones.

Paper 17, by Williams, is a definitive review of Gregory rift volcanism.

Paper 18 is by Baker, Mohr, and Williams, all of whom have long field experience in Kenya or Ethiopia and have authoritative knowledge of those aspects on which they write. The excerpts selected give a brief but comprehensive view of geology and structural relationships. The remainder of the paper is devoted chiefly to volcanism, geophysics, and tectonic theory. There is also a comprehensive bibliography. For a broader treatment of the Gregory Rift Valley and its relationship to the other East African rifts, see Paper 3 and Paper 7.

The Gregory (Kenya) Rift Valley, the associated Kavirondo Rift Valley (Shackleton, 1951) and the Skot trough (Wright, 1966) belong to the Kenya domal uplift of volcano-tectonic origin imposed on the northeastern part of the East African swell. It is the second of the sections into which Paper 18 divides the Eastern Rift System (pp. 25–27), the first being the subject of Paper 17.

King (1978) provided a more recent authoritative account based on detailed and intensive mapping done by the East African Geological Research unit, which he directed over a period of ten years. Six sheets in preparation for publication fill in the areas that remained unmapped by the Kenya Geological Survey.

A paper of a general nature, which summarizes a part of the massive report on the Soviet expedition of the late 1960s (Beloussov et al., 1974), is that by Logatchev et al. (1972) describing the structural and magmatic evolution by stages.

The geomorphology of the rift valley zone of Kenya received early treatment by Gregory (1921) and Sikes (1926). The latter contains illustrations of special interest. Busk (1939) explains an early block diagram of the southern half of the Kenya rift valley. Pulfrey (1960) did the first systematic treatment of Kenya physiography and planation surfaces and clarified the concept "rise to the rift" in a manner that differs radically from the proposals by Cloos (1939) and others. Saggerson and Baker (1965) contains a short section on rift-valley structure and physiography, but deals mostly with eastern Kenya. Both papers give accounts of erosion surfaces in Kenya, relating them to King's (1963) and Dixey's (1946) classifications of African surfaces. Cooke (1957) is also of value. Temperley (1966) provides good illustrations of features of fault and fault-line scarps as they are related to volcanics and sediments and discusses the age of the rift faulting.

Age and nomenclature of planation surfaces and their relations with sediments and volcanics are discussed by Shackleton (1951) as well as by Pulfrey (1960), Kent (1944), and Busk (1945).

The Kenya and Ethiopian domes are separated by a crustal sag or transitional zone of low altitude. Here is the third section of the Eastern Rift System, the Turkana depression of Paper 18 (pp. 28–31). The Sugutu Graben and southern half of Lake Turkana (Fuchs, 1939) terminate the Kenya Rift.

Study of this region is important because it lies between the Gregory and Ethiopian rifts. Results of such study could determine if there is structural continuity. Unless this can be proved, the claim that the Eastern Rift System is a plate boundary fails. In this transitional zone or "saddle" between two major volcano-tectonic domal uplifts, the rift faulting is minimal and en echelon. On the other hand, for the Kenya and Ethiopia rifts, the rift faulting of greatest magnitude corresponds with the maxima of the domal uplifts.

Paper 18 (pp. 13–17) contains a brief account of volcanism of the Gregory and Ethiopia-Afar zones. This combined treatment allows parallels and contrasts to be drawn. The Gregory is wholly continental but in the Ethiopian and Afar zones magmatic evolution from continental to oceanic is revealed. The Gregory rift zone is classified as volcano-tectonic (as is the Arabo-Ethiopian) and the study of volcanism is of special importance. Of modern accounts those by Williams, King, and Baker are most authoritative. Fieldwork by Williams is recorded in Kenya Geological Survey reports; Paper 17 is a summary of his work to 1970. Later accounts by Williams are those of 1978a and 1978b. King and Chapman (1972) and King (Paper 3) describe the volcanics.

Age relationship of volcanism, uplift, and rift faulting, and the nature of the latter—whether compressional or tensional—have in the past involved much discussion and speculation. These problems are discussed in the original of Paper 18 as well as in Paper 3.

On the one hand, the subject can be approached as tracing the historical development and, on the other, as relating the secondary effects of a single primary cause, the emanation of energy (rise of isotherms) from the mantle (Baker and Wohlenberg, 1971). Of special interest in relating volcanism and tectonic history of the Gregory and Ethiopian rifts, is the paper by Shackleton (1978) in which the radiometric time scale is applied. The paper by Baker et al. (1970) is an earlier exercise on the same theme. A more recent paper by Baker et al. (1978) traces the tectonic and magmatic history as now interpreted.

For the Gregory and Kavirondo rift valleys, faulting in the crystalline basement is generally obscured by volcanics and the elucidation of structure must be by consideration of models derived from geophysical interpretation of gravity, seismology, heat flow, and magnetic observations. We can pass over earlier models, that is, subsidence of the keystone of the arch (Gregory, 1921), formation of a crestal graben (Cloos, 1939) and other hypotheses. The immediate question to be resolved is whether the rift faulting of this volcano-tectonic domal uplift is the result of, or is accompanied by: (1) crustal or lithospheric separation with the rise of mantle material either along a zone or localized (Khan and Mansfield, 1971; Khan, 1975; Khan and Swain, 1978); or (2) extension by thinning of the lithosphere with intrusion of dense material (Searle, 1970a, 1970b); or (3) doming and faulting as related effects of some subcrustal process (Baker and Wohlenburg, 1971; Paper 18); or (4) spreading along a constructive plate margin, the boundary between Somali and Nubian plates (McKenzie et al., 1970; Darracott et al., 1973 (but see Gilluly, 1971); or (5) crustal extension without separation by an amount equal to the total of horizontal components of movement on normal faults (King, Paper 3, 1978; Logatchev et al., 1972). Burke and Whiteman (1973) regard the meeting of the north Kenya, south Kenya, Kavirondo rifts as a triple junction on the Kenya dome. An interesting variant is the proposal of Oxburgh and Turcotte (1974) and Oxburgh (1978) that rifting is a result of plate movement on the nonspherical surface of the earth.

Gilluly (1971, pp. 2384–2385) has studied the implications of the magmatic evolution and changes in the volcano-chemistry of igneous activity on plate margins and specifically compares continental and oceanic rift vulcanism, especially in regard to African rifts. He states that "any generalization yet made as to the magmatic associations of the oceanic ridges is entirely inapplicable to the continental rifts so far studied." Gilluly also makes other adverse comparisons.

The Main Ethiopian Rift

This is the fourth section of the Eastern Rift System of Africa as it is described by Baker, Mohr, and Williams in Paper 18, pp. 31–33 (see Paper 5). The remainder of Paper 18 describes volcanism and geophysics.

Although there are marked differences, this rift zone can be compared with the Gregory. Apparently they both evolved by marginal faulting and warping (which was accompanied by early volcanic activity) so that graben resulted (Mohr and Potter, 1973). The depressed blocks were faulted longitudinally and there may have been crustal separation or at least thinning of the sialic crust.

There are dextrally offset segments (compare the sinistrally offset rift valleys of the Western Rift System). The map by Merla et al. (1973) reveals this pattern and the continuity of Precambrian basement across the zone at 6°N latitude.

Dainelli (1943) laid the foundation for subsequent work. This has been built on by Mohr whose *The Geology of Ethiopia* (1962a) was the beginning of intensive investigation. Systematic description of the rift valley was initiated by Mohr (1962b); he described the Wonji fault belt, established the relationship of the rift movements to the Arabo-Ethiopian swell, and established the time relationship of volcanic and tectonic episodes. He also suggested the presence of right lateral, transverse, transcurrent faulting accompanying crustal separation. His contribution to the *Tectonics of Africa* (1971) incorporates most of his work to that date. Gibson (1969) and Gibson and Tazieff (1970) describe details of volcanic fields and the presence of longitudinal normal faults in en-echelon pattern but cannot support Mohr's transverse faulting. They interpret the fault pattern as evidence of left-lateral shear. A valuable detailed description of volcanic activity and structures is given by Paola (1973). Meyer et al. (1975) add important conclusions on the stratigraphy of the volcanics and structural evolution. They distinguish early and later sequences initiated by renewed tectonic activity along the Wonji fault-belt.

With regard to structure, as in the case of the Gregory rift the deep-seated faulting is concealed by the younger volcanics and sediments, and recourse must be had to interpretation of geophysical data. Paper 18 contains references to this work.

The pattern of gravity anomalies over the rift can be interpreted as indicating (1) that beneath the low density volcanics there is a graben structure; (2) that there is intrusion of denser basic material into a zone of crustal attenuation; (3) or that there is actual crustal separation with a median zone of intrusive mantle material (Mohr, 1973a). Further problems are whether this last condition applies to the full width of the rift valley or only to the Wonji fault belt (Searle and Gouin, 1972), and whether these questions refer to conditions in one segment only, that between north latitudes 7° and 9°.

The geology of the Ethiopia rift assumes major significance for a number of reasons. Whereas the Gregory rift may be explainable as essentially a crestal graben terminated to both the north and south and therefore neither a plate boundary nor a connected extension of the world rift system, the main Ethiopian rift is in a different category. Of the rift valleys so far considered this is the first that (1) *may* belong to the world rift system; (2) may be the penetration into a continent of the mid-oceanic rift; and/or (3) may be the failed arm of a triple junction. However, it is not, as could have been expected, a simple rift, that is, a crustal separation in the Wegener sense, narrowing as it penetrates the continent. There are no splay faults beyond the southern termination as there are at both ends of the Gregory rift.

Unless the rift can be demonstrated on geological and geophysical evidence to be the result of an actual crustal separation—not merely an attenuation and extension—and unless this condition can be shown to extend to the south along the Eastern Rift System, then Somalia cannot be a plate or a discrete crustal segment. The same arguments can be applied when considering the left-lateral shear hypotheses of Gibson and Tazieff (1970), which would require the southerly extension of such a wrench fault zone.

Structural continuity of the Gregory and Ethiopian rifts has been assumed in some applications of plate tectonics to eastern Africa (McKenzie et al., 1970; Darracott et al., 1973). The main Ethiopian rift terminates at 5°N. It belongs to a different tectonic order from the East African rifts (Pilger and Rosler, 1976a). It is the failed arm of a three-arm rift system that trisects the Arabo-Ethiopian dome (Paper 19). Burke (1977) regards this rift as an example of a failed rift arm striking into a continent from a triple junction and therefore that it is what he regards as an aulacogen. This view is tied to the concept that aulacogens can develop into continental margins, and that the Horn of Africa is in the process of separation from Africa. Milanovsky (1981) points out that the "penetrating" is only one of three types of aulacogen. The subject of aulacogens is not pursued here because its possible relevance to the Afro-Arabian rift system appears to be established only for the Arabo-Ethiopian dome.

The Afar Depression

Baker, Mohr, and Williams (Paper 18, pp. 33-36) include the Afar depression as the fifth and most northerly member of the sections of the Eastern African System. That the Afar should be described as a rift valley is controversial. Discussion is included here because it is one of the succession of tectonic geomorphic forms of the Afro-Arabian Rift System, and within it are found many volcano-tectonic features with one or other of which many rift valleys, continental or oceanic, can find a match. IUCG Scientific Report No. 16 on the recent international symposium on rift valley problems (Pilger and Rösler, 1976b), has the appropriate title *Afar Between Continental and Oceanic Rifting,* and this indeed appears to state the case.

Many of the references on the general geology of Ethiopia have accounts of both the Ethiopian rift and the Afar (see previous references, especially Mohr 1962a, 1967; Gibson and Tazieff, 1970; Merla et al., 1973; and Paper 5). Tazieff (1970), which includes excellent plates, is valuable in giving an overall impression of the depression. IUCG Scientific Report No. 14, *Afar Depression of Ethiopia* (Pilger and Rösler, 1975) of the Bad Bergzabern Symposium brings together the results of detailed geological and geophysical surveys during the period 1966 to 1975. Of general interest are the introduction by Marinelli and Tazieff; the Afro-Arabian dome by Gass and Gibson (pp. 10-18); the structural setting by Mohr (pp. 27–37); the structural evolution by Barberi et al. (pp. 38–54); crustal attenuation by Morton and Black (pp. 55-65); faulting and structure by Christianson et al. (pp. 259-276); and conclusions from the papers on the nature of the crust by Barberi and Varet (pp. 375–378). Tiercelin and Faure (1978) in a study of subsidence and sedimentation use results from the Afar. The second volume (Pilger and Rösler, 1976b) includes papers on the related wider aspects and regional problems of this part of the Afro-Arabian rift. Pilger and Rösler (pp. 1–25) attempt to establish temporal relationships along the length of the Afro-Arabian rift.

Attention is focused on the Afar because of the revelation of crustal structure (Needham et al., 1976) and related geodynamic phenomena that, along rift valleys, are generally concealed by sediments and volcanics and by lake waters or oceans. The nature and structure of the margins, whether or not they are downwarps (Gouin, 1970; Mohr, 1973b), and of the crust underlying the Afar depression are significant in determining: (1) whether any credence should still be given to the possibility of its formation as a graben structure involving vertical tectonics only (Picard, 1970); or (2) whether the floor is an area of continental crustal extension or separation with emphasis on attenuation by thinning or faulting with or without the creation of oceanic crest (Black, 1973; Morton and Black, 1975); or (3) separation has been by faulting without preceding attenuation and, if so, whether evolution has been in stages (Barberi et al., 1975; Mohr, 1975). For a plausible synopsis there must be synchronization of sequences of tectonic events along the Afro-Arabian rift system, as attempted by Pilger and Rösler (1976b, pp. 1–25), as well as geometric consistency in the horizontal translations and rotations of crustal units (Paper 19).

REFERENCES

Anonymous, 1965, Report on the geology and geophysics of the East African Rift System, *I.U.M.P. Sci. Rep. 6,* Nairobi University Press, Nairobi, 265p.

Baker, B. H., and J. Wohlenberg, 1971, Structure and evolution of the Kenya Rift Valley, *Nature* **229:**538–542.

Baker, B. H., L. A. J. Williams, J. A. Miller, and F. J. Fitch, 1970, Sequence and geochronology of the Kenya Rift volcanics, *Tectonophysics* **11:**191-215.

Baker, B. H., R. Crossley, and G. G. Coles, 1978, Tectonic and magmatic evolution of the southern part of the Kenya rift valley, in *Petrology and Geochemistry of Continental Rifts,* E. R. Neumann and I. B. Ramberg, eds., D. Reidel, Dordrecht, pp. 29-50.

Barberi, F., G. Ferrara, R. Santacroce, and J. Varet, 1975, Structural evolution of the Afar triple junction, in *Afar Depression in Ethiopia,* A. Pilger and A. Rösler, eds., Schweizerbart'sche, Stuttgart, pp. 38-54.

Beloussov, V. V., V. I. Gerasimovsky, A. V. Goriachev, V. V. Dobrovolsky, A. P. Kapitsa, N. A. Logachev, E. E. Milanovsky, A. I. Poliakov, L. N. Rykunov, and V. V. Sedov, 1974, *East African Rift System, Major Features of Structure, Stratigraphy,* vol. 1; *Hypergenic Formations, Geomorphology, Neotectonics,* vol. 2; *Geochemistry of Volcanics, Seismological Investigations, Main Results,* vol. 3,[In Russian], Soviet Geophysical Committee, Nauka, Moscow, 775p.

Black, R., 1973, Structures of the Afar Floor (Ethiopia) in *Implications of Continental Drift to the Earth Sciences,* D. H. Tarling and S. K. Runcorn, eds., Academic Press, London, p. 777.

Burke, K., 1977, Aulacogens and continental breakup, *Ann. Rev. Earth Planet Sci.* **5:**371-96.

Burke, K., and A. J. Whiteman, 1973, Uplifting, rifting, and the break-up of Africa, in *Implications of Continental Drift to the Earth Sciences,* D. H. Tarling and S. K. Runcorn, eds., Academic Press, London, pp. 735-755.

Busk, H. G., 1939, Explanatory note on the block diagram of the Great Rift Valley from Nakuru to Lake Magadi, *Geol. Soc. London Quart. Jour.* **45:**231-233.

Busk, H. G., 1945, On the normal faulting of rift valley structure, *Geol. Mag.* **82:**37-44.

Cloos, H., 1939, Der Nubisch-Arabische Schild, in Hebung, Spaltung, Vulcanismus, *Geol. Rundschau* **30:**434-445.

Cooke, H. B. S., 1957, Observations relating to Quaternary environments in East and Southern Africa, *Geol. Soc. S. Africa Bull.* **60**(annex):1-73.

Dainelli, G., 1943, *Geologia dell-Africa Orientale,* Reale Accad. Italia, Rome, 4 vols.

Darracott, B. W., R. W. Girdler, J. D. Fairhead, and S. A. Hall, 1973, The East African Rift System, in *Implications of Continental Drift to the Earth Sciences,* D. H. Tarling and S. K. Runcorn, eds., Academic Press, London, pp. 757-766.

Dixey, F., 1946, Erosion and tectonics in the East African Rift system, *Geol. Soc. London Quart. Jour.* **102:**339-388.

Dundas, D. L., 1965, Review of rift faulting in Tanzania, in *Report on the geology and geophysics of the East African Rift System,* Nairobi University Press, Nairobi, pp. 95-100.

Fuchs, V. F., 1939, The geological history of the Lake Rudolph Basin, Kenya Colony, *Royal Soc. London Philos. Trans.* **B229:**219-274.

Gibson, I. L., 1969, The structure and volcanic geology of an axial portion of the main Ethiopian Rift, *Tectonophysics* **8:**561-565.

Gibson, I. L., and H. Tazieff, 1970, The structure of Afar and the northern part of the Ethiopian Rift, *Royal Soc. London Philos. Trans.* **A267:**331-338.

Gilluly, J., 1971, Plate tectonics and magmatic evolution, *Geol. Soc. America Bull.* **82:**2383-2396.

Gouin, P., 1970, Seismic and gravity data from Afar in relation to surrounding areas, *Royal Soc. London Philos. Trans.* **A267:**339-358.

Gregory, J. W., 1921, The rift valleys and geology of East Africa, Seeley Service Co., London, 479p.

Kennerley, J. B., 1971, Fault map of Tanzania, in *Tectonics of Africa,* (Earth Sciences, vol. 6), UNESCO, Paris, p. 523.

Kent, P. E., 1944, The age and tectonic relationships of the East African volcanic rocks, *Geol. Mag.* **81:**15-27.

Khan, M. A., 1975, The Afro-Arabian rift system, *Sci. Progress* **62:**207-236.

Khan, M. A., and J. Mansfield, 1971, Gravity measurement in the Gregory Rift, *Nature* **229:**72-75.

Khan, M. A., and C. J. Swain, 1978, Geophysical investigations and the rift valley geology of Kenya, in *Geological Background to Fossil Man: Recent Research in the Gregory Rift Valley, East Africa,* W. W. Bishop, ed., Scottish Academic Press, Edinburgh, pp. 72-83.

King, B. C., 1963, *South African scenery,* 3rd ed., Oliver and Boyd, Edinburgh, 308p.

King, B. C., 1978, Structure and volcanic evolution of the Gregory Rift Valley, in *Geological Background to Fossil Man: Recent Research in the Gregory Rift Valley, East Africa,* W. W. Bishop, ed., Scottish Academic Press, Edinburgh, pp. 29-54.

King, B. C., and G. R. Chapman, 1972, Volcanism of the Kenya rift valley, *Royal Soc. London Philos. Trans.* **A271:**185-208.

Logatchev, N. A., V. V. Beloussov, and E. E. Milanovsky, 1972, East African rift development, *Tectonophysics* **15:**71-81.

McKenzie, D. P., D. Davies, and P. Molnar, 1970, Plate tectonics of the Red Sea and East Africa, *Nature* **226:**243-248.

Merla, G., E. Abbate, P. Canuti, M. Sagri, and P. Tacconi, 1973, *Geological Map of Ethiopia and Somalia,* scale 1:2m, Geol. Palaeontol. Inst., University of Florence, Florence, Italy.

Meyer, W., A. Pilger, A. Rösler, and J. Stets, 1975, Tectonic evolution of the northern part of the Main Ethiopian Rift in southern Ethiopia, in *Afar Depression of Ethiopia,* A. Pilger and A. Rösler, eds., Schweizerbart'sche, Stuttgart, pp. 352-362.

Milanovsky, E. E., 1981, Aulacogens of ancient platforms: problems of their origin and tectonic development, *Tectonophysics* **73:**213-248.

Mohr, P. A., 1962a, *The Geology of Ethiopia,* University College Press, Addis Ababa, 268p.

Mohr, P. A., 1962b, The Ethiopian Rift System, *Addis Ababa Geophys. Observ. Bull.* **3:**33-59.

Mohr, P. A., 1967, The Ethiopian Rift System, *Addis Ababa Geophys. Observ. Bull.* **11:**1-65.

Mohr, P. A., 1971, Outline tectonics of Ethiopia, in *Tectonics of Africa Earth Sciences, vol. 6,* UNESCO, Paris, pp. 447–458.

Mohr, P. A., 1973a, Crustal deformation rate and the evolution of the Ethiopian Rift, in *Implications of Continental Drift to the Earth Sciences,* D. H. Tarling and S. K. Runcorn, eds., Academic Press, London, pp. 767–776.

Mohr, P. A., 1973b, Structural elements of the Afar margins: Data from ERTS-1 imagery, *Addis Ababa Geophys. Observ. Bull.* **15:**83–89.

Mohr, P. A., 1975, Structural setting and evolution of Afar, in *Afar Depression of Ethiopia,* A. Pilger and A. Rösler, eds., Schweizerbart'sche, Stuttgart, pp. 27–37.

Mohr, P. A., and E. C. Potter, 1973, *ERTS-1 Ground Truth Investigation of a Margin Sector of the Ethiopian Rift Valley,* U.S. Dept. of Commerce, NASA CR-132033, 5p.

Morton, W. H., and R. Black, 1975, Crustal attenuation in Afar, in *Afar Depression of Ethiopia,* A. Pilger and A Rösler, eds., Schweizerbart'sche, Stuttgart, pp. 55–65.

Needham, H. D., P. Choukroune, J. L. Cheminee, X. Le Pichon, J. Franchetau, and P. Tapponnier, 1976, The accreting plate boundary: Ardoukoba (Northeast Africa) and the Oceanic Rift Valley, *Earth and Planetary Science Lett.* **28:**439–453.

Oxburgh, E. R., 1978, Rifting in East Africa and large-scale tectonic processes, in *Geological Background to Fossil Man: Recent Research in the Gregory Rift Valley, East Africa,* W. W. Bishop, ed., Scottish Academic Press, Edinburgh, pp. 1–12.

Oxburgh, E. R., and D. L. Turcotte, 1974, Membrane Tectonics and the East African Rift, *Earth and Planetary Science Lett.* **22:**133–140.

Paola, G. M. di, 1973, The Ethiopian Rift Valley (between 7° 00′ and 8° 40′ lat. N), *Bull. Volcanol.* **36:**517–572.

Picard, L., 1970, On Afro-Arabian graben tectonics, *Geol. Rundschau* **5:**337–381.

Pilger, A., and A. Rösler, eds., 1975, *Afar Depression of Ethiopia,* Schweizerbart'sche, Stuttgart, 416p.

Pilger, A., and A. Rösler, 1976a, Temporal relationships in the tectonic evolution of the Afar depression (Ethiopia) and the adjacent Afro-Arabian rift system, in *Afar Between Continental and Oceanic Rifting,* A. Pilger and A. Rösler, eds., Schweizerbart'sche, Stuttgard, pp. 1–25.

Pilger, A., and A. Rösler, eds., 1976b, *Afar Between Continental and Oceanic Rifting,* Schweizerbart'sche, Stuttgart, 210p.

Pulfrey, W., 1960, Shape of the sub-Miocene erosion bevel in Kenya, *Kenya Geol. Survey Bull.* **3:**1–18.

Saggerson,. E. P., and B. H. Baker, 1965, Post-Jurassic erosion surfaces in eastern Kenya and their deformation in relation to rift structure, *Geol. Soc. London Quart. Jour.* **121:**51–72.

Searle, R. C., 1970a, Lateral extension in the East African Rift Valleys, *Nature* **227:**267–268.

Searle, R. C., 1970b, Evidence from gravity anomalies for thinning of the lithosphere beneath the rift valley in Kenya, *Royal Astron. Soc. Geophys. Jour.* **21:**13–31.

Searle, R. C., and P. Gouin, 1972, A gravity survey of the central part of the Ethiopian rift valley, *Tectonophysics* **15:**41–52.

Shackleton, R. M., 1951, A contribution to the geology of the Kavirondo Rift Valley, *Geol. Soc. London Quart. Jour.* **106:**345-392.

Shackleton, R. M., 1978, Structural development of the East African Rift System, in *Geological Background to Fossil Man: Recent Research in the Gregory Rift Valley, East Africa,* W. W. Bishop, ed., Scottish Academic Press, Edinburgh, pp. 19-28.

Sikes, H. L., 1926, The structure of the eastern flank of the rift valley near Nairobi, *Geog. Jour.* **68:**385-402.

Tazieff, H., 1970, The Afar triangle, *Sci. American* **222:**32-51.

Teale, E. O., and E. Harvey, 1933, A physiographical map of Tanganyika Territory, *Geog. Rev.* **23:**403-413.

Temperley, B. N., 1966, The faced scarp structure and the age of the Kenya Rift Valley, *Overseas Geol. Mineral Resour.* **10:**11-29.

Tiercelin, J. J., and H. Faure, 1978, Rates of sedimentation and vertical subsidence in neorifts and palaeorifts, in *Tectonics and Geophysics of Continental Rifts,* I. B. Ramberg and E. R. Neumann, eds., D. Reidel, Dordrecht, pp. 41-47.

Williams, L. A. J., 1978a, Character of Quaternary volcanism in the Gregory Rift Valley, in *Geological Background to Fossil Man: Recent Research in the Gregory Rift Valley, East Africa,* W. W. Bishop, ed., Scottish Academic Press, Edinburgh, pp. 55-70.

Williams, L. A. J., 1978b, The volcanological development of the Kenya Rift, in *Petrology and Geochemistry of Continental Rifts,* E. R. Neumann and I. B. Ramberg, eds., D. Reidel, Dordrecht, pp. 101-121.

Wright, J. B., 1966, Evidence for trough faulting in East Central Kenya, *Overseas Geol. Mineral Resour.* **10:**30-41.

16

Reprinted from *Geol. Soc. London Proc.* **1663**:151–153 (1970)

NEOGENE RIFT TECTONICS AND VOLCANISM IN NORTHERN TANZANIA

J. B. Dawson

The Neogene volcanic province of northern Tanzania is a southern extension of the more widespread volcanic areas of Ethiopia and Kenya. The earliest volcanism was preceded by a major phase of late Tertiary (? Miocene–Pliocene) warping that broke up the Basement terrain, and disrupted the mid-Tertiary erosion surface. The movements gave rise to a tectonic depression, bounded by either faults or warps, that is considerably broader than the trough-shaped depression in Kenya to the north. The depression is bounded to the south by the Masai Block around which the area of tectonic disturbance bifurcates, continuing to the south-west as the Eyasi and Yaida depressions and to the south-east as the Pangani graben. Within this tectonic depression there erupted a group of major shield volcanoes whose lavas infilled the depression and eventually, at several points, overstepped the fault-scarps bounding the depression. The earliest date of eruption of the volcanics is not known but, by analogy with the volcanics in Kenya, is tentatively ascribed to the Pliocene. These first volcanoes produced mainly lavas of the mildly alkaline continental olivine–basalt–trachyte–phonolite association (Oldoinyo Sambu, Elanairobi, Olmoti, Loolmalasin, Ngorongoro, Lemagrut, Oldeani, Tarosero, Ketumbeine, Gelai, and the Mawenzi, Shira and Kibo centres on Kilimanjaro), though the activity at Lemagrut started off with nephelinitic activity and rare nephelinites are known in the Loolmalasin volcanics.

In addition Mosonik, which has been dated as Upper Pliocene in age, is a nephelinite–carbonatite volcano. Some of the basaltic volcanoes developed major calderas in their later stages.

Lava extrusions continued into the upper and middle Pleistocene at some centres, along with sedimentation in the Olduvai, Peninj (L. Natron), Amboseli and Ketumbeine–Tarosero areas. There is some evidence of minor Lower Pleistocene faulting in the Natron–Manyara area but in Middle–Upper Pleistocene times the terrain was again broken up by a major phase of faulting. Some of the boundary faults of the Tertiary depression were reactivated to a minor extent, but the major manifestation of the movements was the major fault which runs from L. Natron southwards to Lake Balangida. This fault roughly bisects the western part of the Tertiary depression and cuts into the Masai block south of Lake Manyara. The faulting was of the 'trap-door' variety, with the hinge being along the eastern boundary of the Tertiary depression and the major movement along the major fault whose escarpment is the present-day western boundary of the Rift Valley. To the west of this fault, minor faulting was not extensive except on the Mbulu Plateau; but to the east of the fault, on the floor of the present-day Rift Valley, there was extensive minor faulting, causing minor horsts, grabens and tilted fault blocks, most of which are limited to the confines of the earlier Tertiary depression.

Following the Upper Pleistocene faulting there was another major phase of volcanic activity, which in eruptive style, relatively small volume of extruded material, and magma type contrasted with the earlier, relatively quite massive extrusions from the basaltic shield volcanoes. The activity was of a highly explosive type giving rise to features varying from major steep cones consisting of dominantly pyroclastic material (Meru, Monduli, Oldoinyo Lengai, Kerimasi, Burko, Essimingor, Kwaraha and Hanang) to areas of minor tuff cones and explosion craters. Some of the major volcanoes lie very close to the Rift fault (e.g. Oldoinyo Lengai, Kerimasi, Hanang) and Meru lies at the south-east end of a minor Pleistocene graben. Oldoinyo Lengai and Kerimasi erupted on a horst block separating the Natron and Engaruka basins, and Essimingor and Burko erupted along a swell structure separating the Engaruka and Manyara Basins. In addition the tuff-cone clusters coincide with areas of intense minor faulting.

The magma type at most of these later centres was ultrabasic–ultra-alkaline, and carbonatite is present at many volcanoes either as intrusive bodies or as extrusives. Modern carbonatite lava extrusives have occurred at the active volcano Oldoinyo Lengai. However at the Monduli volcano the main extrusives were trachybasalt pyroclastics and on Meru a wide variety of rock types have been found in the pyroclastics or as minor flows on the main cone (sodic trachyte, phonolite, nephelinite and nepheline tephrite) and nephelinites occur late in the sequence forming the inner cone in the explosion caldera.

In addition to these Upper Pleistocene–Recent centres, activity at some of the older centres persisted after the Middle Pleistocene.

Minor faulting has occurred in Upper Pleistocene–Recent times, the pyroclastics of Kerimasi being faulted prior to the initial eruption of Oldoinyo Lengai, the youngest centre.

In summary, the Neogene tectonics and volcanism in the rift area of northern Tanzania are intimately related. A major phase of late Tertiary faulting was followed by extrusion of large amounts of basaltic–trachyte magma from large shield volcanoes. This was separated by a second major phase of faulting from a later Upper Pleistocene–Recent phase of volcanism which contrasted with the earlier phase in its volume, dominant magma type and eruptive style. The location of the later volcanoes was closely connected with the rift faults.

17

Reprinted from pp. 439–441, 444–445, and 459–465 of *Bull. Volcanol.*
34:439–465 (1970)

The Volcanics of the Gregory Rift Valley, East Africa *

L. A. J. WILLIAMS

(University College Nairobi, Kenya)

Abstract

The paper reviews the stratigraphy, style of activity and some aspects of the petrology of Tertiary to Recent sodic alkaline volcanic rocks in Kenya, eastern Uganda and northern Tanzania.

Repeated extrusions of basaltic and nephelinitic volcanics occurred from Miocene times onwards, confirming indications from chemical data that magmas of these compositions were parental. At some central volcanoes, a basalt-trachyte-phonolite series evidently arose by fractional crystallization of basaltic magma, whereas various courses of crystallization from a nephelinitic parent led to the production of phonolites, tephrites and basanites as well as olivine- and melilite-bearing nephelinites and melanephelinites.

Phonolitic and trachytic volcanics which dominate an area of repeated upwarping (the Kenya dome) probably originated by processes of partial melting rather than by differentiation of basaltic magma. The basalt-trachyte association which characterizes many central volcanoes north and south of the dome can perhaps best be explained by postulating independent sources for the basic and salic volcanics.

Introduction

Over the last two or three decades systematic sheet mapping by the Geological Surveys of Kenya, Tanzania and Uganda has resulted in the accumulation of a great deal of information on the Tertiary to Recent volcanics associated with the Eastern Rift. The Kenyan part of this rift zone is often referred to as the Gregory Rift Valley in recognition of the pioneer work carried out by J. W. GREGORY (1896;

---

* Paper read at the International Symposium « Volcanoes and Their Roots », Oxford, England, Sept., 1969.

1921). No strict limits have been attached to the name and it is used here to include the section of the rift in northern Tanzania. The volcanic centres of eastern Uganda are also covered by this review for they belong to the same sodic alkaline province as the volcanics farther east.

Despite the quantity of data available, few reviews have appeared in print. Some aspects of the volcanic stratigraphy are summarized in regional accounts by KENT (1944), SHACKLETON (1951), WILCOCKSON (1964), WILLIAMS (1964; 1967) and BAKER (1965). Among the early petrological accounts, those by SMITH (1931; 1938) deserve special mention. In more recent years, general comments on the volcanic petrology are given in papers by KING and SUTHERLAND (1960), WRIGHT (1963; 1965), SAGGERSON and WILLIAMS (1964), WILLIAMS (1965; 1969b), KING (1965), NIXON and CLARK (1967) and VARNE (1968). The writer (WILLIAMS, 1969a) recently described the chief volcanic associations in the Gregory Rift Valley.

Stratigraphy and Petrography

The stratigraphy of the Miocene to Recent volcanics in Kenya, northern Tanzania and eastern Uganda is summarized in Fig. 1. The formations considered to have originated by fissure and multi-centre eruptions (for convenience they are referred to in this account as the plateau volcanics) are separated from the products of major central volcanoes, for it has already been noted (WILLIAMS, 1969a) that these two environments are characterized by distinct volcanic associations. The plateau volcanics are readily subdivided for descriptive purposes on an age and composition basis, but a classification of the central volcanoes is more complex since many of them have representatives of two distinct alkaline series.

The more important central volcanoes are distinguished from the plateau volcanics in Fig. 2 which shows the distribution of the main volcanic associations in the region. The occurrence of both nephelinite-phonolite and basalt-trachyte-phonolite associations at many of the central volcanoes is indicated by superimposed ornaments, the relative densities of the symbols being a measure of the abundance of volcanics of each suite. In general, the ages of the plateau volcanics are known with greater certainty than the time ranges represented by many of the central volcanoes.

[*Editor's Note:* Material has been omitted at this point.]

231

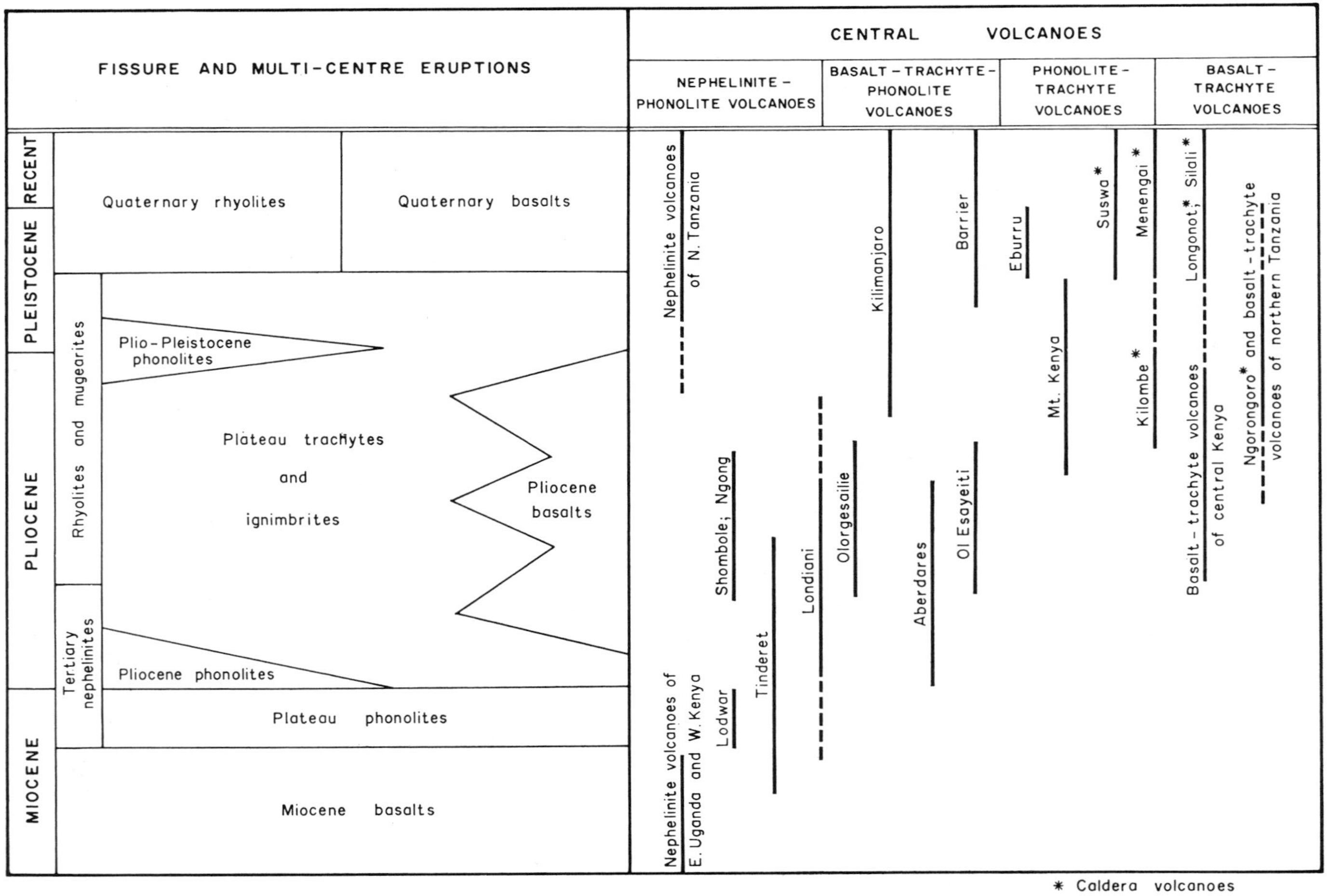

FIG. 1 - Diagram summarizing the stratigraphy of the Miocene to Recent volcanic rocks associated with Gregory Rift Valley.

(2) Distribution of the Volcanics in Space and Time

Previous reviews which have touched on some aspects of the petrogenesis of the rift volcanics in East Africa have largely failed to take into account the character of the eruptions and the fundamental matter of the distribution in space and time of contrasting rock types. Some significant changes in time in the composition of the volcanics and in the style and focus of activity have, therefore, gone unnoticed or unreported. To illustrate some of these changes, an attempt is made in Fig. 3 to show the compositions and extent of plateau volcanics in Miocene, Pliocene and Quaternary times, and to indicate the dominant associations at major central volcanoes which were active during these periods.

The pattern of volcanicity during the Miocene was relatively simple. The extensive flood basalts of north-western Kenya were erupted at about the same time as the build-up of the large nephelinite-phonolite volcanoes of eastern Uganda and western Kenya. One of the central volcanoes (Moroto) is exceptional in having mildly alkaline rocks as wel as nephelinites. Upwarping to produce the « Kenya dome » was followed towards the end of the Miocene by extrusion of the plateau phonolites which flowed down the flanks of the dome from an axial region destined later to become part of the rift floor. Nephelinitic fissure eruptions occurred on a very minor scale in areas both north and south of the dome.

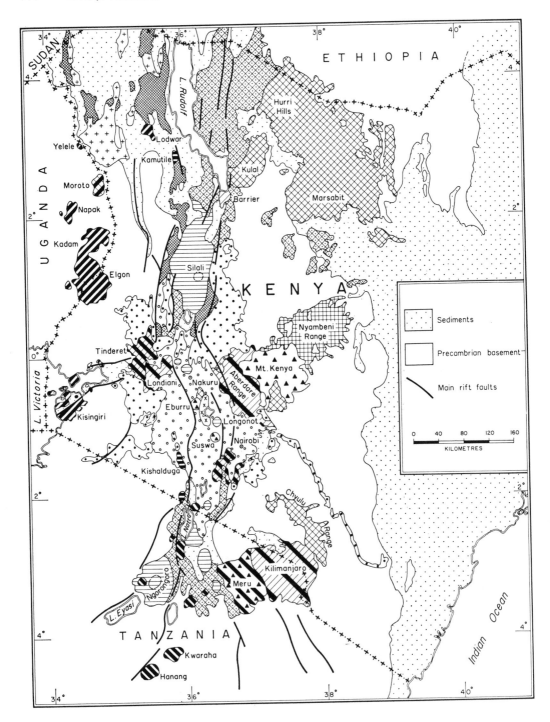

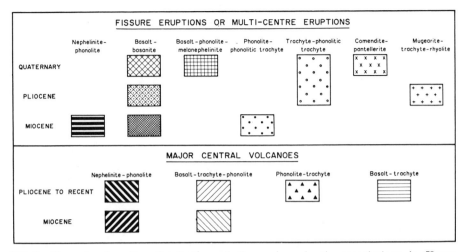

F𝐈G. 2 - Map showing the distribution of the main volcanic associations in Kenya, eastern Uganda and northern Tanzania (*After* W𝐈LL𝐈AMS, 1969*a*, Fig. 1, p. 63; modified to distinguish Miocene, Pliocene and Quaternary volcanics).

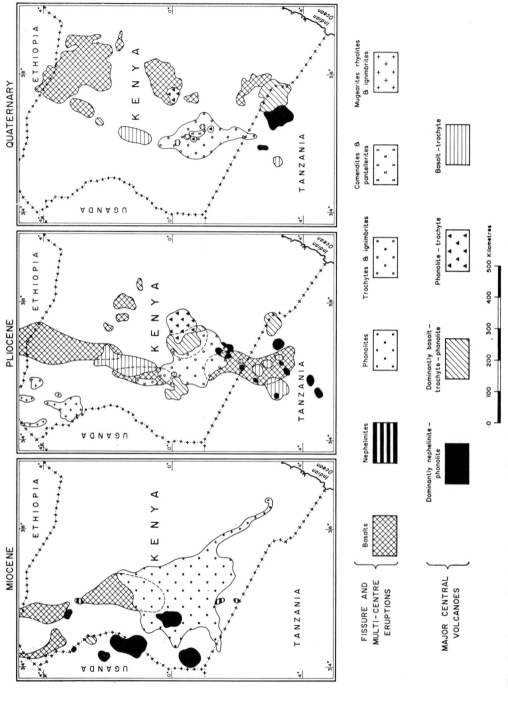

Fig. 3 - The distribution in space and time of the volcanics. Faults are not shown but the Gregory Rift Valley corresponds closely to the meridional zone occupied by the Pliocene volcanics.

Perhaps the most striking aspect of volcanicity in Pliocene times was the tendency for many of the eruptions to correspond closely to the line of the developing rift valley. The initiation of major faulting along the crest of the Kenya dome was accompanied by a marked change there in the character of plateau volcanics, trachytes and ignimbrites taking the place of the earlier phonolites. Flood basalts were erupted repeatedly along the rift floor north and south of the dome. Fissure and multi-centre eruptions occurred in only two areas beyond the meridional volcanic belt which marked the newly formed rift. Extrusion of the ignimbrites, rhyolites and mugearites of north-western Kenya took place in an area of broad downwarping. The sheets of Pliocene basalt in central Kenya were merely the forerunners of much more extensive Quaternary basaltic activity east of the rift valley.

The Pliocene central volcanoes display a great diversity of rock types and stand in strong contrast to the almost exclusively nephelinitic Miocene centres. Nephelinite-phonolite associations are certainly well represented among Pliocene volcanoes but, by the end of the Miocene, the focus of strongly alkaline volcanicity had moved from eastern Uganda to northern Tanzania. Several basalt-trachyte-phonolite volcanoes grew up on the flanks of the Kenya dome and were active before or during extrusion of the intervening plateau trachytes and ignimbrites. Another major centre (Kilimanjaro) which is dominated by lavas of a basalt-trachyte-phonolite association is far removed from those fringing the dome. It is also much younger than the others, activity extending through the Pleistocene up to Recent times. Large areas of the rift floor both north and south of the Kenya dome are marked by numerous basalt-trachyte volcanoes. To the east of the dome is a major phonolite-trachyte volcano (Mt Kenya) at which the main activity had ceased in late Pliocene times, though flank eruptions continued during the Pleistocene.

Quaternary fissure and multi-centre eruptions resulted in a spectacular separation of basaltic and trachytic volcanics. Basic fields are situated well to the east of the rift valley, whereas plateau trachytes occupy the rift floor and shoulders in southern Kenya and represent a continuation of Pliocene trachytic activity. A small area of acid volcanics, belonging to a comendite-pantellerite association, lies within the plateau trachyte field.

The Quaternary central volcanoes are confined mainly to areas in the rift floor. Basalt-trachyte volcanicity occurred predominantly in the central part of the Kenyan rift, but some of the caldera volcanoes situated within the plateau trachyte field to the south are composed of trachytes and ignimbrites with or without basalts. Other centres in the same region are characterized by a phonolite-trachyte association. The most recent strongly alkaline centres are in northern Tanzania. One of them (Ol Doinyo Lengai) is the only currently active volcano in this part of East Africa; it is well known for eruptions of soda-carbonatite lavas and pyroclastics. Some recently active centres in the northern part of the Kenyan rift belong to a basalt-trachyte-phonolite caldera volcano. The main phase of activity at Kilimanjaro, another basalt-trachyte-phonolite volcano, took place during the Pleistocene, but at Mt Kenya (phonolite-trachyte) the only Quaternary activity was in the form of late flank eruptions.

Summary and Conclusions

(a) Alkali olivine basalts were erupted repeatedly from Miocene times onwards. They figure prominently at many central volcanoes and are common products of fissure and multi-centre eruptions. The distribution in space and time of the basaltic volcanics confirms indications from chemical data that parental magmas of alkali olivine basalt composition were constantly available.

(b) Miocene and Pliocene basaltic volcanism was closely controlled by the developing rift valley. Quaternary basalts, however, occur chiefly in areas east of the rift and their distribution is evidently related to a pattern of fracture zones radiating outwards from a region of repeated upwarping now bisected by the rift valley.

(c) At some major central volcanoes a well differentiated series (including olivine basalts, basanites, mugearites, hawaiites, trachybasalts, trachyandesites, trachytes and phonolites) evidently evolved by fractional crystallization of basaltic magma. In contrast, the basaltic volcanics of fissure and multi-centre fields seldom exhibit evidence of widespread differentiation.

(d) Extrusions of nephelinitic volcanics occurred from the Miocene onwards but, unlike the basalts, these rocks are virtually confined to major central volcanoes. The nephelinitic activity was centred on eastern Uganda during the Miocene, but subsequent strongly alkaline volcanism occurred in northern Tanzania and southern Kenya close to major rift faults.

(e) Many of the nephelinite volcanoes are characterized by a strongly alkaline series embracing melanephelinites, nephelinites and phonolites. Olivine and/or melilite are often prominent constituents of the feldspar-free rocks. Some tephrites and basanites probably belong to the same series. A parental magma of olivine nephelinite composition is envisaged, subtle differences in the course of crystallization evidently leading to the production of melilite-, plagioclase-, or alkali feldspar-bearing rocks.

(f) The trachytes and some of the phonolites found at nephelinite centres may have arisen by metasomatic processes or by feldspathization of feldspathoidal rocks.

(g) Some central volcanoes are characterized by representatives of both basalt-trachyte-phonolite and nephelinite-phonolite associations, suggesting that parental magmas for the two series were derived from a common source in the mantle.

(h) The sudden appearance in late Miocene times of enormous volumes of plateau phonolites, and the general lack of volcanics intermediate in character between these lavas and the preceding flood basalts, are not consistent with an origin by differentiation of basaltic magma. The location of the phonolites in an area of crustal upwarping might be taken to support the concept that magmas of phonolitic composition have been generated by processes of partial melting in response to relief of pressure. If, however, the upwarping is itself an expression of subcrustal processes, it is likely that the production of phonolitic magma depends more on the geothermal state of the upper mantle than on local reduction of pressure.

(i) An abrupt change in the composition of plateau volcanics from phonolites to trachytes (including ignimbritic types) coincided with a major phase of faulting in Pliocene times. It is considered unlikely that these trachytes evolved by differentiation of pre-existing basaltic or phonolitic magma, and an origin by selective fusion in the mantle or lower part of the crust is preferred.

(j) Quaternary peralkaline rhyolites occur in a relatively small area within the plateau trachyte field and are possibly the ultimate derivatives of trachytic magma. In contrast, Pliocene rhyolites and ignimbrites in north-western Kenya are associated with mugearitic lavas in an area characterized by broad downwarping; trachytes are rare and a magma of this composition is unlikely to represent a source for the rhyolitic volcanics.

(k) At some centres (including caldera volcanoes) north and south of the Kenya dome alkali trachytes and/or phonolitic trachytes seem to be the only volcanics that accompany olivine basalts, the apparent lack or scarcity of rocks of intermediate composition suggesting that the salic products are not merely fractional derivatives of basaltic magma. Trachytes and phonolites become increasingly abundant at centres situated along the downfaulted and downwarped axial region of the dome, and at some caldera volcanoes they are the only exposed volcanics. It is difficult to escape from the notion that the basalt-trachyte central volcanoes occur in areas marked by the intermingling of two contrasting magma types: a salic magma derived by partial melting processes beneath the Kenya dome, and a primary basalt magma which is well represented by the plateau volcanics erupted some distance away from the dome.

(l) Whereas both basaltic and trachytic volcanics are present at many of the centres north and south of the Kenya dome, a more marked segregation of basic and salic types tends to occur at volcanoes on the eastern and western flanks. The Aberdare volcano is predominantly basaltic, Mt Kenya is a phonolite-trachyte centre, and Londiani is mainly trachytic.

References

BAKER, B. H., 1965, *An outline of the geology of the Kenya Rift Valley.* In *East African Rift System: UMC/UNESCO Seminar Nairobi, April 1965. II. Report on the geology and geophysics of the East African Rift System.* pp. 1-19. University College, Nairobi.

GREGORY, J. W., 1896, *The Great Rift Valley.* John Murray, London, 422 pp.

————, 1921, *Rift valleys and geology of East Africa.* Seeley Service, London, 479 pp.

KENT, P. E., 1944, *The age and tectonic relationships of East African volcanic rocks.* Geol. Mag. Vol. 81, pp. 15-27.

KING, B. C., 1965, *Petrogenesis of the alkaline igneous rock suites of the volcanic and intrusive centres of eastern Uganda.* J. Petrology, Vol. 6, pp. 67-100.

————, and SUTHERLAND, D. S., 1960, *Alkaline rocks of eastern and southern Africa.* Sci. Progress. Vol. 48, pp. 298-321, 504-524, 709-720.

McCALL, G. J. H., 1964, *Froth flows in Kenya.* Geol. Rdsch. Vol. 54, pp. 1148- 1195.

NIXON, P. H., and CLARK, L., 1967, *The alkaline centre of Yelele and its bearing on the petrogenesis of other eastern Uganda volcanoes.* Geol. Mag. Vol. 104, pp. 455-472.

SAGGERSON, E. P., 1968, *Eclogite nodules associated with alkaline olivine basalts, Kenya.* Geol. Rdsch. Vol. 57, pp. 890-903.

————, and WILLIAMS, L. A. J., 1964, *Ngurumanite from southern Kenya and its bearing on the origin of rocks in the northern Tanganyika alkaline district.* J. Petrology Vol. 5, pp. 40-81.

SHACKLETON, R. M., 1951, *A contribution to the geology of the Kavirondo rift valley.* Quart. J. geol. Soc. Lond. Vol. 106, pp. 345-392.

SMITH, W. C., 1931, *A classification of some rhyolites, trachytes and phonolites from part of Kenya Colony, with a note on some associated basaltic rocks.* Quart. J. geol. Soc. Lond. Vol. 87, pp. 212-258.

————, 1938, *Petrographic description of volcanic rocks from Turkana, Kenya Colony, with notes on their field occurrence from the manuscript of Mr. A. M. Champion.* Quart. J. geol. Soc. Lond. Vol. 94, pp. 507-553.

VARNE, R., 1968, *The petrology of Moroto Mountain, eastern Uganda, and the origin of nephelinites.* J. Petrology Vol. 9, pp. 169-190.

WILCOCKSON, W. H., 1964, *Some aspects of East African vulcanology.* Advmt. Sci. Lond. Vol. 21, pp. 400-412.

WILKINSON, P., 1966, *The Kilimanjaro-Meru region.* Proc. geol. Soc. Lond. no. 1929, pp. 28-30.

WILLIAMS, L. A. J., 1964, *The geology of the Narok District.* Proc. E. Afr. Acad. Vol. 1, pp. 37-49.

————, 1965, *Petrology of volcanic rocks associated with the rift system in Kenya.* In *East African Rift System: UMC/UNESCO Seminar Nairobi, April 1965. II. Report on the geology and geophysics of the East African Rift System.* pp. 33-39. University College, Nairobi.

————, 1967, *Geology* (of the Nairobi region). In *Nairobi: City and Region* (ed. W. T. W. Morgan), pp. 1-13. Oxford Univ. Press, 154 pp.

————, 1969a, *Volcanic associations in the Gregory Rift Valley, East Africa.* Nature Lond. Vol. 224, pp. 61-64.

————, 1969b, *Geochemistry and petrogenesis of the Kilimanjaro volcanic rocks of the Amboseli area, Kenya.* Bull. Volcan. Vol. 33-3, pp. 862-888.

WRIGHT, J. B., 1963, *A note on possible differentiation trends in Tertiary to Recent lavas of Kenya.* Geol. Mag. Vol. 100, pp. 164-180.

————, 1965, *Petrographic sub-provinces in the Tertiary to Recent volcanics of Kenya.* Geol. Mag. Vol. 102, pp. 541-557.

241

18

Reprinted from pp. 1, 9–17, 25–36, 40–41, 43–45, and 53–54 of *Geol. Soc. America Spec. Paper 136,* 1972, 67p.

GEOLOGY OF THE EASTERN RIFT SYSTEM OF AFRICA

B. H. Baker, P. A. Mohr, and L. A. J. Williams

Abstract

The eastern rift of Africa is a zone of normal faults separating the Horn of Africa from the remainder of the continent. The zone is typically troughlike, 40 to 65 km wide, and traverses two broad, elongated domal uplifts in Ethiopia and Kenya.

The foundation rocks of the region are metasediments and intrusives of the late Precambrian orogenic belt, which has a meridional trend. The Paleozoic was dominantly an era of denudation in eastern Africa, but late Paleozoic continental sediments (Karroo System) are locally preserved. Mesozoic marine sediments represent an epicontinental marine transgression and regression. Severe coastal warping occurred along the Indian Ocean margin, and in the early Tertiary such warping initiated the Red Sea, Gulf of Aden, and Afar depressions.

Uplift of the Ethiopian and Kenyan domes has been synchronous in three major pulses of late Eocene, mid-Miocene, and Plio-Pleistocene age. Volcanism of intermediate and silicic type shows some relation to uplift in time and space and to the onset of graben faulting, but major flood basalt extrusions in the early Tertiary in Ethiopia were related to massive crustal warping along the future rift margins. The volcanism associated with the eastern rift is overwhelmingly alkaline, and at some volcanoes a strongly alkaline fractionation series is distinguished from a more mildly alkaline series. The flood phonolites, trachytes, rhyolites, and ignimbrites of Kenya, and the pantelleritic ignimbrites of Ethiopia, could have resulted from anatexis of a mantle-derived accreted layer at the base of the crust.

The eastern rift began as a chain of marginally warped depressions which were accentuated as domal uplift proceeded, until, in mid-Miocene to early Pliocene times, faulting produced asymmetrical grabens. The final uplift phase in the early Pleistocene was accompanied by major graben faulting, and subsequent faulting has intensely fractured the floor of the rift along an axial zone marked by caldera volcanoes. The evolution and nature of the faulting, the evidence from the distribution and ages of volcanoes, and seismic and gravity data all indicate that the eastern rift lies along a zone of progressive crustal thinning with local crustal disruption.

The eastern rift can be considered as a plate boundary which meets the Red Sea and Gulf of Aden spreading axes at the Afar triple junction. Plate analysis suggests that the eastern rift marks a line of very slow crustal spreading, which helps account for many of the peculiar or unique features of this continental rift.

[*Editor's Note:* Material has been omitted at this point.]

Domal Uplift and Erosion Surfaces

The evolution of the Kenyan and Ethiopian domes is evidenced in preserved planar erosion surfaces (or bevels). In eastern Kenya, Saggerson
and Baker (1965) mapped three major erosion surfaces whose ages are ascribed,
on the basis of overlying strata and ensuing deformation, to the end-
Cretaceous, sub-Miocene (mid-Tertiary), and end-Tertiary. The same authors
suggest that uplift movements were sporadic and separated by long periods
of crustal stability and erosion.

In central Kenya an end-Cretaceous uplift of more than 400 m can be
demonstrated, decreasing in magnitude eastward until beyond long. 39½° E.
there was subsidence. In the Miocene a further uplift of 300 m occurred
in central Kenya, decreasing to 150 to 100 m in the east and south, while
the coastal region was again a zone of flexuring and subsidence (Fig. 4).
At about this time, the sub-Miocene surface on the northwest flank of the
Kenya dome began to be warped down, initiating a broad, partly faulted
depression. More extensive rift faulting followed at the end of the Miocene.
The partly developed Kenyan and Ethiopian domes were now separated by a
broad depression, extending from northeastern Kenya via Lake Rudolf to
southern Sudan (Saggerson and Baker, 1965; Berry and Whiteman, 1968).

Major uplift of the Kenyan dome occurred near the end of the Tertiary
and was of the order of 1,500 m in central Kenya (Fig. 4), again decreasing
in magnitude eastward and involving downflexing along the coast. The
margins of the rift trough were warped down and strongly faulted, to produce
a true graben for the first time. Arching and flexing within the Kenyan
dome have continued on a smaller scale at intervals throughout the Quaternary, as indicated by the 300-m uplift of the Plio-Pleistocene sediments of
the upper Tana River basin (Saggerson and Baker, 1965) and by large displacements of lower Pleistocene sediments and volcanics at the rift margins
(Isaac, 1967; McCall and others, 1967).

No quantitative survey of individual erosion surfaces in Ethiopia has
yet been made. A regional unconformity separates the end Jurassic-Early
Cretaceous Upper Sandstone from the Eocene Trap Series flood basalts and
has been interpreted as a late Mesozoic erosion surface (Merla and Minucci,
1938). This surface cuts progressively across older Mesozoic marine formations from east to west, suggesting a greater end-Cretaceous uplift in
western than in central Ethiopia, contrary to the pattern of the later
movements.

Major uplifts of the Ethiopian dome occurred in the upper Eocene,
lower-middle Miocene, and Pleistocene. Dainelli (1943) attributed the
formation of the whole of the Ethiopian dome to a single upper Eocene
uplift, but the three major episodes listed above are now clearly recognized (Desio, 1940; Azzaroli, 1958; Merla, 1963). Merla (1963) has estimated an Oligocene-early Miocene uplift of 500 m in central Ethiopia and
attributes the termination of planation on the Ethiopian plateau to this
episode. Other workers (for example, Dainelli, 1943) consider that the
planar surface of the Ethiopian plateau merely represents the early
Tertiary flood-basalt surface, although where these basalts are tilted,
as in the middle Omo Valley and Afar margins, they tend to be roughly
beveled.

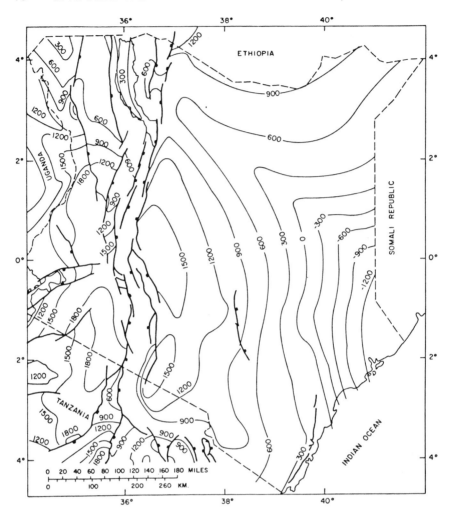

Figure 4. Isobases of the sub-Miocene erosion surface in Kenya. Present-day elevation of the surface given in meters.

    The mid-Tertiary rise of the Ethiopian dome (Azzaroli, 1958; Mohr, 1962b, 1967a) was accompanied by further downwarping along the margins of the rift trough, and in Afar there was massive boundary faulting of at least 1,000 m magnitude. The largest uplift of the Ethiopian dome has been, as in Kenya, the latest one. This post-Pliocene volcanics uplift, intimately associated with major graben faulting in the main Ethiopian rift, took place in the lower and middle Pleistocene, and notably affected drainage patterns on the rift shoulders (Mohr, 1962a).
    The uplift of the Kenyan and Ethiopian domes has therefore occurred in broadly synchronous pulses, a fact which emphasizes the general unity of the African rift system and its genesis. Figure 5 shows the total effect of

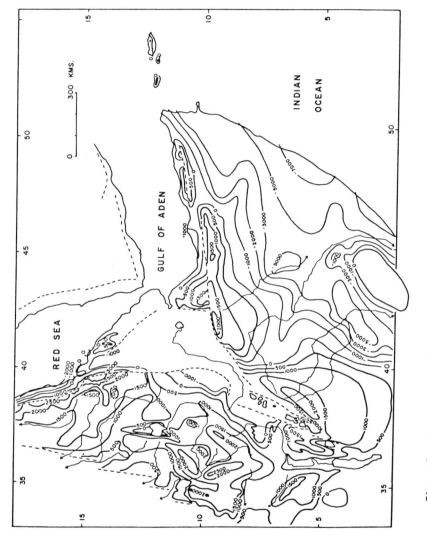

Figure 5. Isohypsals of the Precambrian basement in Ethiopia. Elevation in meters.

Ethiopian domal uplift in terms of contours of the present Precambrian basement surface. For the Kenyan dome the small extent of Mesozoic and Paleo-
gene sedimentary cover necessitates use of data expressing the uplift of the
mid-Tertiary surface (Fig. 4). Approximate volumetric data for uplift and
subsidence in Kenya and Ethiopia are listed in Table 1. The implications of
these figures in terms of the cause of domal uplift are discussed later.

TABLE 1.   APPROXIMATE VOLUMETRIC DATA FOR
UPLIFT AND SUBSIDENCE IN KENYA AND ETHIOPIA

		Ethiopia-Somalia	Kenya
Neogene:	domal uplift	0.9	0.3
	coastal subsidence	-0.2	-0.15
Paleogene:	domal uplift	0.3	0.5
	coastal subsidence	-0.3	-0.7
Late Mesozoic:	domal uplift	?	0.15
	coastal subsidence	-2.1	-0.6

Figures are $km^3 \times 10^6$.

Volcanism

The extensive occurrence of alkaline volcanics along parts of the African rift system became well known from the pioneer field studies of Blanford (1870) and Gregory (1896, 1921) and from early petrographic work of Prior (1900, 1903), Weber (1906), and Manasse (1909). Smith (1931, 1938) and Bowen (1937, 1938) stressed the importance of phonolites, trachytes, and alkaline rhyolites in East Africa. Comucci (numerous publications—*see especially* 1950) and Hieke-Merlin (1950, 1953) emphasized the abundance in Ethiopia of alkali basalts and representatives of a comendite-pantellerite suite.

VOLCANIC HISTORY

The stratigraphy of the eastern rift volcanics is very complex, with extensive formations derived from fissure and multicenter eruptions inter-calated with localized but often very thick accumulations of lavas and pyroclastics of central volcanoes. The volcanic history of eastern Africa is based mainly on correlation of the better known and less differentiated flood and multicenter formations, the ages of which are fairly well estab-lished in Kenya (Bishop and others, 1969; Williams, 1970; Baker and others, 1971), but are less well known in Ethiopia (Mohr, 1968). A tentative corre-lation between the two regions is given in Table 2.

Small, isolated alkali basalt and trachybasalt flows are interbedded in late Mesozoic sediments of eastern Ethiopia (Ogaden), North Somalia, and the Danakil horst, but they are probably not related to the evolution of the rift.

The largest major volcanic episode in the history of the eastern rift produced the Trap Series fissure basalts, which cover much of the Ethiopian and Somalian plateaus, and also the Yemen highlands of southwestern Arabia (Fig. 6). They are Eocene-Oligocene in the north and Oligocene in the south (Grasty and others, 1963; Merla, 1963). Flows thicken and increase in num-ber from within successions a few hundred meters thick on the plateaus to over 2,000 m at the downwarped rift margins. Dike-swarm feeders are now exposed along these warped margins, but some also are exposed in the plateau interiors. There was no equivalent early Tertiary volcanism in Kenya.

The Shield Group basalts pass up from the Trap Series with only local unconformity (Mohr and Rogers, 1966; Abbate and others, 1968) and are large low-angled shields cut by basaltic and silicic dikes and stocks. The group is up to 4,000 m thick in northern Ethiopia where the age has been established as lower-middle Miocene (Mohr, 1967b; Brown, 1970).

The southern and central part of Afar is floored by a series of flood basalts—the Afar Series—which was formerly considered to be a downfaulted extension of the Trap Series of the plateaus and rift margins (Gortani and Bianchi, 1941; Mohr, 1962a). This correlation is now discounted (Mohr, 1967a, 1968), and while the lower age limit of the series remains unknown (but must be post Lower Cretaceous), radiometric ages for the upper part of the series are 24 m.y. and 8 to 3.7 m.y. (Bannert and others, 1970). The total thickness of the Afar Series is suspected to be much greater than the 400 m exposed in eastern Afar. No dike feeders have yet been discovered.

TABLE 2. VOLCANIC SUCCESSIONS OF THE EASTERN RIFT

	Kenya		Ethiopia	
	Fissure/multicenter eruptions	Central volcanoes	Fissure/multicenter eruptions	Central volcanoes
QUATERNARY	Basalts (east of rift)	Trachytes, ignimbrites, phonolites, basalts (caldera volcanoes, rift floor) Nephelinites, phonolites, carbonatites (North Tanzania) Basalts, trachytes, phonolites (North Tanzania)	Basalts, trachytes, rare comendites (North Afar) Basalts Aden Series (rift floor)	Trachytes, pantelleritic ignimbrites and obsidian lavas, basalts Aden Series (caldera volcanoes on rift floor) Basanites, phonolites, trachytes Aden Series (plateaus and Tana rift)
PLIOCENE	Trachytes, rhyolites, ignimbrites, rare phonolites (rift floor) Basalts (rift floor) Phonolites, trachytes (rift floor) Rhyolites, ignimbrites, mugearites (northwestern Kenya)	Basalts, trachytes (rift floor) Phonolites, trachytes (east of rift) Basalts, trachytes, phonolites (rift margins) Nephelinites, phonolites (rift floor and margins)	Trachyte-pantellerite ignimbrites and lavas (rift margins, extending across rift and plateaus in southern and central Ethiopia) Basalts (central Afar) Basalts, mugearites Afar Series (Afar; lower age limit at least pre-middle Miocene)	Basalts, basanites, melanephelinites (rift margins and plateaus)
MIOCENE	Phonolites (central Kenya plateaus) Basalts (northwestern Kenya)	Nephelinites, phonolites, carbonatites (Kenya-Uganda border)		Basalts, trachybasalts (Red Sea and Afar margins) Rare nephelinites, phonolites (plateaus) Basalts, minor comendites Shield Group (plateaus)
EOCENE-OLIGOCENE			Basalts Trap Series (plateaus and rift margins)	

248

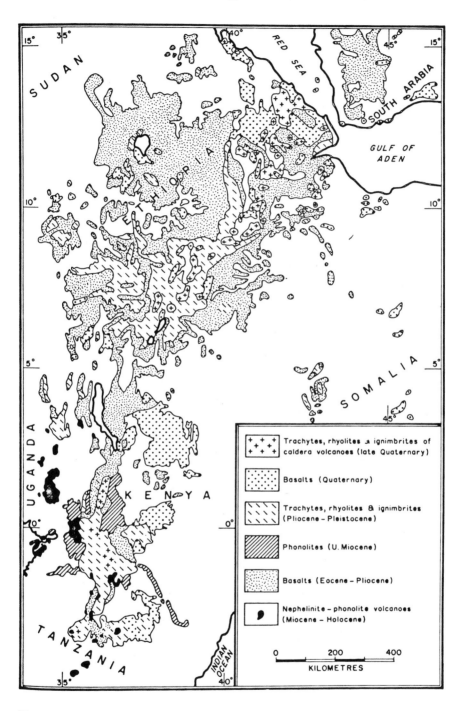

Figure 6. Distribution of the main volcanic groups of the eastern rift zone.

Volcanism in East Africa commenced with explosive activity at early Miocene nephelinite volcanoes, mainly in the Kenya-Uganda border region (King, 1965, 1970; McCall, 1958). Contemporaneous widespread basaltic eruptions took place in the pre-rift Turkana depression of northwestern Kenya, where the lavas are at least 1,000 m thick (Walsh and Dodson, 1969; Rhemtulla, 1970). The Turkana basalts are broadly equivalent to the later part of the Snield Group of northern Ethiopia, although the Trap Series—Shield Group basalts are considered to become progressively younger southward across the Ethiopian plateau. The great Miocene nephelinite eruptions in East Africa have only minor counterparts farther north in localized nephelinitic and phonolitic volcanics of the Ethiopian and Somalian plateaus (LeBas and Mohr, 1968).

The Miocene basalts were followed by extensive and reliably dated flood phonolites erupted from the crest of the Kenya dome in the late Miocene and early Pliocene (Baker and others, 1971). Thicknesses of up to 900 m are revealed by faulting at the rift margins, but the phonolites are probably thicker in the rift floor where they are mostly concealed by younger formations. No phonolitic volcanism of comparable scale took place in Ethiopia.

An important phase of silicic volcanism in northwestern Kenya is provisionally assigned to the lower Pliocene. Rhyolites, ignimbrites, and mugearites derived from fissures are about 900 m thick and rest unconformably on Miocene basalts (Walsh and Dodson, 1969).

After the first phase of rift faulting, which was mid-Miocene in Afar but early Pliocene in Kenya and south-central Ethiopia, volcanic activity was largely confined to the rift floor and margins. Pliocene basalts were erupted from fissures or scattered cones along the Kenyan rift and are frequently banked against pre-existing fault scarps, reaching a thickness of at least 450 m. Approximately contemporaneous phonolites and trachytes in central Kenya locally overflowed onto the eastern plateau (Baker and others, 1971).

Numerous central volcanoes were active along the Gregory rift during the Pliocene (Williams, 1970). Nephelinite centers were confined to the floor and shoulders of the southern section of this rift and to the eastern end of the Kavirondo trough. Basalt-trachyte volcanoes erupted on the rift floor in central Kenya and North Tanzania, and basalt-trachyte-phonolite centers were active along its eastern margin (Shackleton, 1945). At about this time a broad phonolitic shield volcano was built up at the Mt. Kenya center about 75 km east of the rift (Baker, 1967).

In Ethiopia, the earliest basalts of the Pliocene-Quaternary Aden Series are best developed in central Afar, where a distinction from the Afar Series is somewhat arbitrary. Widely scattered Pliocene basalts on the rift margins and plateau interiors were erupted from small vents which occasionally produced basanites and melanephelinites (LeBas and Mohr, 1968).

In the upper Pliocene, very voluminous fissure and central eruptions of trachyte-pantellerite ignimbrites covered most of the southern Ethiopian plateau and filled the main Ethiopian rift (Mohr, 1968). The sources were situated chiefly at the rift margins where thicknesses of 300 to 500 m have been measured, but the thickness in the rift floor is probably much greater. Comparable extensive eruptions of late Pliocene—early Pleistocene trachytic lavas and ignimbrites, probably derived mainly from fissures, filled the central part of the Kenyan rift and locally flowed onto the plateaus. These volcanics are more than 900 m thick in the rift floor but are much thinner on the adjacent plateaus (Williams, 1970).

Middle Pleistocene and later volcanism formed a chain of trachyte caldera volcanoes along the eastern rift from North Tanzania to Afar. Basalts are sometimes associated with these centers, but more typical, particularly in the Aden Series of Ethiopia, are silicic volcanics including ignimbrites/froth flows; phonolites occur locally in Kenya. The caldera volcanoes tend to be situated on the very young axial fracturing of the rift floor, and in the main Ethiopian rift and southern Afar are related to en echelon offsets of the Wonji fault belt (Mohr, 1967a, 1967c). Concurrent basaltic activity on a very large scale took place in Kenya in multicenter fields 150 to 250 km east of the rift (Williams, 1970), but in Ethiopia the younger basalts of the Aden Series were erupted in relatively minor volumes both on the rift floor and on the plateaus, including the Tana rift.

Quaternary nephelinite-carbonatite volcanism was largely restricted to North Tanzania, where the currently active volcano Ol Doinyo Lengai periodically erupts soda carbonatite lava and ash (Dawson and others, 1968). Nephelinites were erupted from Meru during the Holocene, but late Pleistocene—Holocene flows from Kilimanjaro are all phonolitic and followed a predominantly basaltic phase in which nephelinites are sparsely represented (Williams, 1969a).

In the Salt Plain of northern Afar the multicenter Erta-ali basalt range has shown a variable but ever-present activity since its discovery (Richard and Neumann van Padang, 1957; Tazieff and Varet, 1969), and basalt centers in both the Kenyan and the main Ethiopian rifts have erupted within the last few hundred years, a period which probably also includes the youngest pantellerite obsidian lavas of some Ethiopian volcanoes (Mohr, 1962c, 1966a; Gibson, 1970).

[*Editor's Note:* Material has been omitted at this point.]

Tectonics

The eastern rift traverses Ethiopia and Kenya and extends nearly to the North Tanzania coast (Fig. 9). In East Tanzania, a belt of widely spaced faults of Neogene age extends southwest to link the south section of the eastern rift with the western rift, but is not included in this account.

The eastern rift consists of normal grabens, asymmetric grabens, and monoclinally flexed depressions which separate the East Kenya–Somalia crustal block from the remainder of Africa. Well-defined grabens are found only on the central parts and south slopes of the Kenyan and Ethiopian uplifts, namely the Gregory and main Ethiopian rifts. In North Tanzania and North Kenya, the less uplifted areas are characterized by splay faults of antithetic type traversing broad depressions. At the north end of the eastern rift the main faults diverge at the junction of the Ethiopian, Red Sea, and Gulf of Aden rifts, and frame the Afar depression.

The eastern rift is divisible into five main sections, each with a distinctive tectonic style: the North Tanzania divergence, the Gregory and Kavirondo rifts, the Turkana depression, the main Ethiopian rift, and the Afar depression.

NORTH TANZANIA DIVERGENCE

The transition from the south flank of the Kenya domal uplift to the less high interior plateau of North Tanzania takes place between lat. 2° and 3° S. and is marked by the disappearance of the main eastern fault of the Gregory rift and splitting of its western faults (Fig. 10). Major faults separate the Mbulu, Kondoa, and Lelatema structural blocks; each of these is upfaulted at its eastern margin and slopes generally northwest (Fig. 11). The deepest part of the rift zone extends southeast from Lake Natron through the Engaruka basin to the Arusha area, and probably continues under the south flank of the Meru and Kilimanjaro volcanoes to connect with the Pangani graben. East of the Pangani graben the Pare and Usambara horsts extend southeast nearly to the Indian Ocean (Dundas, 1965).

The oldest faults in this area are those on the west side of the Pare and Usambara horsts and may be mid-Tertiary (E. E. Milanovsky, 1970, oral commun.). The Sonjo–Lake Eyasi fault is early Pliocene, and the faults in the central part of the divergence zone are late Pliocene and Pleistocene.

GREGORY AND KAVIRONDO RIFTS

The Gregory rift is the type continental rift (Gregory, 1921); it extends from the Magadi-Natron basin in the south to the Baringo and Suguta grabens in the north and is a complex graben bisecting the Kenya domal uplift (Figs. 4 and 10). Neglecting high volcanoes, the adjacent plateaus stand at heights of 1,600 to 3,200 m, while the rift floor descends from 2,000 m in the central area to 650 m in the south.

The main fault scarps range from 300 to 1,600 m in height, are en echelon in plan, and form a complex graben 60 to 70 km wide. The east side of the graben is stepped, producing the Kinangop and Bahati platforms, and on both sides there are sloping "ramp" structures at major fault offsets (Fig. 10). These marginal structures reduce the width of the inner

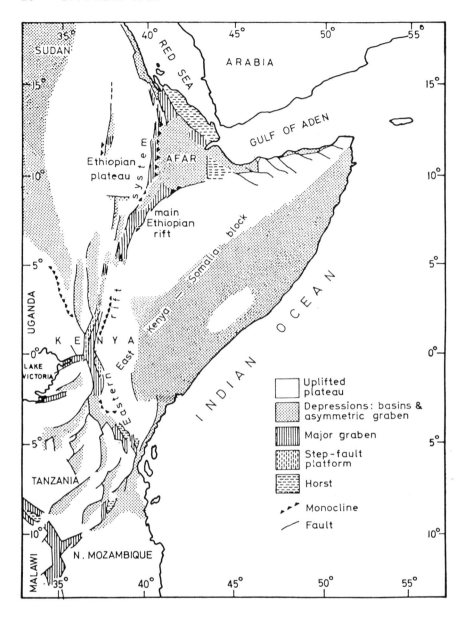

Figure 9. Structural elements of the eastern rift system.

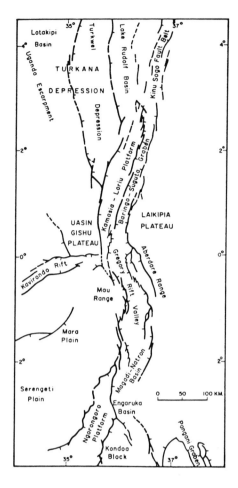

Figure 10. Major faults of the Gregory rift valley.

graben floor to 17 to 35 km. The
fault fractures are rarely exposed,
but the available evidence indicates normal faulting (Baker,
1958; Shackleton, 1955). Escarpment heights and measurable displacements of stratigraphic units
indicate minimum fault throws of
1,600 to 2,200 m on the west side
of the graben and up to 1,000 m
on the east side. Since it can be
proved that some of the faults
moved more than once, with intervening volcanism in the rift floor
(Baker, 1963a; McCall, 1967), the
total downthrow of the older marginal faults is considerably
greater than the height of the
escarpments. The volcanic succession in the flanks of the rift
valley is over 2,000 m thick in
places, and the succession in the
rift floor is certainly thicker
than this, suggesting that the sub-
volcanic surface within the rift is
below sea level and that the total
throw of the marginal faults may
reach as much as 3,000 to 4,000 m.
   The graben floor and step-
fault platforms are composed of
Plio-Pleistocene volcanics cut by
swarms of closely spaced young
faults which give rise to nearly
perfectly preserved small scarps
(Figs. 11 and 12). These faults
are subparallel, reach densities
of two or three faults per kilometer, and trend along the rift
parallel to the main faults. The
fault throws are rarely more than
150 m, and they give rise to horst
and graben structure and to step
and antithetic fault patterns
(Baker, 1958, 1963a; Randel, 1971;
McCall, 1967; Walsh, 1969). The
faults of the rift floor are locally obscured by late Quaternary volcanic
piles (McCall, 1968), and the depressions are partly filled by lower and
middle Pleistocene sediments (McCall and others, 1967).
   The Kavirondo rift branches from the Gregory rift at the center of the
Kenya uplift and trends west and southwest, bisecting the highest part of
the western plateau and descending westward into Lake Victoria (Fig. 11).
It is 15 to 25 km wide and consists of two graben sectors separated by a
section along which gentle monoclinal downflexing is more important than
faulting (Shackleton, 1951). The fault throws range up to 700 m in the
central graben sector, but in the east, near the junction with the Gregory
rift, the structure is obscured by long-active central volcanoes which
partly filled the graben and locally overflowed its flanks. Unlike the
Gregory rift there was no Quaternary faulting nor volcanism in the floor of
the Kavirondo rift; it seems to have formed in the early Pliocene.

TURKANA DEPRESSION

North of lat. 1° N. the Kenya rift gradually widens and its structural
style is nearly a mirror image of that in North Tanzania (Figs. 10 and 12).
The axial graben of the Gregory rift extends north through the Baringo
trough and continues to the south end of Lake Rudolf as the Suguta graben,
which is 20 km wide, with flanking escarpments rarely more than 400 m high
(Dodson, 1963). The northward continuation of the Suguta graben is the
Kinu Sogo fault belt east of Lake Rudolf and is a zone of Pleistocene faults

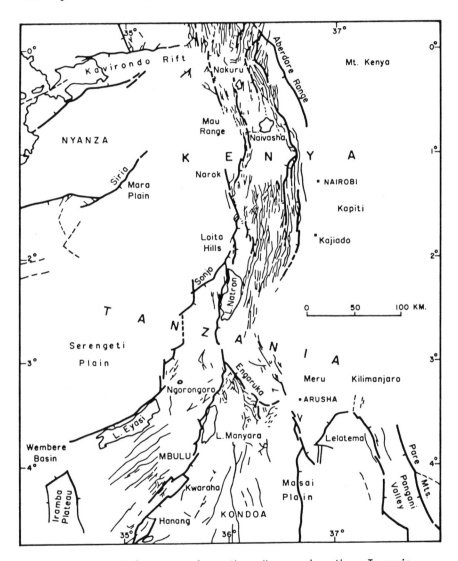

Figure 11.   Fault pattern in southern Kenya and northern Tanzania.

255

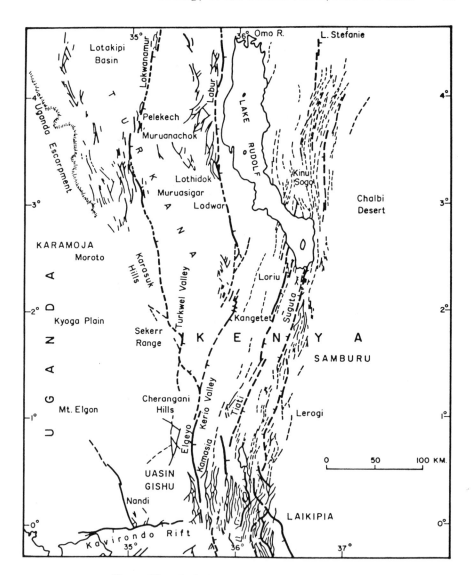

Figure 12.   Fault pattern in northern Kenya.

of small throw crossing the crest of a very shallow rise (Fig. 10). At the north end of this belt a distinct graben is again developed on the Ethiopian border and is occupied by Lake Stefanie (Fig. 13a).

Along part of the eastern side of the Baringo-Suguta graben, the lower part of the volcanic succession of the plateau dips west toward the graben at angles of 20° to 30°, showing that monoclinal flexuring took place in late Miocene and Pliocene times, accompanied by volcanism and succeeded by graben faulting (Fig. 13b; Shackleton, 1946; Baker, 1963b).

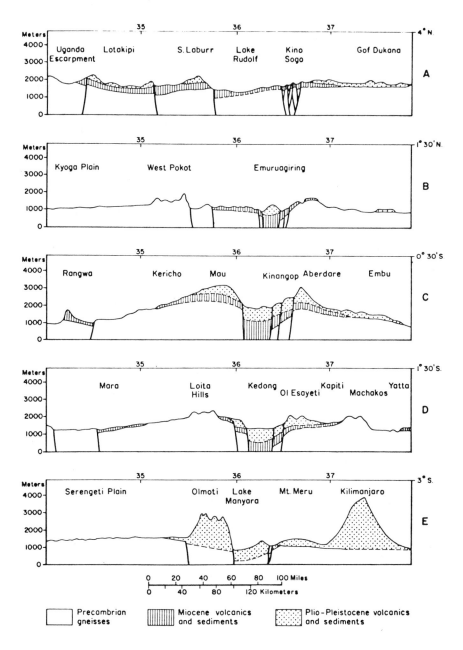

Figure 13. Sections across the Kenya rift valley.

The Baringo-Suguta graben is bounded on the west by the Tiati-Loriu platform (Figs. 10 and 12), which is 2,200 m high at its south end and slopes down to 900 m in the north. At its elevated southern end there is a dominant westerly dip toward the foot of the Elgeyo escarpment, but farther north the formations of the platform are nearly flat-lying. The Kamasia-Loriu platform and the Baringo-Suguta graben together constitute the direct northward extension of the Gregory rift.

The remainder of the Turkana depression is a triangular lowland 300 to 1,000 m in elevation, bordered on the west by the Turkwel and Uganda escarpments (Figs. 10 and 12). The Turkwel escarpment is the highest, oldest, and most dissected of the rift fault escarpments in Kenya. Its mountainous crest reaches heights of 2,500 to 3,500 m, and it descends abruptly to its foot at 1,000 m. The Uganda escarpment farther north is much less steep and is sinuous in plan, and has been interpreted as an eroded monoclinal flexure (Walsh and Dodson, 1969). Diverse views have been ex-pressed on the structural and geomorphological history of these escarp-ments (Pulfrey, 1960; McCall, 1964; Bishop and Trendall, 1967). Re-examina-tion of this area suggests that the Turkwel fault and the Uganda monocline began to develop during the lower Miocene, forming a depression over much of Turkana within which the Miocene basalts were erupted. In early Pliocene times major submeridional antithetic faults cut the floor of the Turkana depression, dividing it into west-tilted blocks and extending the Turkwel fault southward as the Elgeyo fault.

The Turkana depression is thus a triangular lowland between the Kenyan and Ethiopian domal uplifts. It is 200 km wide in the north and narrows southward, its margins being increasingly strongly faulted until it merges into the Gregory graben.

MAIN ETHIOPIAN RIFT

In the transitional zone between the Kenyan and Ethiopian domes, the deepest rift depressions are disconnected structures offset en echelon to the northeast. The Kenyan and Ethiopian rifts are connected through the Kinu Sogo fault belt, the Stefanie graben, and the fault zone of the Galana Dulay Valley. The last dies out, however, north of lat. 5½° N. and the main rift valley resumes 50 km farther east. Possibly there is a transverse fault along the east-west, linear Sagan valley related to this offset.

West of the Kinu Sogo—Stefanie fault zone, the Lake Rudolf basin extends north up the lower Omo Valley (Howell, 1968), which is formed by north—north-northeast faults with westerly upthrows and some east-west faults (Butzer and Thurber, 1969). The valley is a tectonic depression extending into the southwestern flank of the Ethiopian plateau, and Precam-brian rocks are locally exposed in its floor.

The main Ethiopian rift begins north of the Sagan offset in a zone in which the Amaro horst separates the Ruspoli (Lake Chamo) graben on the west from the Galana graben to the east (Fig. 14). The Amaro horst, formed of Precambrian gneisses and Tertiary volcanics rising 2,000 m above the neigh-boring grabens, shows evidence of repeated uplifts, mainly on its eastern step-faulted side, giving it a westward dip slope. The 35-km-wide Ruspoli graben to the west of the horst contains an axial swarm of young faults which continues with varying intensity along the entire length of the main Ethiopian rift floor and is known as the Wonji fault belt (Mohr, 1960).

The Amaro horst declines abruptly at its northern end, the Galana and Ruspoli grabens merging into the Lake Margherita basin to form a rift 60 km wide. The rift is bounded to the east by a broad zone of locally rejuven-ated faults that narrows northward to a double or single escarpment formed by late Pleistocene faulting in the Lake Awasa basin, the form of which is partly determined by east-west faults associated with Quaternary caldera volcanism. East of Lake Awasa and Shashamanne, a strongly denuded older fault scarp evidences a late Miocene—Pliocene downwarping of the Somalian

plateau into the rift (Merla, 1963).   Farther north, in the Galla Lakes
region (Lakes Shala, Abyata, Langana, and Zway), the eastern escarpment is
strongly upwarped riftward from the Katar basin of the Somalian plateau
(Mohr, 1966b).   This strongly stepped escarpment is formed by a complex of
north-northwest- and north-northeast-trending late Pleistocene faults (Fig.
14), coinciding with part of a huge craterlike feature centered on the rift
at lat. 7½° N., diameter 150 km (NASA Gemini astronaut photograph S-65-63162).

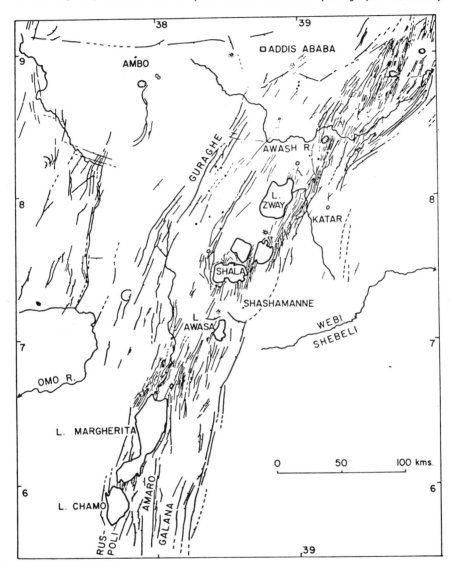

Figure 14. Major faults of the southern Ethiopian rift zone (boundary
faults indicated by thicker line trace).

Along the western margin of the Ethiopian rift there is often a close association between silicic volcanism and faulting. West of Lake Margherita the flank of the massive Gughe volcano is upwarped at the rift margin and cut by early Pleistocene north-northeast-trending step faults. In the Soddu region farther north the structure of the rift margin is obscured by a line of small trachyte and rhyolite centers which separate the main rift floor from the lower lying Omo Valley to the west (Figs. 14 and 17b). The north-ward trend of the rift margin in this area is parallel to the major faults of the middle Omo Valley and results in widening of the rift to over 70 km at lat. 8° N. West of the Galla Lakes, major faults reappear and curve back to the typical north-northeast—northeast rift trend (Fig. 14).

Between lat. 8° and 8½° N., the Guraghe escarpment forms the west margin of the rift and rises more than 1,000 m above its floor (Fig. 14). This escarpment is faced by a prominent antithetic fault 6 to 10 km within the rift floor, forming a "marginal graben" (Gouin and Mohr, 1964) of a kind typical of the Afar-Ethiopian plateau margin farther north. The Ethiopian rift is notably different from the Gregory rift in lacking platform and related ramp structures (*see* Fig. 9).

The rift floor in the Galla Lakes basin is occupied by Pliocene-Quaternary sediments and volcanics, all but the youngest of which are cut by the Wonji fault belt. The Wonji fault belt is a 2-to-5-km-wide zone of intense faulting (Figs. 14 and 17b) which tends to be axial to the rift floor (Mohr, 1960, 1962b). The faults are normal, short, and sinuous, and in places are associated with open tensional fissures (Gibson, 1967; Gouin and Mohr, 1967). The faults comprising the Wonji belt tend to be disposed en echelon along the rift (Fig. 14), with the zones of offset being charac-terized by young silicic caldera volcanoes (Mohr, 1967a, 1967c; Cole, 1969; Gibson, 1969b). These offsets appear to match those in the adjacent rift margin faulting and may be related to possible transverse gravity anomalies on the rift floor (Mohr and Gouin, 1968). North of the Galla Lakes basin, major offsetting of the Wonji fault belt brings it into juxta-position with the east boundary faults of the rift, so that the greater part of the rift floor is not significantly faulted.

The main Ethiopian rift reaches its highest elevation (1,700 m) at the watershed between Lake Zway and the Awash River, southeast of Addis Ababa. Here the eastern margin of the rift is formed by 800 m total magnitude of well-preserved, strongly stepped fault scarps. The western margin of the rift in the Addis Ababa area is a broad, gentle downwarp cut by minor faults with associated Quaternary cinder cones and maars (Mohr, 1961). The northern side of the resulting rift embayment at Addis Ababa (Fig. 14) is a zone of west—west-northwest Pliocene trachytic volcanism, superimposed by Quaternary faulting which extends westward for at least 300 km across the central Ethio-pian plateau. This faulting divides the Ethiopian plateau into two con-trasting subprovinces on the basis of structure and stratigraphy (Figs. 14 and 17a).

AFAR DEPRESSION

At the latitude of Addis Ababa the main Ethiopian rift begins to widen out into the Afar depression, which occupies the area of junction of the Eastern, Red Sea, and Gulf of Aden rifts (Mohr, 1967a, 1970c; Bannert and others, 1970). Afar is almost completely surrounded by continental crustal blocks: to the west the Ethiopian plateau; to the south the Somalian pla-teau; to the east the Aisha (Ali Sabiet) horst; and to the north the Danakil horst (Figs. 9, 15, and 16). Mesozoic marine sediments and early Tertiary Trap Series basalts of the Ethiopian and Somalian plateaus dip and thicken toward the southern and central parts of the Afar depression, dips increas-ing with depth in the succession (Jepsen and Athearn, 1961; Teilhard de Chardin, 1930). The surface dips largely flatten out 40 to 80 km east of the Ethiopian plateau where the older plateau formations are cut by major

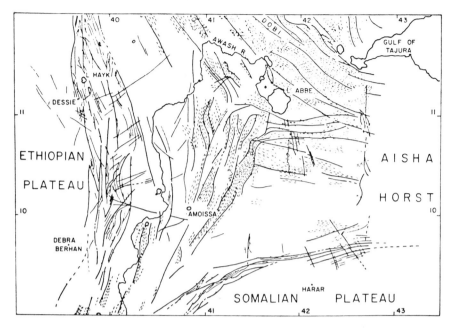

Figure 15. Major faults of the central Ethiopian rift zone: Afar fault belts indicated schematically.

Miocene faults (Mohr, 1962b; Brinckmann and Kürsten, 1969a). These features and the orientations of the dike swarms which fed the Trap Series show that the plateaus were progressively downwarped toward Afar in early Tertiary times, but locally upwarped in the late Tertiary toward lines of rift faulting. The original Afar downwarp is now a zone of intense dissection and recession from older fault scarps and was cut across in the Pleistocene by characteristic discontinuous marginal grabens and belts of antithetic faults (Figs. 15 and 17c). These grabens are offset right en echelon, possibly in relation to traverse structures on the adjacent plateau rim. The eastern sides of the grabens are strongly upwarped or tilted in places and remain seismically active (Gouin, 1970). Along the southern, upwarped margin of Afar, the boundary faults are younger than those of the Ethiopian plateau. Immediately to the north of the boundary faulting, and west of lat. 41° E., antithetic faults cut north-dipping formations. There are no marginal grabens along the south side of the Afar depression.

    A variety of structures trend perpendicular to the margins of the Afar depression, including normal faults, tight monoclines, basalt dike swarms, and even mild compressional folds ranging from Jurassic to Quaternary in age (Merla and Minucci, 1938; Mohr and Rogers, 1966; Abbate and others, 1968; Knetsch, 1970). Although Mohr (1967c) exaggerated the importance of cross-rift structures in Ethiopia, some workers question the existence of any significant traverse faulting (Tazieff and Varet, 1969; Tazieff and others, 1969; Gibson and Tazieff, 1970). However, recent seismic evidence, together with detailed mapping of sectors of the Ethiopian plateau—Afar margin, seems to confirm crustal movement related to such trends (Fairhead and Girdler, 1970; Dakin and others, 1971; Mohr, 1971b).

The floor of the Afar depression is dominated by subparallel belts of closely spaced mid-Pleistocene faults (Figs. 15 and 16), which tend to persist over long distances and reach densities of two or three faults per kilometer. Horst and graben patterns occur, but, as in the Wonji fault belt, the structure is commonly antithetic. The fault throws rarely exceed 100 m and in some areas they are associated with gaping fissures (Tazieff and Varet, 1969). The fault belts tend to be oriented parallel to the nearest margin of Afar (Mohr, 1967a), but there are many deviations, in particular south of Lake Julietti (Fig. 16) and in the annular fault zones of northeast Afar and the Lake Abbe region (Fig. 15).

The Wonji fault belt and its associated silicic volcanoes can be traced north-northeast from the main Ethiopian rift across southern Afar to Lake Abbe. It lies a little west of the median line between the divergent Ethiopian and Somalian plateau escarpments, and locally contains a very young, narrow graben 1 to 2 km wide. It shows the usual en echelon pattern with 35-to-50-km offsets in the Amoissa and Lake Abbe areas (Fig. 15). North of Lake Abbe the Wonji fault belt cannot be identified with certainty.

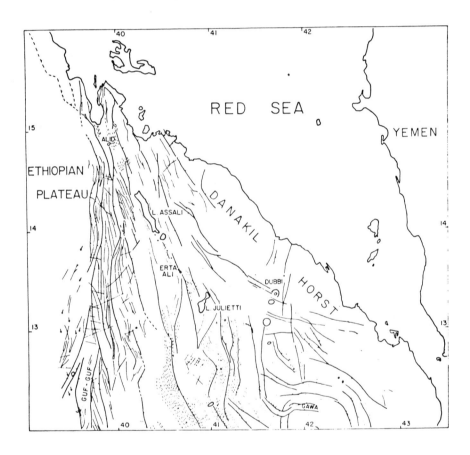

Figure. 16. Major faults of the northern Ethiopian rift zone: Afar fault belts indicated schematically.

It may extend to the Gulf of Tajura (Mohr, 1962b), but a zone of very recent faulting trends also north across the Danakil horst to reach the Red Sea coast at Edd (Figs. 16 and 17c). Along this fault zone there is a crude alignment of Quaternary trachyte-pantellerite caldera volcanoes with associated Holocene fissure basalts such as typify the Wonji fault belt farther south; it also coincides with a possible line of seismic epicenters (Gouin, 1970).

A barely perceptible topographic rise divides north from south Afar along lat. 12° N. North of this line the floor of the depression declines gently to below sea level in the Lake Assali area, which is underlain by thick Neogene evaporites similar to those of the Red Sea shelves (Holwerda and Hutchinson, 1968; Hutchinson and Engels, 1970). It is a graben-in-graben structure and contains an axial northwest-southeast alignment of active basalt shield volcanoes (Tazieff and others, 1969; Bannert and others, 1970).

[*Editor's Note:* Material has been omitted at this point.]

STRUCTURAL EVOLUTION OF THE EASTERN RIFT

The widespread late Eocene uplift of the Ethiopian-Arabian region, evidenced by sudden marine regression, may be taken as the first well-established stage of the formation of the eastern rift system. Uplift and axial downwarping in Ethiopia were preceded and accompanied by outpouring of the Eocene–Oligocene Trap Series fissure basalts. By late Oligocene—early Miocene time, embryonic troughs had formed along the present lines of the Red Sea and Gulf of Aden (Swartz and Arden, 1960; Azzaroli, 1968). This mid-Tertiary phase of development extended southward through Afar and Ethiopia and was manifested by downflexing and faulting of the Turkana depression in northwestern Kenya.

A second uplift phase has been variously estimated to be of middle to late Miocene age, ending the sub-Miocene erosion cycle in Kenya (Saggerson and Baker, 1965) and in Ethiopia (Filjak and others, 1959; Merla, 1963).

The Miocene uplift in Kenya was approximately 700 m on the western and 300 m on the eastern side of the raised area and was accompanied by eruption of fissure phonolites of late Miocene age. The pre-existing Turkwel fault extended southward, and an asymmetrical graben, faulted on its western side, was created along the curving line extending from west Turkana to Lake Eyasi in northern Tanzania (Fig. 10). Subsequent mid-Pliocene dominantly trachytic and basaltic volcanism partly filled the rift depression and locally built high volcanic ranges on its eastern flank. During most of the early and middle Pliocene the marginal plateau was comparatively stable, for it was during this time that the end-Tertiary erosion surface of East and North Kenya was beveled.

During late Pliocene times massive eruptions of ignimbrites began in the central part of the Kenya rift, locally filling it and overflowing its flanks. At about the same time, the last and major uplift phase began, which raised central Kenya by a further 1,500 m. It was accompanied by major graben faulting along most of the Gregory rift at the end of the Pliocene (2.5 m.y.) and was followed by trachyte volcanism that flooded most of the graben floor (Baker and others, 1971). During the early Pleistocene the floor of the graben was broken by dense swarms of minor faults, and local rejuvenations of the marginal faults took place. Since the later mid-Pleistocene, only minor faulting has occurred, affecting the Paleolithic artifact-bearing sediments that accumulated in faulted depressions in the rift floor (McCall and others, 1967).

The structural evolution of the Ethiopian rift was similar to that in Kenya but is less well known. The pre-Pliocene history of the rift is obscured by thick ignimbritic cover. On the plateau west of the rift, the Omo Valley was downwarped in the Oligocene, uplifted and faulted in the early Miocene and late Pliocene, partially filled with upper Pliocene ignimbrites, and substantially uplifted and faulted in the early Pleistocene. By the Pliocene, the main Ethiopian rift was a topographically shallow trough with deep infilling of silicic volcanics erupted from centers close to the rift margins. Major uplift and graben faulting took place in early Pleistocene times producing the high escarpments overlooking the Galla Lakes and Lake Margherita. Parallel tectonism occurred on the neighboring plateaus (Mohr, 1966b) and in the Tana rift. This episode was followed by rejuvenation of some of the faults east of the Galla Lakes and by late Pleistocene—Holocene fragmentation of the rift floor along a narrow axial belt (the Wonji fault belt).

The tectonic history of the Afar depression differs from that of the main Ethiopian rift in several respects. Downwarping and associated faulting of the margins of the depression probably began in the late Mesozoic and was accentuated during the Eocene-Oligocene, with accompanying mild north-south compression. Faulting of the Afar-Ethiopian plateau boundary occurred in the middle? Miocene, and those of the Afar-Somalian plateau in the Pliocene. Major early Pleistocene uplift of the plateau areas was accompanied by the formation (still continuing) of the marginal grabens of the Ethiopian plateau boundary, by renewed faulting along the Afar-Danakil horst and Aisha horst boundaries, and by intense fracturing of the floor of Afar in relation to crustal spreading (Mohr, 1970c). Late Quaternary volcano-tectonic activity has been concentrated along the Wonji fault belt and the northern apex of Afar. In the late Pliocene, northern Afar was flooded by a marine incursion from the Red Sea, but isolation and evaporation ensued within the Salt Plain in the early Pleistocene (Dainelli, 1943).

In summary, the eastern rift was formed by repeated warpings of the crust, preceded and accompanied by volcanism and faulting. During the early stages monoclinal flexuring and fissure volcanism were important, creating broad depressions that were subsequently converted into narrow grabens in the regions of greatest uplift. These grabens have most recently been subject to intense crustal fracturing and dilatation. The peripheral parts of the domal uplifts are characterized by divergent fault systems crossing broad depressions, as in Turkana between the Kenyan and Ethiopian uplifts and in North Tanzania.

Discussion

INFLUENCE OF OLDER STRUCTURES ON THE PATTERN OF THE EASTERN RIFT

Local parallelism between the structural grain of Precambrian metamorphic rocks, the boundaries of Paleozoic and Mesozoic fault troughs, and the trend of African Cenozoic rift faults has led some writers to the conclusion that there is a genetic connection between the generations of structures (McConnell, 1951, 1967, 1969, 1970; Dixey, 1956, 1959). Examples are the parallelism of rift faults and Precambrian foliation trends in the Upemba and Rukwa grabens of southwestern Tanzania (McConnell, 1951), in the Pare and Usambara horsts of northeastern Tanzania and the Turkwel escarpment of west-central Kenya (Dixey, 1956), and in the western rift (McConnell, 1959). Other examples could be quoted—the Siria fault of southwestern Kenya (Williams, 1964) and the Nandi fault branching from the Kavirondo rift (Sanders, 1965) represent late Tertiary rejuvenations of Precambrian faults.

These observations have led to proposals of a concept of "resurgent tectonics" in the belief that much of the African rift system has a long structural history of block movements embracing Precambrian orogenesis, late Paleozoic (Karroo) warping and faulting, Cretaceous rifting in Malawi, and the Cenozoic rift faulting. The validity of the concept is a question of interest in geotectonics for, if it can be inferred that some major structures have a very long history and that their cause lies deep in the mantle, then relative horizontal motions between lithosphere and asthenosphere become improbable (Beloussov, 1969).

Advances in regional geological mapping and in understanding of the structure of the Precambrian foundation in eastern Africa make it possible to re-examine this question. From the north end of Lake Malawi the Rukwa-Urema-Luama grabens of the western rift trend northwest parallel to the Ubendian and Ruzizian fold belts (1,800 m.y.; Cahen and Snelling, 1966), but the much deeper Lake Tanganyika rift crosses this zone obliquely. The northern part of the western rift crosses the Burundian–Karagwe Ankolean (1,100 m.y.) fold belt obliquely and the Kibali-Toro (2,100 m.y.) fold belt at right angles (Pallister and Hepworth, 1956). In Tanzania, the southwest-trending branches of the eastern rift cut obliquely across the structures of the Mozambique belt (about 650 m.y.) and cross its western front onto the central shield (2,600 m.y.) to the west (Hepworth and others, 1967). In Kenya and Ethiopia the relation between Precambrian foundation structure and rift faults is often obscured by the Cenozoic volcanic cover, and although there is a gross similarity of trend of the rift and Mozambique fold belt there are many cross-cutting relations.

The nature of the relation between late Paleozoic (Karroo) and Jurassic-Cretaceous episodes of block faulting and igneous activity and the Cenozoic development of the southern part of the African rifts is beyond the scope of this paper (*see* Vail, 1968; Gair, 1959; Lambert, 1956; de Swardt, 1965; K. Bloomfield and F. Hapgood, 1966, unpub. rept.), but neither the tectonic style nor the nature of the igneous activity accompanying the earlier tectonism justifies regarding it as a period of Mesozoic rifting.

In the eastern rift (excluding Afar) there are no recognizable pre-Tertiary rift structures, and the concept of a genetic connection between Precambrian orogenesis and Cenozoic taphrogenesis is untenable. The evidence indicates that the rift faults have followed Precambrian structures in a few places where the older structures happened to be favorably located and oriented, but the rift system as a whole was formed by imposition of new structural lines across the heterogeneous pre-existing fabric of eastern Africa. Similar conclusions have been drawn about the relation of the Rhine graben to its Hercynian foundation (Illies, 1967; Knetsch, 1967) and for the Baikal rift system (Artemjev and Artyushkov, 1971).

RELATION BETWEEN DOMAL UPLIFTS, VOLCANISM, AND FAULTING

Most of the volcanism of the eastern rift has been centered on the Kenyan and Ethiopian domes, which have a history of uplift dating from the early Cenozoic. The early, middle, and late Cenozoic uplift phases seem to represent local culminations of a series of epeirogenic uplifts that affected much of Africa (Cooke, 1957; King, 1962). Despite the close spatial relation between uplift and volcanism, time relations between uplift, volcanism, and faulting need further study.

In Kenya, Miocene basalt and nephelinite volcanism occurred north and west of the area destined to become the region of greatest domal uplift. The main phase of uplift in the late Miocene was accompanied, however, by large-scale eruptions of flood phonolites from the crest of the dome. Most of the Pliocene flood volcanism took place along the developing rift. Basalts were erupted along the greater part of the floor, and trachytic and ignimbritic flows became progressively more important than phonolites, culminating in great outpourings in central Kenya during late Pliocene and early Pleistocene uplift and faulting. Central volcanoes situated on the floor and margins of the Pliocene rift were petrologically more complex than the earlier Miocene centers. Trachytes and phonolites are most abundant at volcanoes in close proximity to the axial region of the dome, whereas nephelinites become more prominent on its southern flanks toward a focus of strongly alkaline volcanism in North Tanzania (Williams, 1969a, 1969b, 1970). Trachytic activity continued along the southern part of the rift floor in the Quaternary, and rhyolitic volcanism occurred locally. Some basalts are associated with trachytic caldera volcanoes in the floor, but the most spectacular Quaternary basaltic eruptions took place well east of the rift along zones radiating outward from the main area of uplift (Williams, 1970).

In Ethiopia there is a suggestion that major volcanism immediately preceded and accompanied domal uplift (Mohr, 1967a). Fissure basalts are mainly associated with the warped margins of broad structural depressions, these margins being subsequently emphasized by faulting. Silicic volcanism has similar structural relations but is more intimately associated with uplift in both time and space (Mohr, 1970b); nevertheless, not all uplift phases were everywhere accompanied by coeval volcanism; for example, the northern regions of the Somalian and Ethiopian plateaus have only minor Pliocene-Quaternary volcanic rocks.

The large subsidence that resulted in the deposition of great thicknesses of sediments in the coastal and hinterland areas of East Kenya and Somalia might appear to be complementary to the uplifts of the interior (*see* Figs. 3 to 5; Table 1). Most of the subsidence of the continental margin took place in the late Mesozoic and early Tertiary, however, while the uplift of the interior was late Tertiary and Quaternary. Thus, the movements were not genetically connected in such a way that lateral transfer of materials in the upper mantle could be invoked.

Despite difficulties of detailed correlation, a broad association of volcanism and domal uplift is undeniably present, and there is some evident connection between uplift and faulting in both time and location. The relations suggest that these crustal phenomena together are the result of some fundamental subcrustal thermal process rather than effects of independent external stresses.

SEISMICITY AND GRAVITY DATA

The regional seismicity of the eastern rift has been reviewed by Gutenberg and Richter (1954), Sykes and Landisman (1964), Loupekine (1965), Wohlenberg (1968), Fairhead and Girdler (1970), Sykes (1970), and Gouin (1970). Travel-time and surface-wave-dispersion studies have yielded preliminary results on gross crustal structure (Press and others, 1956; Ocal, 1965; Jones, 1968; Molnar and Oliver, 1969; Gumper and Pomeroy, 1970; Searle and Gouin, 1971).

[*Editor's Note:* Material has been omitted at this point.]

Summary and Conclusions

The eastern rift extends southward from its junction with the Red Sea and Gulf of Aden and crosses the crests of two elongated domes formed by periodic uplift which raised the subvolcanic surface to maximum elevations of 3,000 m in Ethiopia and 1,800 m in Kenya.

Voluminous alkaline volcanism accompanied the development of the eastern rift. Early Tertiary flood basalts issued from crustal warps and covered much of the plateaus in Ethiopia and southwestern Arabia. Miocene basaltic volcanism extended from Ethiopia into the Turkana depression of northwestern Kenya, and contemporaneous nephelinitic central volcanoes built up in the Kenya-Uganda border area. In the late Miocene, flood phonolites were the first eruptives in central Kenya. All these volcanics preceded the first main period of rift faulting, which occurred in the mid-Miocene in the north and early Pliocene in the south. This faulting was accompanied by a remarkable change in the composition of the volcanics, there being a complete reversal of the proportions of basaltic and silicic lavas in Ethiopia and a change from phonolitic to trachytic flood volcanism in Kenya. In the later Pliocene and Quaternary, major domal uplift and formation of deep grabens were accompanied by extensive trachyte-pantellerite volcanism, much of it ignimbritic, which locally filled and even overflowed from the rift. The graben subsidence may well have been partly due to the rapid transfer of enormous volumes of silicic magma to the surface.

The cause of the change in the composition of the volcanics and its undoubted connection with episodes of graben faulting is one of the major problems of eastern rift evolution. The large volumes (nearly 50,000 km³ in Ethiopia) of trachyte-rhyolite could not have been produced by fractional crystallization or direct partial melting of the mantle alone. One possibility is to consider development of higher-than-normal temperatures in the mantle beneath the rift zone, with crustal arching and thinning, intense dike injection, and fracturing and collapse of the rift floor, accompanied by derivation of partial melts from the mantle and their contamination with sialic crust. More strontium-isotope ratio and rare-earth-abundance spectrum work is needed to test whether sialic crust was involved in the genesis of the silicic volcanics. Another possibility is long-term accumulation of a mantle-derived salic-silicic layer at the base of the crust, from which flood trachytes and rhyolites, and possibly phonolites, could be obtained in bulk by rapid remelting.

The growth of the domal uplifts was accompanied by increasingly clear definition of the crustal rift zones. These rift zones were subsiding from the very beginning, invalidating the "fallen keystone of the arch" hypothesis of rift valley genesis. Early Tertiary downwarping of Afar and the northern part of the main Ethiopian rift was succeeded by Miocene faulting. In Kenya, early Miocene downwarping and faulting of the Turkana depression was extended south in early Pliocene times to form a shallow asymmetric graben, locally sharply flexed on its eastern flank, across the Kenyan dome. Strong graben faulting occurred in the late Pliocene–early Pleistocene, together with major uplift of the graben shoulders, and in the mid-Pleisto-

cene the floors of the grabens and of Afar were shattered by swarms of
minor faults and pierced by caldera volcanoes. Such a structural sequence
could result from progressive crustal doming and thinning and upward intrusion of a wedge-shaped asthenospheric diapir through the lithosphere and
into the crust. Geophysical data suggest that the areas of domal uplift
are underlain by mantle of anomalously low density and high temperature,
these conditions being most marked under the rift itself.

In longitudinal profile the elevation of the rift floors reflects the
uparched shape of the adjacent plateaus. Considering the greater depth
of sedimentary and volcanic infilling of some graben sectors, however, it
seems that the subvolcanic surface under the rift floor is generally
deepest where the adjacent plateau was most uplifted, and the surface is
below sea level along much of the rift. It cannot be maintained that the
rift floors merely lagged behind the uplift of the plateaus; they definitely
dropped downward, as is evident in some sectors of the western rift. The
problem of why volcanism is absent from "deep" sectors of the African rift
system (for example, the Lake Tanganyika basin) now becomes less acute:
the Gregory and Ethiopian rift floors are probably even deeper, beneath
their volcanic fill.

The geomorphic and geological connection with the Red Sea and Gulf of
Aden through Afar links the eastern rift to sea-floor-spreading processes.
Plate-tectonics analysis indicates that crustal extension across the eastern rift has been very slow by oceanic standards. Geological data suggest
the total extension is unlikely to have been more than 30 km in Ethiopia
and 10 km in Kenya, and in Turkana and north Tanzania the upper limit is
2 to 3 km. The maximum spreading rates are thus 0.4 to 1.0 mm/yr, which
is one or two orders of magnitude lower than typical oceanic spreading
rates. This slow rate of extension of a thick crust distinguishes the
eastern rift from the neo-oceanic spreading zones of the Red Sea and Gulf
of Aden and helps account for the lack of a rectilinear rift and transform
fault pattern such as occurs in the ocean, and also permits the localized
structural variety found along the eastern rift. The massive alkaline
volcanism of the eastern rift conforms with slow continental crustal distension, there being smaller-scale melting at greater depths than for
typical oceanic volcanism.

In summary, plate tectonics confirms the genetic connection between
the formation of the eastern rift and the Red Sea and Gulf of Aden troughs.
But the eastern rift is considered to be atypical of the first stage of
continental disruption and drift; its slow spreading rate has permitted a
degree of vertical uplift and associated voluminous alkaline magmatism
rarely found, for example, along the once contiguous continental margins
of the Atlantic Ocean. Furthermore, when it is possible to compare the
structure of the eastern rift with an oceanic rift of similar spreading
rate, it will likely be found that continental crust exerts some significant modifications to oceanic spreading phenomena. Thus, the eastern rift
is an unusual structure marked by a transitional character from continental
block faulting to a quasi-oceanic spreading zone, but the transition is
not regular along its length.

[*Editor's Note:* Only references cited in the preceding excerpts are reprinted here.]

REFERENCES

Abbate, E., Azzaroli, A., Zanettin, B., and Visentin, E. J., 1968, A geologic and petrographic mission of the "Consiglio nazionale delle ricerche" to Ethiopia, 1967–1968. Preliminary results: *Soc. Geol. Italiana Boll.*, v. 87, p. 561–580.

Artemjev, M. E., and Artyushkov, E. V., 1971, Structure and isostasy of the Baikal rift and the mechanism of rifting: *Jour. Geophys. Research*, v. 76, p. 1197–1211.

Azzaroli, A., 1958, L'Oligocene e il Miocene della Somalia. Stratigrafia, tettonica, paleontologia: *Paleont. Ital.*, v. 52, p. 1–142.

Azzaroli, A., 1968, On the evolution of the Gulf of Aden: *Internat. Geol. Cong., 23d, Prague, 1968*, v. 1, p. 125–134.

Baker, B. H., 1958, Geology of the Magadi area: *Kenya Geol. Survey Rept. no. 42*, 81p.

Baker, B. H., 1963a, Geology of the area south of Magadi: *Kenya Geol. Survey Rept. no. 61*, 27p.

Baker, B. H., 1963b, Geology of the Baragoi area: *Kenya Geol. Survey Rept. no. 53*, 74p.

Baker, B. H., 1967, Geology of the Mount Kenya area: *Kenya Geol. Survey Rept. no. 79*, 78p.

Baker, B. H., Williams, L. A. J., Miller, J. A., and Fitch, F. J., 1971, Sequence and geochronology of the Kenya rift volcanics: *Tectonophysics*, v. 11, p. 191–215.

Bannert, D., Brinckmann, J., Käding, K.-Ch., Knetsch, G., Kürsten, M., and Mayrhoffer, H., 1970, Zur Geologie der Danakil Senke: *Geol. Rundschau*, v. 59, p. 409–443.

Beloussov, V. V., 1969, Interrelations between the earth's crust and upper mantle, in Hart, P. J., ed., The earth's crust and upper mantle: *Am. Geophys. Union Geophys. Mon. 13*, p. 698–712.

Berry, L., and Whiteman, A. J., 1968, The Nile in the Sudan: *Geog. Jour.*, v. 134, p. 1–37.

Bishop, W. W., and Trendall, A. F., 1967, Erosion surfaces, tectonics and volcanic activity in Uganda: *Geol. Soc. London Quart. Jour.*, v. 122, p. 385–420.

Bishop, W. W., Miller, J. A., and Fitch, F. J., 1969, New Potassium-argon age determinations relevant to the Miocene fossil mammal sequence in East Africa: *Am. Jour. Sci.*, v. 267, p. 669–699.

Blanford, W. T., 1870, *Observations on the geology and zoology of Abyssinia, made during the progress of the British expedition to that country in 1867–1868*: London, Macmillan, 487p.

Bowen, N. L., 1937, Recent high-temperature research on silicates and its significance in igneous geology: *Am. Jour. Sci.*, v. 33, p. 1–21.

Bowen, N. L., 1938, Lavas of the African rift valleys and their tectonic setting: *Am. Jour. Sci.*, v. 35A, p. 19–33.

Brinckmann, J., and Kürsten, M., 1969a, *Geological sketchmap of the Danakil depression (1:250,000), 4 sheets*: Bundesanstalt für Bodenforschung, Hanover.

Brown, G. F., 1970, Eastern margin of the Red Sea and the coastal structures in Saudi Arabia: *Royal Soc. London Philos. Trans.*, Ser. A, v. 267, p. 75–87.

Butzer, K. W., and Thurber, D. L., 1969, Some late Cenozoic formations of the lower Omo basin: *Nature*, v. 222, p. 1135–1138.

Cahen, L., and Snelling, N. J., 1966, *The geochronology of equatorial Africa:* Amsterdam, North-Holland Pub. Co., 195p.

Cole, J. W., 1969, Gariboldi volcanic complex, Ethiopia: *Bull. Volcanol.*, v. 33, p. 566–578.

Comucci, P., 1950, *Le vulcaniti del Lago Tana (Africa Orientale):* Accad. Naz. Lincei, Roma, 209p.

Cooke, H. B. S., 1957, Observations relating to Quaternary environments in East and Southern African (A. L. du Toit memoir lectures): *Geol. Soc. South Africa Trans. and Proc.*, annex to v. 60, p. 1–73.

Dainelli, G., 1943, *Geologia dell'Africa Orientale:* Reale Accad. Ital., Roma, 3 v. and 1 v. maps.

Dakin, F., Gouin, P., and Searle, R. C., 1971, The 1969 earthquakes in Serdo (Ethiopia): *Bull. Geophys. Obs. Addis Ababa,* no. 13, p. 19–56.

Dawson, J. B., Bowden, P., and Clark, G. C., 1968, Activity of the carbonatite volcano Oldoinyo Lengai, 1966: *Geol. Rundschau,* v. 57, p. 865–879.

Desio, A., 1940, Resti di artiche superfici di degradazione nell'Ethiopia centrale: *Riv. Geog. Ital.*, v. 47, p. 17–24.

de Swardt, A. M. J., 1965, Rift Faulting in Zambia, in *East African rift system:* Nairobi, Univ. Coll., pt. 1, p. 105–114.

Dixey, F., 1956, The East African rift system: *Colonial Geol. Mining Research Bull.*, Supp. no. 1, 71p.

Dixey, F., 1959, Vertical tectonics in the East African rift zone: *Assoc. Serv. Geol. Africa,* v. 20, p. 359–375.

Dodson, R. G., 1963, Geology of the South Horr area: *Kenya Geol. Survey Rept. no. 60,* 53p.

Dundas, D. L., 1965, Review of rift faulting in Tanzania, in *Rept. geol. geophys. East African rift system* — UMC–UNESCO seminar: Nairobi, Univ. Coll., pt. 2, p. 95–103.

Fairhead, J. D., and Girdler, R. W., 1970, Seismicity of the Red Sea, Gulf of Aden and Afar triangle: *Royal Soc. London Philos. Trans.*, Ser. A, v. 267, p. 49–74.

Filjak, R., Glumicic, N., Zagorac, Z., 1959, *Oil possibilities of the Red Sea region in Ethiopia:* Zagreb, Naftaplin, 104p.

Gair, H. S., 1959, The Karroo system and coal resources of the Gwembe district, north-east sections: *Northern Rhodesia, Geol. Survey Bull, 1.*

Gibson, I. L., 1967, Preliminary account of the volcanic geology of Fantale, Shoa: *Bull. Geophys Obs. Addis Ababa,* no. 10, p. 59–67.

Gibson, I. L., 1969b, The structure and volcanic geology of an axial portion of the main Ethiopian rift: *Tectonophysics,* v. 8, p. 561–565.

Gibson, I. L., 1970, *Quaternary pantelleritic volcanism in the main Ethiopian rift:* Univ. Leeds, Dept. Earth Sci., 14th Ann. Rept. Research Inst. Africa Geol. p. 35–38.

Gibson, I. L., and Tazieff, H., 1970, The structure of Afar and the northern part of the Ethiopian rift: *Royal Soc. London Philos. Trans.*, Ser. A, v. 267, p. 331–338.

Gortani, M., and Bianchi, A., 1941, Note illustrative su la carta geologica degli altipiani hararini e della Dancalia meridionale: *Mem. Royal Accad. Sci. Ist. Bologna*, v. 8, p. 89–104.

Gouin, P., 1970, Seismic and gravity data from Afar in relation to surrounding areas: *Royal Soc. London Philos. Trans.*, Ser. A, v. 267, p.339–358.

Gouin, P., and Mohr, P. A., 1964, Gravity traverses in Ethiopia (interim rept.): *Bull. Geophys. Obs. Addis Ababa*, no. 7, p. 185–239.

Gouin, P., and Mohr, P. A., 1967, Recent effects possibly due to tensional separation in the Ethiopian rift system: *Bull. Geophys. Obs. Addis Ababa*, no. 10, p. 69–78.

Grasty, R., Miller, J. A., Mohr, P. A., 1963, Preliminary results of potassium-argon age determinations on some Ethiopian Trap Series basalts: *Bull. Geophys. Obs. Addis Ababa*, no. 6, p. 97–102.

Gregory, J. W., 1896, *The great rift valley*: London, John Murray, 424p.

Gregory, J. W., 1921, *The rift valleys and geology of East Africa*: London, Seeley, Service, 479p.

Gumper, F., and Pomeroy, P. W., 1970, Seismic wave velocities and earth structure on the African continent: *Seismol. Soc. Am. Bull.*, v. 60, p. 651–668.

Gutenberg, B., and Richter, C. F., 1954, *Seismicity of the earth and related phenomena* (2nd ed.): Princeton, New Jersey, Princeton Univ. Press, 310p.

Hepworth, J. V., Kennerly, J. B., and Shackleton, R. M., 1967, Photogeological investigation of the Mozambique front in Tanzania: *Nature*, v. 216, p. 146–147.

Hieke-Merlin, O., 1950, I basalti dell'Africa Orientale: *Mem. Inst. Geol. Min. Univ. Padova*, v. 17, 42p.

Hieke-Merlin, O., 1953, Le vulcaniti acide dell'Africa Orientale: *Mem. Inst. Geol. Min. Univ. Padova*, v. 18, 45p.

Holwerda, J. G., and Hutchinson, R. W., 1968, Potash-bearing evaporites in the Danakil area, Ethiopia: *Econ. Geology*, v. 63, p. 124–150.

Howell, F. C., 1968, Omo research expedition: *Nature*, v. 219, p. 567–572.

Hutchinson, R. W., and Engels, G. G., 1970, Tectonic significance of regional geology and evaporite lithofacies in northeastern Ethiopia: *Royal Soc. London Philos. Trans.*, Ser. A, v. 267, p. 313–329.

Illies, H., 1967, Development and tectonic pattern of the Rhinegraben: *Rhinegraben Prog. Rept. no. 6*, p. 7–9.

Isaac, G. L., 1967, Stratigraphy of the Peninj group — early middle Pleistocene formations west of Lake Natron, Tanzania, in Bishop, W. W., and Clark, J. D., eds., *Background to evolution in Africa*: Chicago, Chicago Univ. Press, p. 229–258.

Jepsen, D. H., and Athearn, M. J., 1961, *General geology map of the Blue Nile basin, Ethiopia (1:1 million)*: U.S. Dept. of State, Washington, D.C.

Jones, P. B., 1968, Surface seismic wave dispersion and crustal structure between the Gulf of Aden and Addis Ababa: *Bull. Geophys. Obs. Addis Ababa*, no. 12, p. 19–26.

King, B. C., 1965, Petrogenesis of the alkaline igneous rock suites of the volcanic and intrusive centres of eastern Uganda: *Jour. Petrology*, v. 6, p. 67–100.

King, B. C., 1970, Vulcanicity and rift tectonics in East Africa, in Clifford, T. N., and Gass, I. G., eds., *African magmatism and tectonics*: Edinburgh, Oliver and Boyd, p. 263–283.

King, L. C., 1962, *The morphology of the earth:* London, Oliver and Boyd, 699p.

Knetsch, G., 1967, Changing roles of the Upper Rhine lineament in the course of geological times and events: *Rhinegraben Prog. Rept. no. 6*, p. 13–15.

Knetsch, G., 1970, Danakil reconnaissance 1968, in Illies, J. H., and Mueller, St., eds., *Graben problems*: Stuttgart, Schweizerbartsche Verlagsbuchhandlung, p. 267–279.

Lambert, H. H. J., 1956, The origin of the mid-Zambezi Valley and its significance in a regional interpretation of the Rhodesias: *Proc. C. C. T. A. East-Central Reg. Comm. Geol. Conf.*, Dar-es-Salaam, p. 129–138.

LeBas, M. J., and Mohr, P. A., 1968, Feldspathoidal rocks from the Cainozoic volcanic province of Ethiopia: *Geol. Rundschau*, v. 58, p. 273–280.

Loupekine, I. S., 1965, Seismology in East Africa, in *Rept. geol. geophys. East African rift system*: UMC–UNESCO Seminar, Nairobi, Univ. Coll., pt. 2, p. 20–32.

Manasse, E., 1909, *Contribuzioni allo Studio Petrografico della Colonia Eritrea*: Nova, Siena, 169p.

McCall, G. J. H., 1958, Geology of the Gwasi area: *Kenya Geol. Survey Rept. no. 45,* 88p.

McCall, G. J. H., 1964, Geology of the Sekerr area: *Kenya Geol. Survey Rept. no. 65,* 84p.

McCall, G. J. H., 1967, Geology of the Nakuru-Thomson's Falls-Lake Hannington area: *Kenya Geol. Survey Rept. no. 78*, 122p.

McCall, G. J. H., 1968 The five caldera volcanoes of the central rift valley, Kenya: *Geol. Soc. London Proc.*, no. 1647, p. 54–59.

McCall, G. J. H., Baker, B. H., and Walsh, J., 1967, Late Tertiary and Quaternary sediments of the Kenya rift valley, in Bishop, W. W., and Clark, J. D., eds., *Background to evolution in Africa*: Chicago, Chicago Univ. Press, p. 191–220.

McConnell, R. B., 1951, *Rift and shield structure in East Africa*: Internat. Geol. Cong., 18th, London 1948, v. 14, p. 199–207.

McConnell, R. B., 1959, Outline of the geology of the Ruwenzori Mountains: *Colonial Geol. Mining Research*, v. 7., p. 245–268.

McConnell, R. B., 1967, The East African rift system: *Nature*, v. 215, p. 578–581.

McConnell, R. B., 1969, East African rift system: *Nature*, v. 224, p. 65.

McConnell, R. B., 1970, The evolution of the rift system of eastern Africa in the light of Wegmann's concept of tectonic levels, in Illies, J. H., and Mueller, St., eds. *Graben problems*: Stuttgart, Schweitzerbartsche Verlagsbuchhandlung, p. 285–290.

Merla, G., 1963, Missione geologica nell'Ethiopia meridionale del C. N. R. 1959-1960. Notizie geomorfologiche: *G. Geol. Bologna*, v. 31, p. 1–56.

Merla, G., and Minucci, E., 1938, *Missione geologica nel Tigrai*: Reale Accad. Ital., Roma, v. 1, 363p.

Mohr, P. A., 1960, Report on a geological excursion through southern Ethiopia: *Bull. Geophys. Obs. Addis Ababa*, no. 3, p. 9–20.

Mohr, P. A., 1961, The geology, structure and origin of the Bishoftu explosion craters: *Bull. Geophys. Obs. Addis Ababa*, no. 4, p. 65–101.

Mohr, P. A., 1962a, *The geology of Ethiopia*: Univ. Coll. Addis Ababa Press, 268p.

Mohr, P. A., 1962b, The Ethiopian rift system: *Bull Geophys. Obs. Addis Ababa*, no. 5, p. 33–62.

Mohr, P. A., 1962c, Surface cauldron subsidence with associated faulting and fissure basalt eruptions at Gariboldi Pass, Shoa, Ethiopia: *Bull. Volcanol.*, v. 24, p. 421–428.

Mohr, P. A., 1966a, Chabbi volcano (Ethiopia): *Bull. Volcanol.*, v. 29, p. 797–816.

Mohr, P. A., 1966b, Geological report on the Lake Langano and adjacent plateau regions: *Bull. Geophys. Obs. Addis Ababa*, no. 9, p. 59–75.

Mohr, P. A., 1967a, The Ethiopian rift system: *Bull. Geophys. Obs. Addis Ababa,* no. 11, p. 1–65.

Mohr, P. A., 1967b, Review of the geology of the Simien Mountains: *Bull. Geophys. Obs. Addis Ababa*, no. 10, p. 79–93.

Mohr, P. A., 1967c, Major volcano-tectonic limeament in the Ethiopian rift system: *Nature,* v. 213, p. 664–665.

Mohr, P. A., 1968, The Cainozoic volcanic succession in Ethiopia: *Bull. Volcanol.,* v. 32, p. 5–14.

Mohr, P. A.,1970b, Volcanic composition in relation to tectonics in the Ethiopian rift system: A preliminary investigation: *Bull Volcanol.,* v. 34, p. 141–157.

Mohr, P. A., 1970c, The Afar triple junction and sea-floor spreading: *Jour. Geophys, Research,* v. 75, p. 7340–7352.

Mohr, P. A., 1971b, Ethiopian Tertiary dike swarms: *Smithsonian Astrophys. Obs. Spec. Rept.* (in press).

Mohr, P. A., and Gouin, P., 1968, Gravity traversés in Ethiopia (4th interim rept.): *Bull. Geophys. Obs. Addis Ababa,* no. 12, p. 27–56.

Mohr, P. A., and Rogers, A. S., 1966, Gravity traverses in Ethiopia (2d interim rept.): *Bull, Geophys. Obs. Addis Ababa.* no. 9, p. 7–58.

Molnar, P., and Oliver, J., 1969, Lateral variations of attenuation in the upper mantle and discontinuities in the lithosphere: *Jour. Geophys. Research,* v. 74, p. 2648–2682.

Ocal, N., 1965, The dispersion of surface waves and crustal structure in the African continent: *Pure and Appl. Geophys.,* v. 60, p. 74–79.

Pallister, J. W., and Hepworth, J. V., 1956, Notes on mylonite and rift faulting in Uganda: *C.C.T.A. East-Central Reg. Com. Geol. Conf.,* Dar-es-Salam, p. 95–97.

Press, F., Ewing, M., and Oliver, J., 1956, Crustal structure and surface wave dispersion in Africa: *Seismol. Soc. Am. Bull.,* v. 46, p. 97–103.

Prior, G. T., 1900, On aegirine and riebeckite anortheoclase rocks related to the "grorudite-tinguaite" series, from the neighborhood of Adowa and Axum, Abyssinia: *Mineralog. Mag.,* v. 12, p. 255–273.

Prior, G. T., 1903, Contributions to the petrology of British East Africa: *Mineralog. Mag.,* v. 13, p. 228–263.

Pulfrey, W. P., 1960, Shape of the sub-Miocene erosion level in Kenya: *Kenya Geol. Survey Bull. no. 3,* 18p.

Randel, R., 1971, Geology of the Suswa area: *Kenya Geol. Survey Rept. no. 97,* (in press).

Rhemtulla, S., 1970, A geological reconnaissance of south Turkana: *Geog. Jour.,* v. 136, p. 61–73.

Richard, J., and Neumann van Padang, M., 1957, *Catalogue of the active volcanoes of the world including solfatara fields, 4, Africa and the Red Sea:* Internat. Volcanol. Assoc., Naples, Italy.

Saggerson, E. P., and Baker, B. H., 1965, Post-Jurassic erosion surfaces in eastern Kenya and their deformation in relation to rift structure: *Geol. Soc. London Quart. Jour.,* v. 121, p. 51–72.

Sanders, L. D., 1965, Geology of the contact between the Nyanza shield and the Mozambique belt in western Kenya: *Kenya Geol. Survey Bull. no. 7,* 45p.

Searle, R. C., and Gouin, P., 1971, An analysis of some local earthquake phases originating near the Afar triple-junction: *Seismol. Soc. America Bull.,* (in press).

Shackleton, R. M., 1945, Geology of the Nyeri area: *Kenya Geol. Survey Rept. no. 12,* 26p.

Shackleton, R. M., 1946, Geology of the country between Nanyuki and Maralal: *Kenya Geol. Survey Rept. no 11,* 54p.

Shackleton, R. M., 1951, A contribution to the geology of the Kavirondo rift valley: *Geol. Soc. London Quart. Jour.,* v. 106, p. 345–392.

Shackleton, R. M., 1955, Pleistocene movements in the Gregory rift valley: *Geol. Rundschau,* v. 43, p. 257–263.

Smith, W. C., 1931, A classification of some rhyolites, trachytes and phonolites from part of Kenya Colony, with a note on some associated basaltic rocks: *Geol. Soc. London Quart. Jour.,* v. 87, p. 212–258.

Smith, W.C., 1938, Petrographic description of volcanic rocks from Turkana, Kenya Colony, with notes on their field occurrence from the manuscript of Mr. A. M. Champion: *Geol. Soc. London Quart. Jour.,* v. 94, p. 507–553.

Swartz, D. H., and Arden, D. D., 1960, Geologic history of Red Sea area: *Am. Assoc. Petroleum Geologists Bull.,* v. 44, p. 1621–1637.

Sykes, L. R., 1970, Seismicity of the Indian Ocean and a possible nascent island arc between Ceylon and Australia: *Jour. Geophys. Research,* v. 75, p. 5041–5055.

Sykes, L. R., and Landisman, M., 1964, The seismicity of East Africa, the Gulf of Aden and the Arabian and Red Seas: *Seismol. Soc. America Bull.,* v. 54, p. 1927–1940.

Tazieff, H., and Varet, J., 1969, Signification tectonique et magmatique de l'Afar septentrional (Éthiopie): *Rev. Géographie Phys. et Géologie Dynam.,* v. 11, p. 429–450.

Tazieff, H., Marinelli, G., Barberi, F., and Varet, J., 1969, Géologie de l'Afar septentrional: *Bull. Volcanol.,* v., 33, p. 1039–1072.

Teilhard de Chardin, P., 1930, Observations géologiques en Somalie Francaise et au Harrar: *Soc. Géol. France Mem.,* v. 6, p. 5–12.

Vail, J. R., 1968, The southern extension of the East African rift system and related igneous activity: *Geol. Rundschau,* v. 57, p. 601–614.

Walsh, J., 1969, Geology of the Eldama ravine—Kabarnet area: *Kenya Geol. Survey Rept. no. 83,* 48p.

Walsh, J., and Dodson, R. G., 1969, Geology of northern Turkana: *Kenya Geol. Survey Rept. no. 82,* 42p.

Weber, M., 1906, Die Petrographische Ausbeute der Expeditionen O. Neumann-v. Erlanger nach Östafrika und Abessynien 1900–1901: *Geog. Gessel. München Mitt.,* v. 1, p. 637–660.

Williams, L. A. J., 1964, Geology of the Mara River-Siana area: *Kenya Geol. Survey Rept. no. 66,* 47p.

Williams, L. A. J., 1969a, Geochemistry and petrogenesis of the Kilimanjaro volcanic rocks of the Amboseli area, Kenya: *Bull. Volcanol.,* v. 33, p. 862–888.

Williams, L. A. J., 1969b, Volcanic associations in the Gregory rift valley, East Africa: *Nature,* v. 224, p. 61–64.

Williams, L. A. J., 1970, The volcanics of the Gregory rift valley, East Africa: *Bull. Volcanol.,* v. 34, p. 439–465.

Wohlenberg, J., 1968, Seismizität der Östafrikanischen Grabenzonen zwischen 4°N und 12°S sowie 23°E und 40°E: *Bayerischen Akad. Wiss.,* v. 23, 95p.

Part VII

THE AFAR RIFT JUNCTION

Editor's Comments
on Paper 19

19   TAZIEFF et al.
Tectonic Significance of the Afar (or Danakil) Depression

THE AFAR RIFT JUNCTION

Paper 19, by the CNRS-CNR research team led by Tazieff, was published early in the research project that began in 1967. It presents a brief but clear account of the "junction area of three most important structures." The geology of the Afar depression has been described and references given in Paper 18. An important area in the south and east that contains the controversial Aisha Horst is not described in Paper 19. The value of Paper 19 lies in its arguments for the crustal separation that resulted in the Red Sea and Gulf of Aden. Their geology and tectonics are described at some length.

The Afar Depression, the Red Sea, and the Gulf of Aden are not, strictly speaking, rift valleys, although they are tectonic landforms initiated by rift faulting, rifting, and subsidence. The Gulf of Aden and the Red Sea differ from rift valleys in order of magnitude and ratio of subsidence to width. These considerations are overlooked by, among others, Cloos (1939) and Picard (1970), who maintain that they are graben. In any case the Gulf of Aden and the Red Sea are important members of the Afro-Arabian rift system.

The Gulf of Aden

The Gulf of Aden is described by Girdler (1978, p. 331) as "a fully developed oceanic structure on a small scale." Its median zone is a spreading ridge—a transform rift system (Laughton, 1966; Laughton et al., 1970)—that, except for the absence of the development of a continuous mid-oceanic ridge, can be classed with oceanic rift systems.

Geology of the flanking continental blocks of the Gulf of Aden comprises: (1) the stratigraphy of sedimentary rocks and the volcanics and igneous intrusives of the marginal blocks, Arabia and Somalia; (2) the geomorphology of the margins, to what extent they are downwarped, modified by scarp retreat and pedimentation, or by subsequent faulting; or, on the other hand, are unflexed but faulted with fresh or eroded and retreated scarps; (3) the detailed structure within the marginal blocks and any relationship with structures of the gulf floor.

Beydoun (1964, pp. 99–102) summarizes this information for the Arabian margin and later (1970) describes and compares the geology of the Somalia coast, with reference to the spreading hypotheses for the formation of the gulf. Abdel-Gawad (1970) has made a special study of the interpretation of structure of the flanks using satellite photographs. The map by Merla et al. (1973) is of special interest for the geology of the Somalia coastal zone. The topography is shown on the map by Laughton et al. (1970).

With regard to sedimentation, there has been deposition of terrigenous sediments beyond the continental slopes. Thickness may have been considerably greater at the western end.

Members of the Deep Sea Drilling Project (JOIDES) have drilled at site 231 (Fisher et al., 1974), in about 2152 m depth on the gulf floor seaward of the foot of the continental slope at about 46.5°E longitude. The hole penetrated 584 m of mid-Miocene sediments and entered chalk with basalt. At site 232A on the northwest of the Alula-Fartak Trench at the entrance to the gulf and also on the floor, the hole was drilled in 1743 m depth of water and penetrated 396 m of Upper Miocene sandstone. Site 233A is on the opposite side of the trench and penetrated Upper Pliocene sediments, finally entering diabase.

Magnetic and seismic survey results were used by Laughton (1966), Laughton et al. (1970), and Girdler and Styles (1976). The latter made comparisons with magnetic results on profiles of the Red Sea. More recent magnetic profiles were used by Noy (1978) to make additional comparisons. Girdler (1978) gives a more complete account of seismic gravity and magnetic shipborne surveys over the west end of the gulf and relates the results to those from the Red Sea.

The Red Sea

The Red Sea geology appears to be much the same as that of the Gulf of Aden. This includes geomorphology and morphotectonics of the margins and of the floor, structural history, and volcanology of the flanks and of the median zone. They differ appreciably only in the sedimentation and sedimentary history. In the case of the Red Sea, much greater thicknesses of sediments and evaporites, with volcanics, were laid down.

The symposium edited by Falcon et al. (1970) for the Royal Society, London, contains a valuable collection of papers on geology, structure, geophysics, and volcanology, principally of the Red Sea. Plate tectonics influenced most of these papers. Another symposium publication containing papers on geology and geophysics is that by Degens and Ross (1969).

Geologic maps of the Red Sea include: Laughton (1970), the bathymetric contour map; the map compiled by Coleman (1973), gives summaries of terrestrial geology, petroleum exploration wells, and Deep Sea Drilling Project locations. For the bordering countries there are the following maps: Egypt (Egyptian Geological Survey, 1937); Sudan (Vail, 1978); Ethiopia (Merla et al., 1973); Arabia (U.S. Geol. Surv., 1963). These maps are at scale 1:2 m. Other references are given in Coleman (1973).

Geophysical surveys to 1969, chiefly magnetic and seismic, are summarized by Girdler (1969). For seismic surveys and seismicity see the following accounts: Fairhead and Girdler (1970); Phillips and Ross (1970); Allan (1970); Kabbani (1970); Qureshi (1971). Patterns of magnetic anomalies symmetrical about a central axis and models derived from gravity surveys are described in Drake and Girdler (1964) and Noy (1978). These papers correlate results in the Gulf of Aden with those in the Red Sea.

Descriptions of the geology of the adjacent countries, accompanied by both small-scale and detailed maps and sections, include the following: Egypt (Said, 1962; Hume, 1925, vol. I); Sudan (Whiteman, 1968, 1971; Vail, 1978); Ethiopia (Mohr, 1962); Arabia (Brown, 1970); Yemen (Geukens, 1966).

A general account of the geology—the stratigraphy related to structural history—is that by Coleman (1974a, 1974b). Both have the same text but 1974a is accompanied by the map on scale 1:2 m (Coleman, 1973), whereas 1974b has a map on a reduced scale of 1:8 m.

Evaporites of middle to upper Miocene age were deposited in the succession when there was an incursion from the Mediterranean Sea along the Clysmic Gulf and as far as the southern end of the Red Sea (Hume, 1921; Montanaro, 1941; Swartz and Arden, 1960; Heybroek, 1965; Girdler and Whitmarsh, 1974). Sediments older than Miocene are rarely encountered (Coleman, 1974a). Girdler (1969), from applied geophysics and on structural and marine topographical grounds, divides the Red Sea into three parts—northern, central, and southern. In the central part, drilling by the members of the Deep Sea Drilling Project (Coleman, 1973; Whitmarsh et al., 1974) was undertaken in the hope of reaching basement underlying the axial trough. Sediments and evaporites of Miocene age were found to have infilled (by flowage) the central trough, which is supposedly widening by spreading (Girdler and Styles, 1974). One hole reached basalt (oceanic crust?). Petroleum exploration holes drilled near the coasts may in some cases have reached either basic igneous rock, which is either oceanic crust or basalt flow in the sedimentary sequence. The seafloor between the axial zone and the coasts has not been explored by drilling. In the southern part Deep Sea Drilling Project holes did not reach basement. Petroleum exploratory holes drilled on Farsan and Dahlan islands, revealed the presence of evaporites; one hole entered metamorphics.

Coleman (1974a, 1974b) summarizes published information for the whole length of the littoral zone of the Red Sea. This zone consists of a coastal plain of coalescing fans, marine sediments as old as Miocene, and minor volcanic flows, generally backed by erosion scarps and truncated spurs. The plain continues seaward as the continental shelf. There is evidence of emergence in the north and submergence in the south. Faults, unrelated to the erosion of features, do exist, the most easterly being the probable Red Sea marginal fault. They are concealed beneath the surface deposits of the littoral zone.

The landforms of the coast of the Red Sea have a critical bearing on its formation and hence on the geological structure of the depression. Whiteman (1968, p. 235) states "many of the faults shown on general maps do not exist on the ground and have, in fact, been put in on the assumption that most of the escarpments are of fault origin." Auden (1958) states that "the junction of the coastal plain sediments with the base of the Red Sea hills is not primarily a faulted one but is rather of the nature of a marginal overlap across an irregularly indented topography." The most detailed account of the geology of the Red Sea littoral is by Whiteman (1971) who quotes Carella and Scarpa (1962) and others.

The former eastward extent of the Tertiary erosion surface in the Sudan is not known, but if it extended beyond the Red Sea margin, this would give the age of the last stage in the formation of the Red Sea as post mid-Tertiary.

Study of the geomorphology of the adjacent landmass shows that a single uplift zone along the whole length of the sea—the so-called Afro-Arabian swell—does not exist. There is only one area of domal uplift, that which flanks the Red Sea in the south, the Arabo-Ethiopian swell (Mohr, 1962). This has its northwestern margin at 17°N latitude on the west (near the Sudan-Eritrean border), and at 21°N latitude on the east. This uplift began in Eocene times. Beyond this to the northwest on either flank are uplands, falsely described in the literature either as the Arabo-Nubian swell or the Afro-Arabian swell. Hills (1961, pp. 84–85) criticizes this model on which Cloos's (1937) proposition depends—"that the Red Sea is a crestal graben on an up-domed Arabian-Ethiopian massif." From a morphotectonic map, Hills concludes that "it is not possible to conceive of the region as an elliptical upwarp. Instead we find a long narrow uplifted block on the south-west of the Red Sea, and following its trend." Whiteman (1971, Fig. 82) shows the Red Sea hills as residuals rising above the Togni surface (mid-Tertiary) and the postulated later upwarp may be not tectonic but structural, the result of differential erosion. Moreover, the elevation of these uplands cannot be younger than Cretaceous and is probably much older. They were residuals on the mid-Tertiary surface.

A number of hypotheses have been advanced for the structure (and therefore of the origin) of the Red Sea: (1) crestal graben (Suess, 1904, Part II, p. 374: Krenkel, 1957, pp. 123–128); crestal graben with bifurcating ends (Cloos, 1939, pp. 434–445), denied by Hills (1961; see above); (2) graben with splitting and separation along a median zone, exposing emplaced mantle material (Drake and Girdler, 1964, Fig. 14); (3) marginal monoclinal flexures with thinning or attenuation of the crust (accompanying domal uplift and extension) with subsequent separation along a median zone (Whiteman, 1968; Qureshi, 1971); (4) faulting of margins, separation and spreading, thinning being replaced by repeated faulting (Lowell and Genik, 1972; compare this with Morton and Black, 1975, for Afar); (5) faulting of margins adjacent to present coasts and separation (Swarz and Arden, 1960; Wegener, 1966; Holmes, 1965); (6) initial formation of a narrow trough with marginal monoclinal flexures followed by rifting along the axis and crustal separation (Whiteman, 1971, for margins; and Girdler and Styles, 1974, 1976; and Girdler, 1978, for rifting; see also Whiteman, 1971, Fig. 82, note to caption).

The geology of the Red Sea floor is at present known or inferred chiefly from interpretation of geophysical data or from bathymetry and exploratory drilling.

Arabo-Ethiopian Swell (Afar Triple Junction)

The southern Red Sea, the Gulf of Aden, the Ethiopian main rift, and the Afar depression all lie within the area of the Arabo-Ethiopian swell volcano-tectonic domal uplift or arch-volcanic uplift, in Milanovsky's terminology. Paper 18 has described the domal uplifts of Kenya and Ethiopia, relating tectonics and the development of erosion surfaces, the history of volcanism and progressive changes in its nature, generally from continental to oceanic. In general terms, volcanism and tectonism in crustal swells and consequent rifting are the subject of papers by Le Bas (1971), Burke and Whiteman (1973) and Burke (1978). For the Arabo-Ethiopian swell, papers by Gass (1970, 1975) relate magmatism and tectonism. Gass refers to the swell as the "Afro-Arabian dome" (see discussion above).

Paper 19 has served as an introduction to the modern interpretation of the Afar depression and the Arabo-Ethiopian swell. As described in the foregoing, the Afar is not a rift valley, nor is it a "point" triple junction. It is a junction area—a tectonic depression in which is exhibited as nowhere else the unsubmerged and unmasked crust, neither continental nor wholly oceanic, that forms when continental plates first separate.

With realization of its significance in the plate tectonic context, there was a marked increase in interest in the Afar from about 1972 on. Accounts of these investigations appear in Pilger and Rösler (1975). Of these, Black et al. (1972) and Barberi et al. (1975) are important. Later general papers based on this work are found in Pilger and Rösler (1976), Neumann and Ramberg (1978) and Ramberg and Neumann (1978). Paper 19 does not contain an account of the southern Afar, and for this the paper by Christianson et al. (1975, pp. 259–276) should be consulted. The paper by Barberi and Varet (1978, pp. 55–69) is of the same nature and scope but includes additional information on structure and volcanism.

The foregoing descriptions of the structural nature of the three rifts (one rift valley, two rift seas) and of the Afar depression, are incomplete without further reference to originating processes. The vertical tectonics (graben) hypotheses require no further mention. Additional references for plate tectonic hypotheses follow.

Elucidation of structural history in terms of plate tectonics involves four elements: (1) transform faults; (2) magnetic anomaly patterns; (3) fit of plate marginal submarine contours; (4) and from the foregoing, poles of rotation (McKenzie et al., 1970). The transform faults of the Gulf of Aden are described in Laughton (1966), Laughton et al. (1970), and Girdler (1978). For the Red Sea, detection and location of transform faults are made difficult and uncertain because of the sedimentary cover and paucity of paleomagnetic data, which is of such value in this context for the Gulf of Aden. Seismic data provides the most fruitful indicators, the location of earthquake epicenters with fault plane solutions that indicate strike-slip movement on presumed transform faults (Coleman, 1973). However, such locations are possible from events within a time range of only a few decades and on only a few of possible faults. In contrast to the Gulf of Aden, where more than fourteen transforms are recognized (Laughton et al. 1970), only two have been established in the Red Sea and these not on magnetic lineation data but on seismic evidence (Sykes and Landisman, 1964; Fairhead and Girdler, 1970; Whiteman, 1970). These transforms are shown on the map by Coleman (1973) but their azimuths are not as precisely known as suggested by Girdler and as discussed by Whiteman. However, airborne magnetic surveys (Girdler and Styles, 1974; Girdler, 1978) give data on duration and rate of spreading. The fit of the plate margins, at the 500-fathom contour, is not easily determinable for the central and southern Red Sea because of the thick accumulation of evaporites and terrigenous sediments (Coleman, 1974; Girdler and Whitmarsh, 1974); the independent separation of the Danakil Horst (platelet); and the absence for most of the area of magnetic survey data.

For the Gulf of Aden including southern Afar, the outstanding unresolved difficulty remains the presence and nature of the Aisha Horst, whether autochthonous or allochthonous (Mohr, 1970; Christianson et al., 1975; Black et al., 1972).

REFERENCES

Abdel-Gawad, M., 1970, Interpretation of satellite photographs of the Red Sea and Gulf of Aden, *Royal Soc. London Philos. Trans.* **A267**:23–40.

Allan, T. D., 1970, Magnetic and gravity fields over the Red Sea, *Royal Soc. London Philos. Trans.* **A267**:153–180.

Auden, J. B., 1958, Page 22 in discussion of A. M. Quennell, 1958, The structural and geomorphic evolution of the Dead Sea rift, *Geol. Soc. London Quart. Jour.* **114**:1–24.

Barberi, F., G. Ferrara, R. Santacroce, and J. Varet, 1975, Structural evolution of the Afar triple junction, in *Afar Depression of Ethiopia*, A. Pilger and A. Rösler, eds., Schweizerbart'Sche, Stuttgart, pp. 38–54.

Barberi, F., and J. Varet, 1978, The Afar rift junction, in *Petrology and Geochemistry of Continental Rifts*, E. R. Neumann and I. B. Ramberg, eds., D. Reidel, Dordrecht, pp. 55–69.

Beydoun, Z. R., 1964, Stratigraphy and structure of eastern Aden Protectorate, *Overseas Geol. Mineral Resour. Bull., Supplement 5*, 107p.

Beydoun, Z. R., 1970, Southern Arabia and northern Somalia: Comparative geology, *Royal Soc. London Philos. Trans.* **A267**:267–292.

Black, R., W. H. Morton, and J. Varet, 1972, New data on Afar tectonics, *Nature, Phys. Sci.* **240**:170–173.

Brown, G. F., 1970, Eastern margin of the Red Sea and coastal structure in Saudi Arabia, *Royal Soc. London Philos. Trans.* **A267**:75–87.

Burke, K., 1978, Evolution of continental rift systems in the light of plate tectonics, in *Tectonics and Geophysics of Continental Rifts*, I. B. Ramberg and E. R. Neumann, eds., D. Reidel, Dordrecht, pp. 1–10.

Burke, K., and A. J. Whiteman, 1973, Uplifting, rifting and the break-up of Africa, in *Implications of Continental Drift to the Earth Sciences*, D. H. Tarling and S. K. Runcorn, eds., Academic Press, London, pp. 735–755.

Carella, R., and N. Scarpa, 1962, Geological results of exploration in Sudan by AGIP Mineraria Ltd, *4th Arab Petroleum Congr. Proc.*, Beirut, 1962.

Christianson, T. B., H.-U. Schaefer, and M. Schönfield, 1975, Geology of southern and central Afar, Ethiopia, in *Afar Depression of Ethiopia,* A. Pilger and A. Rösler, eds., Schweizerbart'sche, Stuttgart, pp. 259-276.

Cloos, H., 1937, Grosstektonik Hochafrikas und Seiner Umgebung, *Geol. Rundschau* **28:**333-348.

Cloos, H., 1939, Der Nubisch-Arabische Schild, in Hebung, Spaltung, Vulkanismus, *Geol. Rundschau* **30:**434-445.

Coleman, R. G., 1973, Geological Map of the Red Sea, in *Initial Reports of Deep Sea Drilling Project,* vol. 23, U.S. Government Printing Office, Washington, D.C., map. scale 1:2m.

Coleman, R. G., 1974a, Geological background of the Red Sea, in *Initial Reports of Deep Sea Drilling Project,* vol. 23, U.S. Government Printing Office, Washington, D.C., pp. 813-819.

Coleman, R. G., 1974b, Geological background of the Red Sea, in *The Geology of the Continental Margins,* C. A. Banks and C. L. Drake, eds., Springer Verlag, New York, pp. 743-751.

Degens, E. T., and D. A. Ross, eds., 1969, *Hot Brines and Recent Heavy Metal Deposits in the Red Sea,* Springer Verlag, Berlin, 590p.

Degens, E. T., and D. A. Ross, 1970, The Red Sea Hot Brines, *Sci. American* **222:**32-53.

Drake, C. L., and R. W. Girdler, 1964, A geophysical study of the Red Sea, *Royal Astron. Soc. Geophys. Jour.* **8:**473-495.

Egyptian Geological Survey, 1937, *Geological Map of Egypt,* scale 1:2m, Dept. of Survey and Mines, Giza, Egypt.

Fairhead, J. D., and R. W. Girdler, 1970, The seismicity of the Red Sea, Gulf of Aden and Afar triangle, *Royal Soc. London Philos. Trans.* **A267:**49-74.

Falcon, N. L., I. G. Gass, R. W. Girdler, and A. S. Laughton, 1970, A discussion on the structure and evolution of the Red Sea, Gulf of Aden and Ethiopia Rift Junction, *Royal Soc. London Philos. Trans.* **A267:**1-420.

Fisher, R. L., et al., 1974, Gulf of Aden Sites 231-233, in *Initial Reports of the Deep Sea Drilling Project,* vol. 24, R. L. Fisher, E. T. Bunce et al., eds., U. S. Government Printing Office, Washington, D.C., pp. 17-196.

Gass, I. G., 1970, Tectonic and magmatic evolution of the Afro-Arabian dome, in *African Magmatism and Tectonics,* T. N. Clifford and I. G. Gass, eds., Oliver & Boyd, Edinburgh, pp. 285-300.

Gass, I. G., 1975, Magmatic and tectonic processes in the development of the Afro-Arabian dome, in *Afar Depression in Ethiopia,* A. Pilger and A. Rösler, eds., Schweizerbart'sche, Stuttgart, pp. 10-18.

Geukens, F., 1966, Geology of the Arabian Peninsula, Yemen, *U.S. Geol. Survey Prof. Paper 560B,* 23p.

Girdler, R. W., 1969, The Red Sea—a geophysical background, in *Hot Brines and Recent Heavy Metal Deposits in the Red Sea,* E. T. Degens and D. A. Ross, eds., Springer Verlag, Berlin, pp. 38-58.

Girdler, R. W., 1978, Comparison of the East African Rift System and the Permian Oslo rift, in *Tectonics and Geophysics of Continental Rifts,* I. B. Ramberg and E. R. Neumann, eds., D. Reidel, Dordrecht, pp. 329-345.

Girdler, R. W., and P. Styles, 1974, Two-stage Red Sea floor spreading, *Nature* **247:**7-11.

Girdler, R. W., and P. Styles, 1976, The relevance of magnetic anomalies over the southern Red Sea and Gulf of Aden and Afar, in *Afar Between Continental and Oceanic Rifting,* A. Pilger and A. Rösler, eds., Schweizerbart'sche, Stuttgard, pp. 156-170.

285

Girdler, R. W., and R. Whitmarsh, 1974, Miocene evaporites in Red Sea cores and their relevance to the problems of the width and age of oceanic crust beneath the Red Sea, in *Initial Reports of the Deep Sea Drilling Project,* vol. 23, U.S. Government Printing Office, Washington, D.C., pp. 913–921.

Heybroek, F., 1965, The Red Sea Miocene evaporite basin, in *Salt Basins Around Africa,* W. Q. Kennedy, ed., Inst. of Petroleum, London, pp. 17–40.

Hills, E. S., 1961, Morphotectonics and geomorphological sciences with special reference to Australia, *Geol. Soc. London Quart. Jour.* **117:**77–89.

Holmes, A., 1965, *Principles of Physical Geology,* rev. ed., Thomas Nelson, London, 1288p.

Hume, W. F., 1921, Relations of the northern Red Sea and its associated Gulf areas to the "rift" theory, *Geol. Soc. London Quart. Jour.* **77:**97–101.

Hume, W. F., 1925, *Geology of Egypt,* vol. 1, Egyptian Survey Dept., Cairo, 408p.

Kabbani, F. K., 1970, Geophysical and structural aspects of the central Red Sea rift valley, *Royal Soc. London Philos. Trans.* **A267:**89–97.

Krenkel, E., 1957, Der Graben des Roten Meeres: die Erythreis, in *Geologie Afrikas,* vol. 1, rev. ed., Gebrüder Borntraeger, Berlin, pp. 123–128.

Laughton, A. S., 1966, The Gulf of Aden, *Royal Soc. London Philos. Trans.* **A259:**150–171.

Laughton, A. S., 1970, A new bathymetric chart of the Red Sea, *Royal Soc. London Philos. Trans.* **A267:**21–27.

Laughton, A. S., R. B. Whitmarsh, and M. T. Jones, 1970, The evolution of the Gulf of Aden, *Royal Soc. London Philos, Trans.* **A267:**227–266.

Le Bas, M. J., 1971, Peralkaline volcanism, crustal swelling and rifting, *Nature, Phys. Sci.* **230:**85–87.

Lowell, J. D., and G. J. Genik, 1972, Sea-floor spreading and structural evolution of the southern Red Sea, *Am. Assoc. Petroleum Geologists Bull.* **56:**247–259.

McKenzie, D. P., D. Davies, and P. Molnar, 1970, Plate tectonics of the Red Sea and East Africa, *Nature* **226:**243–248.

Merla, G., E. Abbate, P. Canuti, M. Sagri, and P. Tacconi, 1973, *Geological Map of Ethiopia and Somalia,* scale 1:2m, Geol. Palaeontol. Inst., University of Florence, Florence, Italy.

Milanovsky, E. E., 1972, Continental rift zones: Their arrangement and development, *Tectonophysics* **15:**65–70.

Mohr, P. A., 1962, *The geology of Ethiopia,* University College, Addis Ababa, 268p.

Mohr, P. A., 1970, The Afar triple junction and sea-floor spreading, *Jour. Geophys. Research* **75:**7340–7352.

Mohr, P. A., 1972, Surface structure and plate tectonics of Afar, *Tectonophysics* **15:**3–18.

Montanaro, G., 1941, Foraminifera, posizione stratigrafico e facies di un calcare, *Palaeontographica Italica* **40:**67–75.

Morton, W. H., and R. Black, 1975, Crustal attenuation in Afar, in *Afar Depression of Ethiopia,* A. Pilger and A. Rösler, eds., Schweizerbart'sche, Stuttgart, pp. 55–65.

Neumann, E. R., and I. B. Ramberg, eds., 1978, *Petrology and Geochemistry of Continental Rifts,* D. Reidel, Dordrecht, 290p.

Noy, D. J., 1978, A comparison of magnetic anomalies in the Red Sea and Gulf of Aden, in *Tectonics and Geophysics of Continental Rifts,* I. B. Ramberg and E. R. Neumann, eds., D. Reidel, Dordrecht, pp. 278-287.

Phillips, J. D., and D. A. Ross, 1970, Continuous seismic reflexion profiles, *Royal Soc. London Philos. Trans.* **A267:**145-152.

Picard, L., 1970, On Afro-Arabian graben tectonics, *Geol. Rundschau* **59:**337-381.

Pilger, A., and A. Rösler, eds., 1975, *Afar Depression of Ethiopia,* Schweizerbart'sche, Stuttgart, 416p.

Pilger, A., and A. Rösler, eds., 1976, *Afar Between Continental and Oceanic Rifting,* Schweizerbart'sche, Stuttgart, 196p.

Qureshi, I. R., 1971, Gravity measurements in the North Eastern Sudan and Crustal Structure of the Red Sea, *Royal Astron. Soc. Geophys. Jour.* **24:**119-135.

Ramberg, I. B., and E. R. Neumann, eds., 1978, *Tectonics and Geophysics of Continental Rifts,* D. Reidel, Dordrecht, 444p.

Said, R., 1962, *The Geology of Egypt,* Elsevier, Amsterdam, 377p.

Suess, E., 1904, *The Face of the Earth,* vol. I, H. B. C. Sollas, trans., Clarendon Press, Oxford, 604p.

Swartz, D. H., and D. D. Arden, 1960, Geological history of the Red Sea area, *Am. Assoc. Petroleum Geologists Bull.* **44:**1621-1637.

Sykes, L. R., and M. Landisman, 1964, The seismicity of East Africa, the Gulf of Aden and the Arabian and Red seas, *Seismol. Soc. America Bull.* **54:**1927-1940.

U.S. Geological Survey, 1963, *Geological Map of the Arabian Peninsula,* scale 1:2m, Arabian Oil Co., Saudi Arabia, U.S. Department of State.

Vail, J. R., 1978, Outline of the geology and mineral resources of the Democratic Republic of the Sudan and adjacent areas, *Overseas Geol. Mineral Resour.* **49:**1-68.

Wegener, A. 1966, *The Origin of Continents and Oceans,* J. Biram, trans., Methuen, London, 248p.

Whiteman, A. J., 1968, Formation of the Red Sea depression, *Geol. Mag.* **105:**231-246.

Whiteman, A. J., 1970, The existence of transform faults in the Red Sea depression, *Royal Soc. London Philos. Trans.* **A267:**407-408.

Whiteman, A. J., 1971, *The Geology of the Sudan Republic,* Clarendon Press, Oxford, 290p.

Whitmarsh, R. B., et al., 1974, Red Sea Sites 225-230, in *Initial Reports of the Deep Sea Drilling Project,* vol. 23, R. B. Whitmarsh, O. E. Weser, D. A. Ross, et al., eds., U.S. Government Printing Office, Washington, D.C., pp. 595-812.

19

Reprinted from *Nature* **235**:144–147 (1972)

Tectonic Significance of the Afar (or Danakil) Depression

H. TAZIEFF & J. VARET

CNRS RCP 180—Laboratoire de Petrographie et Volcanologie 91 Orsay

F. BARBERI & G. GIGLIA

CNR—Dipartimento di Scienze della Terra, 53 Via S. Maria, Pisa

The Afar depression results from the separation of the Arabian and Nubian plates, with generation of new oceanic crust. Volcano-tectonic spreading axes form part of broad uplifted structures. The Gulf of Aden and Red Sea rifts constitute one and the same tectonic megastructure; but the distinct characteristics of the East African rift system suggest that it does not pertain to the megastructure.

THE Afar depression is the junction area of three most important tectonic structures: the Red Sea, Gulf of Aden, and East African rift systems (Fig. 1). Recent developments in plate tectonics have drawn attention to this region, the significance of which has been interpreted differently by different authors. Controversy exists about stratigraphy, tectonic trends, tectonic structures, presence or absence of oceanic crust, amount of spreading and so on.

The Afar depression being the only emerged portion of the Red Sea–Gulf of Aden structure and a key region for their interpretation, it is rather surprising that no systematic geological research had been undertaken seriously before 1967. Extensive geological field work has been carried out since that date by our CNRS–CNR research team. The results obtained thus far allow a sound reconstruction and permit an elimination of some of the interpretations formerly proposed.

Evidence of Crustal Separation

The amount of spreading has been interpreted either as corresponding to the total separation between the Red Sea

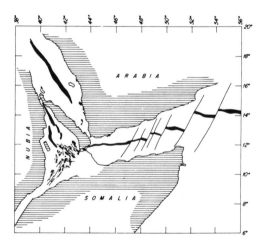

Fig. 1 Location of Afar depression in relation to the Red Sea, Gulf of Aden and East African rift systems. Zones of generation of new oceanic crust are shown in full black. Main faults of the Lake Abhe area are recorded to show the relations between the Afar depression and main Ethiopian rift.

and the Gulf of Aden coastlines[1,2] or as restricted to the width of the median trench[3,4]. The nature of the spreading mechanism has also been questioned[5].

The main tectonic feature of northern and central Afar (north of 11° N, Fig. 2) is a prevailingly NNW–SSE set of distensive faults[6], with typical rift-in-rift structures disposed in an *en échelon* pattern[7,8]. Volcanic rocks of oceanic type characterize the rift-in-rift axes (Fig. 2).

These axial units are built up over NNW–SSE trending fissures. The petrological evolution, from mildly alkalic basalts to peralkaline rhyolites, with numerous intermediate members (such as ferrobasalts and dark trachytes), is clearly related to the volcanological evolution. Differentiation is more developed where more evolved volcanic structures are observed, that is, Erta'Ale[9] or Tat'Ali ranges; in other ranges, such as Alayta[10], N. Loggio or Asal, evolution only reached the dark trachyte member and volcanological characteristics are also of a lower evolutive state. Volcanic rocks of these ranges can be explained in terms of crystal fractionation in low and decreasing fO_2 conditions[11]. Isotopic data confirm the oceanic nature of these rocks, which clearly differ from those located in the volcanic units situated closer to the rift's margins, where crustal influence is observed[10]. All data obtained so far suggest the absence of sialic crust within these axes, which are made up of newly created oceanic rocks.

These Afar axial zones of crustal separation appear at approximately the same latitude as the dying out of the main trough of the Red Sea, close to latitude 15° (Fig. 1); it may thus be concluded that Afar is but an *en échelon* disposed structure pertaining to the main Red Sea system[7,8,10]. This displaced segment of the Red Sea tectonic axis extends over two degrees latitude from Alid to Lake Afrera (13°), over the Salt Plain and the Erta'Ale range. In this segment the amount of crustal separation increases from north–southward, while a corresponding reduction is observed within the Red Sea south from latitude 15°.

South of Lake Afrera the axis of the depression is split into two *en échelon* volcano-tectonic units, both with a Red Sea trend: Alayta and Tat'Ali ranges (Fig. 2). To the south, the geology becomes more complicated: the axes of crustal separation appear less evident and seem to be fragmented into several different short rift-in-rift structures coexisting at the same latitude (11°–12°), as for N. Loggia, Dama Ale, Manda and Asal.

Detailed studies show that important uplifts are affecting Afar rift-in-rift axes. Actual uprising movements sometimes predominate even in collapsed structures. This is demonstrated by the horsts observed along the central axis of northern Afar (Alid-Erta'Ale)[9,12] as well as by upper Pleistocene and Holocene submarine volcanoes of northern and central Afar, which are now exundated and have even reached altitudes several hundred metres above sea level (Mount Asmara)[13,14]; in French Territory, Marinelli (personal communication) has pointed out an obviously arched structure in the Asal graben[15];

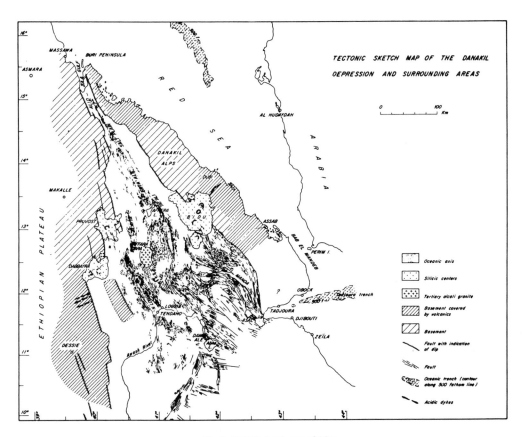

Fig. 2 Tectonic sketch map of Afar.

and the axial trench of the latter is spectacularly uplifted, at a rate that has been calculated on ^{14}C evidence to be of about 90 m during the past $5,800 \pm 150$ yr[12].

These apparently down-faulted grabens, where uprises are actually observed, are marked by normal distensional faults and gaping fissures which allow subcrustal magma to well up. It may accordingly be considered that this arching up is related with both oceanic volcanism and observed separation of tectonic blocks.

Red Sea–Gulf of Aden Megastructure

Magnetic and gravity surveys in the western part of the Gulf of Aden show that the main negative central magnetic anomaly does not pass from the Gulf of Aden into the Red Sea across the straits of Bab-el-Mandeb, but through the Gulf of Tadjoura and the Afar depression. Moreover, recent seismological profiles conducted in the Gulf of Tadjoura and the Asal area have shown the presence of abnormal upper mantle, probably of mid-oceanic ridge nature, similar to that found in the Gulf of Aden[16]. These geophysical data confirm that the main axis of regional expansion actually passes through the Afar depression.

East from longitude 47° E, the Gulf of Aden tectonic trends, as demonstrated by hypsometric and magnetic surveys[17,18], are chiefly ESE–WNW or SE–NW. The Somali highlands bordering the Gulf of Aden to the south exhibit a largely prevailing SE–NW oriented fault system[19]. West from the E–W oriented trench of Tadjoura, these "Sheba" trends resume in central Afar: the magnetic anomalies veer from E–W in southern Afar, to SE–NW in central Afar, and eventually to the SSE–NNW in northern Afar[20]. Tectonic field evidence suggests either that the Sheba trend criss-crosses the NNW Red Sea trend (in the Kyu-Enkebba region or the area north from the Ghoubbet-al-Kharab), or that it progressively bends northward until it merges imperceptibly with the Red Sea direction; this one itself deviates to a NW–SE direction in central Afar[12]. These facts strongly suggest that the Gulf of Aden and the Red Sea rift systems constitute one single megastructure, the inflexion area lying within the Afar depression.

Sheba–Red Sea Megastructure Junction with East African Rift

The NNE–SSW tectonic trend of the African rift system is only observed in the south-western corner of the Afar depression. The SSW–NNE tensional faults of the East African rift system disappear almost immediately once they have crossed the first lineament of the Red Sea–Gulf of Aden megastructure; they are not present in Eastern Afar. Geological evidence is supported by geophysical data which show that neither magnetic nor gravity anomalies exist following NNE trending (that is, African rift trend) in either central or northern Afar. The "Wonji fault belt", allegedly continuing throughout Afar up to the Red Sea[22], disappears north-west of Lake Abhe in our opinion. Consequently, we feel that there is no need to consider further the Afar depression as a "funnelling out" of the East African rift system[1,21,23,24].

Rotation of Danakil Horst

The separation of the Danakil horst from the Nubian block was accompanied by a rotation of the Danakil horst. This can be deduced from a series of arguments such as a study of the faulting directions of the scarps, a comparison of the Danakil block and the Ethiopian plateau geology, the progressive increase of the volcanicity from north southwards in northern and central Afar, the splitting of the single oceanic axial graben into two parallel ones south from 13° N, the presence of crustal rocks splinters in between the oceanic rift-in-rift structures, and palaeomagnetic data[21]. Geological evidence points to a rotation angle of $18° \pm 10°$ with a pole located close to Alid volcano (Fig. 3).

Patterns of the scarps on either side of the depression (Fig. 3) differ: while the western border of the Danakil horst is comparatively straight with a NW–SE direction, the Ethiopian scarp is not simply a N–S oriented tectonic feature, as suggested by many published maps, but actually consists of a series of en échelon normal faults with the characteristic NNW–SSE Red Sea direction. This is the actual trend of the depression's western limits, and consequently reduces by a matter of 12° to 15° the angle of the apparent relative rotation of the Danakil horst, as it might be erroneously deduced from maps showing an alleged N–S direction for the Ethiopian scarp[23,24].

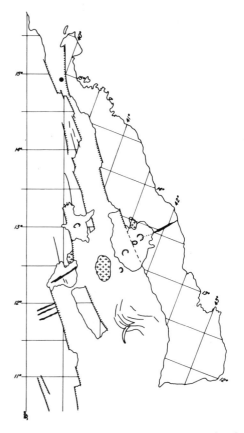

Fig. 3 Position of Danakil block after a clockwise rotation of 18°, with pole located at Alid volcano (full point).

These tectonic features account for continental crust splinters located between the en échelon disposed oceanic axes. The Danakil block can be considered to be such a splinter, and the largest one, because it is located between the main trough of the Red Sea and the longest Afar échelon; smaller splinters of Ethiopian plateau have been mapped closer to the Ethiopian scarp, and the Miocene Affara-Dara alkali granitic body and the surrounding rocks may be considered the exposed portion of a somewhat larger sliver[25].

The presence of still hidden intrusive magmatic bodies, on the other hand, may be responsible for tectonic discrepancies, such as the curvilinear grabens of central Afar. These features have been considered as rotational structures[26], but this hypothesis is in apparent contradiction with field observations, which show only tensional fissures, to the exclusion of any shear-stress features.

Trans-rift Lineaments

Transform or transcurrent faulting could be expected in an area of oceanic nature, and some authors[22] have described such trans-rift lineaments in the Afar depression. Unfortunately, our observations are in formal contradiction with this, as none of these mapped transcurrent faults has been actually observed in the field nor detected on aerial photographs. A few volcano-tectonic features, subperpendicular to the Red Sea trend, are present in both the Ethiopian scarp (Dabbayra) and the Danakil horst (Dubbi)[25]. Being definitely tensional, they cannot be mistaken for shear faults. Moreover, they do not affect the depression proper, and possibly correspond to pre-rift crustal tectonics; they fit perfectly, both in orientation and position, a reconstitution assuming an angle of rotation of about 18 degrees for the Danakil horst (Fig. 3). These tensional fault patterns have been rejuvenated by recent tectonics. The variation of the direction of faulting of the Danakil block north and south of Bidu-Dubbi transverse weak zone may be interpreted assuming a different angle of rotation north and south from this line, the rotation angle of the southern part being a little wider (20°) than the northern one (18°). Therefore, contrary to widespread opinions[22,23], one cannot consider this Bidu-Dubbi structure as the continuation of the East African rift system (Wonji fault belt) further.

Conclusions

The Gulf of Aden and Red Sea rift systems do make up one single megastructure, the tectonics, magmatism and significance of which are different from those of the East African rift.

Crustal separation zones (oceanic rifts) of the Red Sea and Gulf of Aden are connected by the axial volcano-tectonic rift-in-rift structure of the Afar depression (Fig. 1).

The Danakil continental horst is supposed to have moved away from the Nubian plate with a dextral rotation of 18° ± 10°. The exact amount of spreading cannot be deduced from geological basis only, and supplementary geophysical data are required to solve the problem quantitatively and should be required to any sound interpretation of the Afro-Arabian area in terms of plate tectonic mechanisms.

We thank the Ethiopian and French Territorial Authorities for their help. The Shell Company provided fuel for helicopter work.

Received September 28, 1971.

[1] McKenzie, D. P., Davies, D., and Molnar, P., *Nature*, **226**, 243 (1970).
[2] Allan, T. D., *Phil. Trans. Roy. Soc.*, A, **267**, 153 (1970).
[3] Laughton, A. S., Whitmarsh, R. B., and Jones, M. T., *Phil. Trans. Roy. Soc.*, A, **267**, 227 (1970).
[4] Freund, R., *Nature*, **228**, 453 (1970).
[5] Mohr, P. A., *J. Geophys. Res.*, **75**, 7340 (1970); *Nature*, **228**, 547 (1970).
[6] Tazieff, H., *Bull. Soc. Géol. France*, **10**, 468 (1968).
[7] Gibson, I. L., and Tazieff, H., *Phil. Trans. Roy. Soc.*, A, **267**, 331 (1970).
[8] Tazieff, H., *CR Acad. Sci.*, **268**, 2030 (1969).
[9] Barberi, F., and Varet, J., *Bull. Volc.*, **34**, 848 (1970).
[10] Barberi, F., Borsi, S., Ferrara, G., Marinelli, G., and Varet, J., *Phil. Trans. Roy. Soc.*, A, **267**, 293 (1970).
[11] Treuil, M., Varet, J., Billhot, M., and Barberi, F., *Contr. Mineral. and Petrol.*, **30**, 84 (1971).
[12] Tazieff, H., *CR Acad. Sci.*, **272**, 1055 (1971).
[13] Bonatti, E., and Tazieff, H., *Science*, **168**, 1087 (1970).
[14] Tazieff, H., *Geol. Rundschau* (in the press).
[15] Stieltjes, L., *BRGM Paris* (1970).
[16] Laughton, A. S., and Tramontini, C., *Tectonophysics*, **8**, 359 (1969).
[17] Roberts, D. G., and Whitmarsh, R. B., *Earth Planetary Sci. Lett.*, **5**, 253 (1969).
[18] Peter, G., and DeWald, O. E., *Geol. Soc. Amer. Bull.*, **80**, 2313 (1969).
[19] Azzaroli, A., and Fois, V., *Proceedings Twenty-second Intern. Geol. Cong.*, **4**, 293 (1964).
[20] Girdler, R. W., *Phil. Trans. Roy. Soc.*, A, **267**, 359 (1970).
[21] Burek, P. J., *Trans. Amer. Geophys. Union*, **51**, 271 (1970).
[22] Mohr, P. A., *Nature*, **218**, 938 (1968); Baker, B. M., *Phil. Trans. Roy. Soc.*, A, **267**, 383 (1970).
[23] Gouin, P., *Phil. Trans. Roy. Soc.*, A, **267**, 339 (1970).
[24] Gass, I. G., *Phil. Trans. Roy. Soc.*, A, **276**, 369 (1970).
[25] *Geological Map of Danakil Depression*, 1/500,000, CNRS–CNR, 1 (1971).
[26] Mohr, P. A., *Bull. Geophys. Observ. Addis Ababa*, **12**, 1 (1969).

Part VIII

KEY MAPS AND SECTIONS

20A
RIFT VALLEYS AND SYSTEMS OF EASTERN AFRICA
Key Map and Sections

A. M. Quennell

This map was prepared expressly for this Benchmark volume.

Cenozoic: A–A = Red Sea, southern section; B–B = Gulf of Aden; C–C = The Afar; D–D = Ethiopian system (and Wonji fault belt); E–E = Gregory system, Kenya, and north Tanzania; F–F = Eastern system; G–G = Western system, Albert — Kivu zone; H–H = Western system, Tanganyika zone; J–J = Western system, Rukwa—Malawi zone; K–K = Upemba.

Pre-Cenozoic: a–a = Lukuga; b–b = Songwe; c–c = Luhombero— Kidodi; d–d = Ruhuhu; e–e = Luangwa; f–f = Zambezi; g–g = Shire— Urema.

Sections: A–A = Gulf of Aden; B–B = Red Sea, southern section; C–C = Ethiopian rift valley (Lake Abaya); D–D = Gregory rift system, southern Kenya; E–E = Gregory rift system, southern zone; F–F = Western rift system, Lake Albert zone; G–G = Western rift system, Tanganyika, and Rukwa rift valleys and Ufipa plateau (horst); H–H = Western rift system, Malawi (north).

Geographic features: a = Indian Ocean; b = Arabia; c = Africa; d = Horn of Africa; e = Gulf of Tadjours; f = Danakil Horst; g = Lake Abaya; h = Lake Rudolf; i = Lake Naivasha; j = Lake Victoria; k = Kavirondo Gulf; l = Kedong scarp; m = Lake Eyasi; n = Lake Balangida; o = Nile; p = Lake Albert; q = Mt. Ruwenzori; r = Lake Kivu; s = Lake Tanganyika; t = Congo River; u = Ufipa Plateau; v = Lake Rukwa; w = Mbeya Mountain; x = Lake Malawi; y = Shire River; z = Zambezi River; aa = Alula-Fartak "trench."

Symbols: 1 = Cenozoic faults; 2 = Pre-Cenozoic faults and sediments; 3 = Post-faulting sediments in sections; 4 = Volcanics and intrusives in sections.

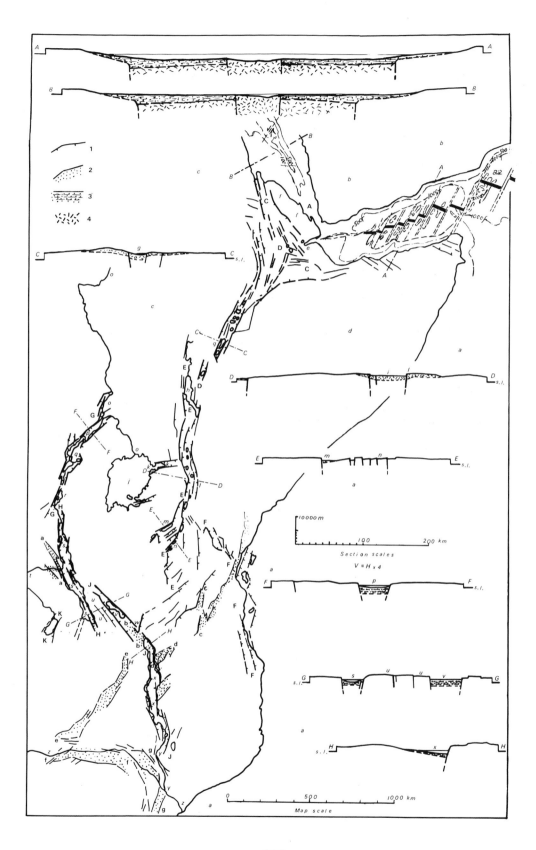

Section scales

V = H × 4

Map scale

20B

RIFT VALLEYS AND SYSTEMS OF THE MIDDLE EAST
Key Map and Sections

A. M. Quennell

This map was prepared expressly for this Benchmark volume.

Cenozoic: A–B = Syrian zone; B–C = Lebanon-Bekaa zone; C–D = Jordan-Dead Sea-Wadi Araba zone; D–E = Aqaba zone; F–G = Gulf of Suez (Clysmic) rift valley; G–H = Red Sea, northern section.

Sections: A–A = Yammouné Fault and Bekaa Valley; *B–B* = Jordan Valley; C–C = Dead Sea; *D–D* = Wadi Araba; *E–E* = Gulf of Aqaba; *F–F* = Gulf of Suez; *G–G* = Red Sea, northern section (shortened by 50 k).

Structural features: J = Bitlis—Zagros overthrust; K= East Anatolian fault; L = Amanos fault; M = Gharb depression (rhombochasm); N = Yammouné fault; O = Bekaa valley (synform); P = Palmyra folds; Q = Sirhan alignment; R = Emeq—Gilboa graben; S = Jordan Valley (synform); T = Dead Sea (rhombochasm); U = Araba-Aqaba Rift; V = Shadwan Island submarine slope.

Geographic features; a = Mediterranean Sea; b = Cyprus; c = Border Folds; d = Toros fold-belt; e = Lebanon Range; f = Anti-Lebanon Range; g = Beyrouth; h = Haifa; j = Lake Tiberias; k = Dead Sea; m = Sinai; n = Arabia; o = Gulf of Aqaba; p = Gulf of Suez; q = Red Sea; r = Africa.

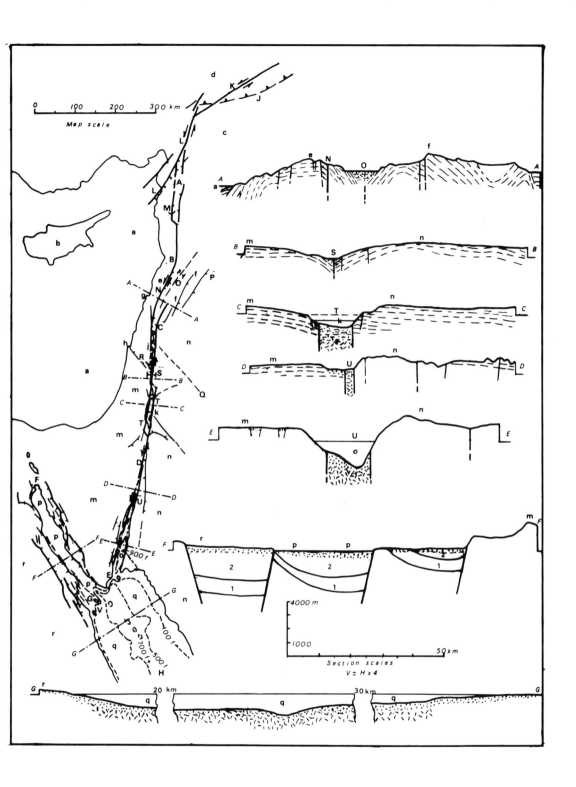

Part IX

GULF OF SUEZ

Editor's Comments
on Papers 21, 22, and 23

21A SAID
Excerpt from *Tectonic Framework of Egypt*

21B SAID
Excerpts from *Geology of the Gulf of Suez Region*

22 ROBSON
Excerpts from *The Structure of the Gulf of Suez (Clysmic) Rift, with Special Reference to the Eastern Side.*

23 THIÉBAUD and ROBSON
A Discussion of the Structure and Evolution of the Red Sea and the Nature of the Red Sea, Gulf of Aden and Ethiopia Rift Junction

Papers 21A and 21B are excerpts from the standard work on the geology of Egypt by Said. The Gulf of Suez is described as a taphro-geosyncline, an active zone of subsidence bordered by normal faults from Paleozoic times, in which, Said states, it differs from the Red Sea, which took shape from mid-Tertiary times. However, as stated above, a narrow graben, an embryonic Red Sea, may have initially extended the Gulf of Suez southward (Whiteman, 1968).

The stratigraphic history is related to the structural history in another chapter of *Geology of Egypt*. The tectonic map, Figure 25, together with the diagrammatic sections, Nos. 1, 8 and 9, and the excerpted text, outline the structure of the Gulf of Suez and of Aqaba.

An account contrasting in important matters is given by Robson in Paper 22, and a summary of the views of Thiebaud and Robson is the subject of Paper 23. Thiebaud and Robson claim that Paleozoic to Tertiary sediments were deposited uniformly over the entire region and not within a confined trough. The older rocks were infaulted, not infilled (note that this situation also applies to the Rhinegraben). Briefly, the Cenozoic rift faulting began in the Oligocene, with the disturbance of a peneplain surface that had been cut across all earlier formations, whether these had been deposited or infaulted within troughs or graben (see also Quennell, 1958). In Paper 22, Robson makes much of supposed parallelism of Cenozoic faulting with Precambrian trends as read from satellite photography, but this is not completely accepted.

References to the geology and structure of the Gulf of Suez, in addition to Said (Papers 21A and 21B), include some already cited for the Red Sea. For submarine topography Coleman (1973) and Laughton (1970) are not sufficiently detailed, but Admiralty Chart (*Suez to the Brothers*) reveals the even floor (20 fathoms in the north and 45 fathoms in the south) and the descent to the floor of the Red Sea (700 fathoms) across the mouth of the gulf. For geology and structural history, an additional reference is Tromp (1950), which deals chiefly with the gulf. Abdel-Gawad (1970) is a valuable review, which contains some controversial matter (see above). Schürmann (1966, 1974) deals with other aspects of the geology. Picard (1970) and Dubertret (1970) mention the gulf briefly.

Information, largely the result of petroleum exploration drilling, on the geological history and of specific interest is given in Heybroek (1965); the deposition of evaporites is linked with that in the central and southern Red Sea.

A point made in Paper 23 is that there is no evidence of either any lateral extension beyond that resulting from normal faulting (contrasting with the Red Sea), or any major horizontal movement (strike slip) on the rift faults. On the latter point, Van de Ploeg (1953) and Abdel-Gawad (1970) and others do favor some strike-slip movement. However, on these two matters, it is doubtful if solutions can be determined on either field or subsurface geological observation alone; recourse must be had to geophysical, seismic, and probably paleomagnetic records and to the application of plate-tectonic theory in identifying plates and their poles of rotation (McKenzie et al., 1970).

The paper by Garfunkel and Barton (1977) was not available when the excerpts for Papers 21A, 21B, 22, and 23 were selected. It is a comprehensive account of the geology and structure of the Gulf of Suez. Generally it reinforces the information and views expressed in the excerpted papers. Two more papers by Thiebaud and Robson (1979, 1981) have appeared recently.

REFERENCES

Abdel-Gawad, M., 1970, The Gulf of Suez: A brief review of stratigraphy and structure, *Royal Soc. London Philos. Trans.* **A267:**41–48.

Coleman, R. G., 1973, Geologic Map of the Red Sea, scale 1:2m, in *Initial; Reports of the Deep Sea Drilling Project,* vol. 23, U.S. Government Printing Office, Washington, D.C. p. 1180.

Dubertret, L., 1970, Review of structural geology of the Red Sea and surrounding areas, *Royal Soc. London Philos. Trans.* **A267**:9-20.

Garfunkel, Z., and Y. Bartov, 1977, The tectonics of the Suez rift, *Geological Survey of Israel, Bull.* **1**:1-44.

Heybroek, F., 1965, The Red Sea Miocene evaporite basin, in *Salt Basins Around Africa,* W. Q. Kennedy, ed., Inst. of Petroleum, London, pp. 17-40.

Laughton, A. S., 1970, A new bathymetric chart of the Red Sea, *Royal Soc. London Philos. Trans.* **A267**:21.

McKenzie, D. P., D. Davies, and P. Molnar, 1970, Plate tectonics of the Red Sea and East Africa, *Nature* **226**:243-248.

Picard, L., 1970, On Afro-Arabian graben tectonics, *Geol. Rundschau* **59**:337-381.

Quennell, A. M., 1958, The structural and geomorphic evolution of the Dead Sea Rift, *Geol. Soc. London, Quart. Jour.* **114**:1-24.

Schürmann, H. M. E., 1966, *The Precambrian Along the Gulf of Suez and Northern Part of the Red Sea,* E. J. Brill, Leiden, 404p.

Schürmann, H. M. E., 1974, *The Precambrian of North Africa,* E. J. Brill, Leiden, 351p.

Thiebaud, C. E., and D. A. Robson, 1979, The geology of the area between Wadi Wardan and Wadi Gharandal, East Clysmic rift, Sinai, Egypt, *Jour. Pet. Geol.* **1**:63-75.

Thiebaud, C. E., and D. A. Robson, 1981, The geology of the Asla oilfield, Western Sinai, Egypt, *Jour. Pet. Geol.* **4**:77-87.

Tromp, S. W., 1950, The age and origin of the Red Sea graben, *Geol. Mag.* **87**:385-392.

Van de Ploeg, P., 1953, Egypt, in *The World's Oilfields: The Eastern Hemisphere,* V. C. Illing, ed., Oxford University Press, Oxford, pp. 151-157.

Whiteman, A. J., 1968, Formation of the Red Sea depression, *Geol. Mag.* **105**:231-246.

21A

TECTONIC FRAMEWORK OF EGYPT

R. Said

[*Editor's Note:* In the original, material precedes this excerpt.]

FAULTING

Many of these folds, particularly in the north, are completely or partially bounded by impressive faults. These often lie parallel to the folds. Tensional adjustment in the south, however, is less noticeable on the surface although several north-south (East African) trending faults have been recorded and many are assumed to exist not far beneath.

North–south (East African) faulting

Among the numerous fault systems known in the Stable Shelf the north–south faults are the most noticeable. The north-south fault of Kharga oasis runs along the center of this depression for almost 60 km and seems to extend even farther beyond (Beadnell, 1909). In the area to the east of the Nile, at Aswan, Attia (1955) describes several faults running mainly in a north–south direction. Beadnell (1900) notes the difference in elevation between the formations of the Libyan plateau to the south of Garra and Kurkur and the formations exposed along the Nile in the immediate vicinity. He ascribes this difference to an inferred fault running in a north–south direction more or less parallel to the Nile. To the north there are slipped Eocene blocks tilted toward the cliffs that border the Nile. These extend from south of Luxor in Gebel Rakhamiya northward past the shattered little hill of El-Qurn at the mouth of Wadi Matuli and further through Gebel Serai to Gebel Abu Had at the mouth of Wadi Qena. This would lead one to infer a great north–south fault along the Nile Valley extending to Wadi Qena. Toward the north, at Gebel southern Galala, and almost at the same longitude of this last-mentioned fault, a series of north–south faults dissect this mountain into a complex pattern of grabens and horsts along the planes of which many basalt dikes and flows are noted. The area separating these two sets of faults (that of southern Galala and that of Luxor – Gebel Abu Had) is occupied mainly by Wadi Qena. This is an ancient tributary of the Nile which runs in a north–south direction. It may thus be inferred that Wadi Qena was excavated along a fault line which connected the faults of southern Galala with those at the mouth of Wadi Qena and Luxor. The many basalt occurrences and the alignment of basement rocks substantiate this hypothesis. It is perhaps along this line that the mechanical boundary between the Red Sea graben frame and the more stable portion of the Stable Shelf may be sought. Yallouze and Knetsch (1954) suggest that this mechanical boundary may be found in a northwest line that runs from Baharia to Wadi Natash and Foul Bay; but while parts of this line can be shown in any structural analysis a great deal of inference is needed to visualize its path and extent.

The north–south (East African) trend is probably that of the old grain of Egypt. It is evident as an old trend in the dike directions, in the strike and dip of foliation, in the trend of faults, and in the schistosity in many of the areas of the foundation crystalline rocks recently mapped by the Geological Survey. Later movements with this alignment may have determined many deflections of present and ancient shore lines. Many deflections of the present day Gulf of Suez and Red Sea have this trend. As shown by bathymetric contours, the submarine slope at the entrance of the Gulf of Suez also has this trend. A similarly-trending fault seems to cut across the Gulf of Suez, and produces the outcropping of basement rocks at Gebel Araba in Sinai and Gebel Zeit and Esh Mellaha ranges in the Eastern Desert. The extension of this latter fault inland to Abu Shaar plateau (near Hurgada) may be inferred from the presence of tilted coral reefs and from the continuation of the fringe of limestone at the base of the Red Sea hills to the south of Abu Shaar plateau.

Northwest (Erythrean or African) faulting

Even more spectacular than the north–south trend is the Erythrean or African (northwest–southeast) trend cutting through the Stable Shelf in a most pronounced way. Although this trend seems to be old, its greatest manifestations are obvious in mid-Tertiary time when it apparently governed the extent and direction of the Red Sea and the Gulf of Suez and many of the topographic features of present-day Egypt. Ball (1909, 1920) suggests the idea that the Gulf of Suez was due to erosion. Most authorities today consider the Gulf to be a part of the great African Rift Valley whose tectonic origin is evident: Walther (1888), Blanckenhorn (1893), Hume (1901), Gregory (1920, 1921, 1923), etc. Gregory formulates the hypothesis that a long wedge-shaped block, tapering downward, sank between normal boundary faults as a result of reduction of lateral pressure on an original arch-like structure. This sunken wedge displaced at depth material which broke out as volcanoes along the cracks.

Since the publication of the classical papers by Suess and Gregory, controversy as to the mechanics of the formation of this great graben has continued – a discussion to which many eminent geologists have subscribed. Willis (1936) advocates that the Rift Valley was produced by deep-seated compression. This hypothesis would explain: (a) the fact that in many areas volcanic activity had ceased altogether while rift movements were still in progress; and (b) the fact that the floor of the Rift Valley is shown by a gravity survey to have abnormally low values (Bullard, 1936). The drawback of this theory is that deeply-seated lateral compression would have produced boundary faults in the nature of steep upthrusts, a conclusion which is not corroborated by field evidence.

Bogolepov (1930) holds that the Red Sea is not a rift, but a broad crack resulting from the 'rotation rift' of the upper layers of the lithosphere. Shalem (1954) terms the depression caused by the moving apart of crustal blocks 'a paar'. Swartz and Arden (1960) seem to subscribe to this view. These authors suggest that such movements took place among four blocks that bounded the Red Sea: Block I, northeast Africa, west of Suez and the Red Sea and north of the Ethiopian Rift Valley; Block II, the Arabian Peninsula; Block III, the Sinai Peninsula; and Block IV, the horn of Africa

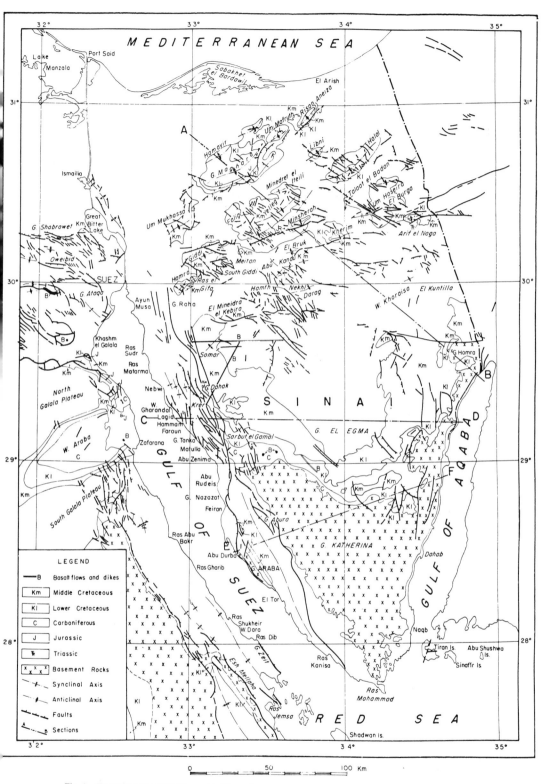

Fig. 4. Tectonic map of Sinai (assembled from the maps prepared by the Survey officers: Ball, Barron, Beadnell, Hume, Madgwick, Moon, Sadek, and the oil company geologists: Bowles, Iskandar, Jones, Shata, and Sourial).

305

east of the rift valleys. Further discussion of the tectonics of the Gulf of Suez and the Red Sea will be found in other chapters of this book.

Whatever the mechanics of the origin of the Red Sea may be, the fact remains that normal faults control almost the entirety of its coasts. In many areas there occur several step faults affecting different formations, a phenomenon which may suggest that the formation of these grabens must have continued for a long time after their inception in lower Tertiary time. African-trending faults bound the east-facing scarp of the Red Sea range as well as that of the smaller fault blocks of the Esh-Mellaha range and Gebel Zeit. They also bound the west-facing scarp of the south Sinai mountain ranges and the smaller range of Gebel Araba, as well as the escarpments of Gebel Ataqa and the Galalas. Faults of similar trend (African), though less spectacular, are recorded from all over the country: in the Helwan-Beni Suef reach of the Nile Valley region, in the Wadi el-Rayan region to the south of Fayum, and in El-Hadahid region to its west. They are also inferred along the western escarpments of the Red Sea range further south. Faults of this trend seem to have controlled the path of the Nile Valley between Qena and Assiut. Within this part of the Nile Valley, these faults are reported in Akhmim and are inferred in other localities from the presence of slipped limestone blocks.

East–west (Tethyan) faulting

Faults having an east-west direction (Tethyan) are also known in the Stable Shelf. Notable among these is the dike of Raqabet el-Naam cutting across central Sinai, and the major fault bounding Gebel Homra in east central Sinai. To this trend belong also many of the faults of the Cairo–Suez road, and they determine the fault-block mountains and horsts of this area, such as Gebel Oweibid, Gafra, Anqabia and Nasuri. Fault movements along this direction occurred as early as the middle Cretaceous and the same trend seems to have continued more intensively during later time.

Northeast–southwest (Aualitic) faults

Aualitic faults are notably developed around the Gulf of Aqaba rift. Though less pronounced, they are recorded from all around the Stable Shelf and bound many of the folds in its northern part. They also govern many of the Red Sea and Nile deflections and traverse Baharia oasis. The movement is probably as old as the Erythrean movement.

In conclusion, it may be said that the trend of most of the structures of the Stable Shelf is normal to the axis of the geanticlinal welt and follows to a large extent the axis of the marginal trough developed along the craton. This trend is probably more notable in the south where most of the structures are either east African, Erythrean, or more rarely Tethyan. As one proceeds away from the craton towards the Unstable Shelf, some aualitic elements begin to be of significance; most of the domes and anticlines recorded in the north are aualitic.

III. GULF OF SUEZ TAPHROGEOSYNCLINE

Situated within the stable belt of Egypt the Gulf of Suez has been an active zone of subsidence throughout its geological history. Great thicknesses of sediments accumulated in this trough, and practically all the stages from the Paleozoic to the Recent are represented.

306

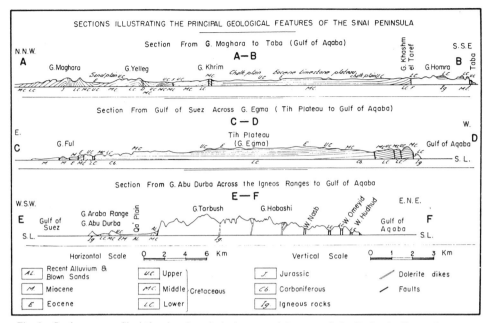

Fig. 5. Sections across Sinai showing the principal structural features of the Peninsula (for section paths see Fig. 4) (after Sadek, 1928).

In these respects the Gulf of Suez is unique. It differs not only from other stable areas but also from the Red Sea and the other parts of the great Rift Valley of which it is to-day a part. These have a comparatively younger history having taken shape only in mid-Tertiary times.

The movements that affected this Gulf were tensional. The Gulf is bordered by, and made up of, a large number of blocks that were continuously rising and sinking with varying magnitudes and intensities. The rectilinearity of these blocks and the fact that they show no warping or tangential effects suggest that they are major fractures due primarily to vertical movements. That these blocks have been active since at least the Paleozoic, as can be deduced from a study of the paleogeographic reconstructions and the stratigraphic sections exposed in the different blocks, may support Picard (1943), Tromp (1950) and Henson (1951) in their conclusions that these vertical movements are probably due to infracrustal fault movements. In fact, normal faults cut the basement and other older formations in many parts of the Gulf.

[*Editor's Note:* In the original, material follows this excerpt. Only references cited in this excerpt are reprinted here.]

REFERENCES

Attia, M. I., 1955. *Topography, geology and iron-ore deposits of the district east of Aswan.* Geol. Survey Egypt, Cairo, 262pp.

Ball, J., 1901. On the origin of the Nile Valley and the Gulf of Suez. *Cairo Sci. J.* **3:**250-252.

Ball, J., 1920 [Reference omitted in original publication.]

Beadnell, H. J. L., 1900. The geological survey of Egypt. *Geol. Mag.* **7**(Decade 4):46-48.

Beadnell, H. J. L., 1909. *An Egyptian Oasis: An account of the oasis of Kharga in the Libyan Desert.* Murray, London, 248pp.

Blanckenhorn, M., 1893. Die Sturkturlinien Syriens und des Roten Meeres. Ein Geotektonische Studie. *Festschr. Ferd. Richtofen,* Berlin, pp. 115-180.

Bogolepov, M., 1930. Die Dehnung der Lithosphäre. *Z. deutsch. geol. Ges.,* **82:**206-228.

Bullard, E. C., 1936. Gravity measurements in East Africa. *Phil. Trans. Roy. Soc. London, Ser. A.,* **235:**445-531.

Gregory, J. W., 1920, The African Rift Valleys. *Geograph. J.,* **56:**13-47; 327-328.

Gregory, J. W., 1921. *The Rift Valleys and geology of East Africa. An account of the origin and history of the Rift Valleys of East Africa and the relation to contemporary earth movements which transformed the geology of the world.* Seeley Service and Co., London, 479pp.

Gregory, J. W., 1923. The structure of the great Rift Valley. *Nature,* **112:**514-516.

Henson, F. R. S., 1951. Observations on the geology and petroleum occurrences in the Middle East. *Proc. Third World Petrol. Congr., Sect. I,* E. J. Brill, Leiden, pp. 118-140.

Hume, W. F., 1901. Sur les 'Rift valleys' de l'est du Sinai. *Compt. rend. 8e congr. géol. intern. Paris, 1900,* fasc. **2:**900-912.

Picard, L., 1943. Structure and evolution of Palestine. *Bull. Geol. Dept. Hebrew Univ. Jerusalem,* 4, No. 2-4: 134pp.

Shalem, N., 1954. The Red Sea and the Erythrean disturbances. *Compt. rend. 19e Congr. geol. intern., Algiers, 1952,* Sect. 15, pt. **17:**223-231.

Swartz, D. H., and Arden, D. D., 1960. Geologic history of Red Sea area. *Bull. Am. Assoc. Petrol. Geologists,* **44:**1621-1637.

Tromp, S. W., 1950. The age and origin of the Red Sea graben. *Geol. Mag.* **87:**385-392.

Walther, J. K., 1888. Die Korallenriffe der Sinaihalbinsel: Geologische und Biologische Beobachtungen. *Abhandl. sächs. Akad. Wiss. Leipzig, Math-Naturw. K1.,* **14:**439-505.

Willis, B., 1936. East African plateaus and Rift Valleys. *Carnegie Inst. Wash. Publ.* 470.

Yallouze, M., and Knetsch, G., 1954. Linear structures in and around the Nile Basin. *Bull. soc. géograph. Égypte,* **27:**153-207.

21B

GEOLOGY OF GULF OF SUEZ REGION

R. Said

The Gulf of Suez region forms a distinct structural unit that was involved throughout time in movements that brought it under the sea for almost the entire length of its geological history. In addition to the great Tertiary movements that brought the Gulf into its present shape the Gulf area was subjected to Paleozoic and Mesozoic movements that led to the accumulation of a great thickness of sediments in this continuously subsiding area.

Schürmann (1949), Said and Shukri (1955), Shata (1955) and Kostandi (1959) give evidence for the subsidence of the Gulf during Carboniferous time. Said and Shukri (1955) note that the trend of the reconstructed Carboniferous shore line may be a rejuvenation of a similar trend known in the foundation crystalline rocks. Kostandi's paleogeographic reconstructions of the Mesozoic as well as the distribution of the Cenozoic sediments show that the Gulf of Suez constituted a zone of subsidence throughout these eras.

The movements that shaped this region were tensional affecting at any one time the whole or parts of the region; the basin is essentially a taphrogeosyncline.

There is indication that the different parts of the Gulf had different geological histories. The succession, facies changes and the relationship of the different blocks that border or build the Gulf are quite variable, so much so that no single area in the Gulf can be wholly representative of the stratigraphy or the structure of the entire region. In fact, the Gulf can be looked upon as a region composed of a large number of blocks that were continuously rising and sinking at different times and with different magnitudes and intensities on their sides.

Viewed as a whole, it can be said that the formation of the Gulf started toward its northwestern side. Bubnoff diagrams as well as isopachous contours show marked thinning of the sediments of all ages toward the southeast, indicating that the center of the basin may be toward the northwest in the vicinity of Ayun Musa. The blocks that border the Gulf on its west were probably more active in the earlier part of geological history than those that border it on its east; and the Gulf remained, in all probability, as a half-graben for part of its history. The eastern blocks became exceptionally active in later time. In fact, there is evidence that the eastern blocks are still subsiding to this day at a rate which far supercedes that of their counterparts on the western side.

The limits of this basin that has been continuously flooded, with minor intervals of erosion, since the Paleozoic, cannot be exactly defined; but paleogeographic reconstructions show that this basin did not extend farther south than Hurgada. Only during the Miocene did this basin extend to the south to form part of the great Red Sea rift.

Viewed as such, the Gulf extends from latitude 30° north to latitude 27° north, takes a general northwest direction, and measures about 350 km from end to end. The northwest end of the depression extends from the foot of the Ataqa scarp across the Suez canal to Gebel Raha plateau in Sinai. Its southeast end extends from the southern tip of the Sinai Peninsula across Hurgada to Abu Shaar plateau. A description of the topography of the Gulf is given in Sadek (1959) [1].

REMARKS ON REGIONAL GEOLOGY AND MAJOR FRACTURE SYSTEMS

The Gulf of Suez may be viewed as a great elongated depression separating the massifs of central Sinai from those forming the backbone of the Eastern Desert. While the center of this depression is occupied by the waters of the Gulf, the comparatively low country on either side is chiefly made up of rocks of Miocene and younger age. The relief is broken by elongated ridges, usually of fair heights, composed of Eocene or Cretaceous strata with occasional outcrops of basement rocks. Outside the depression the rocks forming the high ranges are crystalline rocks in the south followed to the north by Paleozoic to Cenozoic sediments. Fig. 20, 21 and 22 give the areal geology of three selected regions from the Gulf of Suez. Fig. 20 shows the areal geology of the western scarp and the coastal plain between Gebel Ataqa and Gebel northern Galala (El-Galala el-Baharia). Fig. 21 gives the geology of the eastern (Sinai) coastal stretch between Suez and Abu Zeneima. Fig. 22 shows the distribution of strata in the three ranges of Esh-Mellaha, Gebel Zeit, and Gebel Araba, that run parallel to the Gulf of Suez and bound it, at its southern end, on both its western and eastern sides.

The Gulf of Suez represents one of the most intensively-faulted areas on the face of the earth (Fig. 25, 26). Bordering the depression on both sides are two major marginal faults usually marked by lines of high vertical escarpments on the upthrown sides; these faults separate the much-disturbed Gulf region from the almost undisturbed massifs of central Sinai and the central plateau of the Eastern Desert. These two major lines of fracture determine, to a large extent, the configuration of the present Gulf. With the exception of some irregularities in the northwestern side of the depression, the course of both these lines of fracture is parallel to the shores of the Gulf along which they run on either side at an almost equal distance from the shore. The throw of these faults varies greatly when traced from north to south. Fig. 25 gives the main structural elements of the Gulf of Suez and shows the course of these major marginal faults. The marginal faults bound a complex of blocks, each of which displays a unique succession of strata.

---

[1] Written in the early twenties and published posthumously in 1959.

[*Editor's Note:* Material has been omitted at this point.]

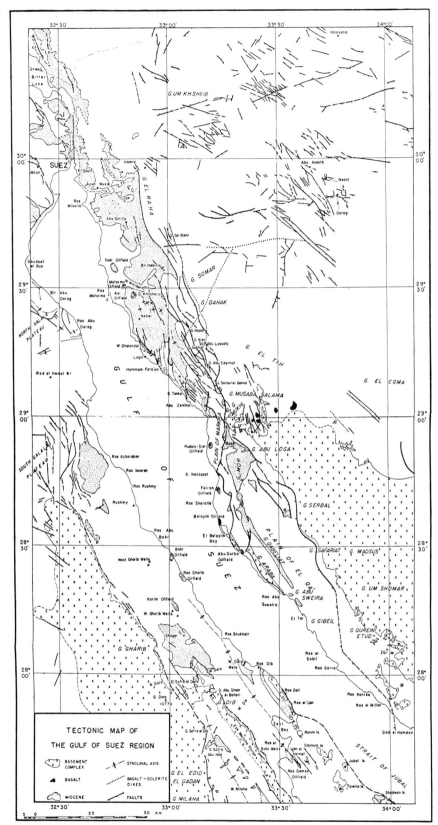

Fig. 25. Tectonic map of the Gulf of Suez region (compiled from the works of Ball, Barron, Hume, Madgwick,
Moon, Sadek, Iskandar and Jones).

311

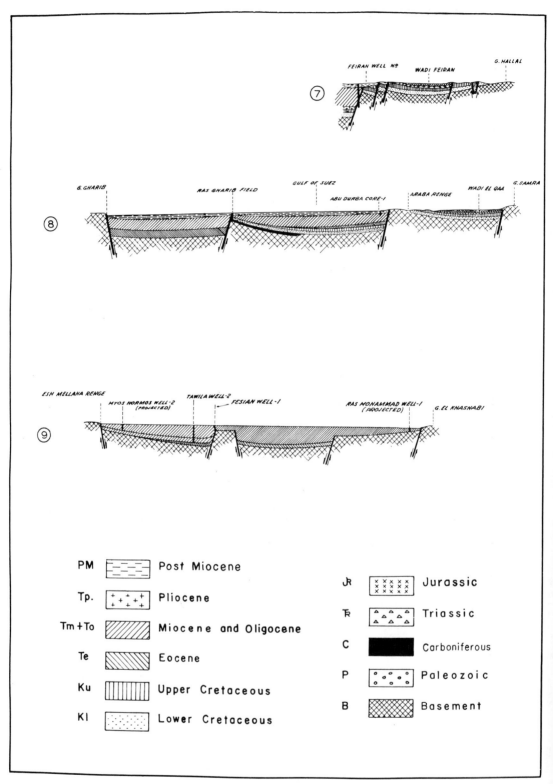

Fig. 26.   Cross sections across the Gulf of Suez (after Jones, 1946, Iskandar, 1946, and Shata, 1956).

312

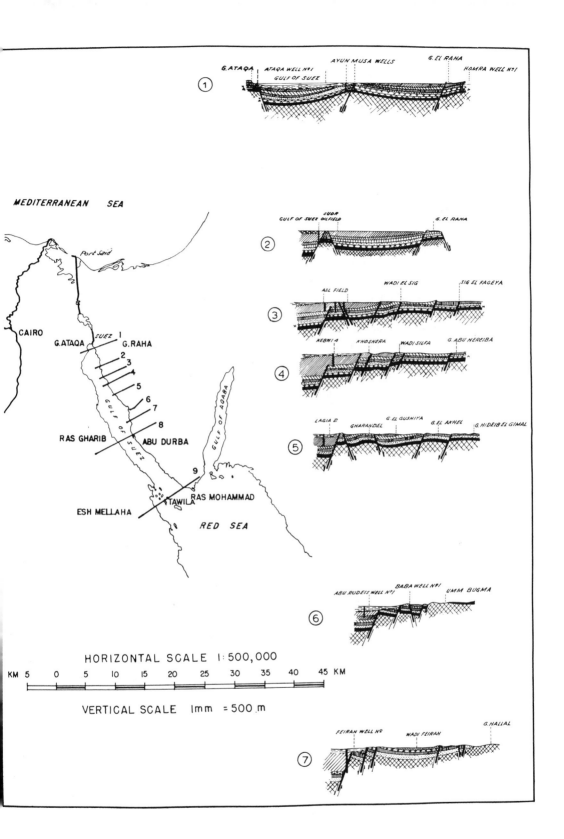

MEDITERRANEAN SEA

Port Said

CAIRO

G.ATAQA SUEZ
G. RAHA

RAS GHARIB ABU DURBA

ESH MELLAHA

RAS MOHAMMAD
TAWILA

RED SEA

GULF OF SUEZ

GULF OF AQABA

HORIZONTAL SCALE 1:500,000

KM 5 0 5 10 15 20 25 30 35 40 45 KM

VERTICAL SCALE 1mm = 500 m

313

TECTONICS

The Gulf of Suez depression represents an intensively faulted area which owes its origin to tensional forces that have been active since early geological time. Folding played a minor role, if any, in determining the structure of the Gulf; all the folds noted were produced either by the bending of the strata before breaking or by movements that caused the less rigid sediments (especially the Miocene) to bend in anticlinal or synclinal folds.

The faulting movements that affected the Gulf of Suez region are old. Reconstructed shore lines of the Carboniferous suggest that the shore lines were fault-determined along lines that have the same trend as those found in the crystalline basement complex. The succession of strata in different parts of the Gulf indicate that the movements were not of the same magnitude in all parts of the Gulf at any one time; and many of the blocks show varying degrees of activity on their different sides. In fact, the history of the Gulf of Suez area can be looked upon as the history of the sinking and rising of a large number of blocks that border or build the Gulf of Suez. Some of these blocks are of sizable dimensions and have been active since early geological time (e.g., Ataqa, northern Galala, and Raha), while others are splinter blocks that seem to be younger. The blocks were especially active during the middle Cretaceous, late Cretaceous, early Eocene, late lower Eocene or early middle Eocene, late Eocene, and finally, with even greater intensity during the Oligocene and later time. The movement at the end of the lower Eocene or the beginning of the middle Eocene seems to have affected most of the blocks, and there is indication that it was during this episode that the differentiation of the Sinai massif and the Red Sea ranges started.

[*Editor's Note:*   Material has been omitted at this point.]

It is not quite clear what forces determined the differentiation of the blocks that bound or build the Gulf of Suez. Several theories have been advanced to account for the formation of the Gulf of Suez graben, such as those by Tromp (1950), Henson (1951) and Swartz and Arden (1960). Any satisfactory theory must explain the following peculiar features: (*1*) that the western blocks were differentiated at an earlier date than the eastern blocks; (*2*) that the differentiation of blocks started from the north, the oldest being to the north; (*3*) that the rising of some blocks was accompanied quite often by the sinking of others, so that when Gebel Somar block rose the blocks to the west and to the north sank; and (*4*) that the intervening blocks were activated at rates and times that were different from those of the bounding blocks (*e.g.*, Ayun Musa block).

[*Editor's Note:* In the original, material follows this excerpt. Only references cited in these excerpts are reprinted here.]

REFERENCES

Henson, F. R. S., 1951. Observations on the geology and petroleum occurrences in the Middle East. *Proc. Third World Petrol. Congr., Sect. I,* E. J. Brill, Leiden, pp. 118-140.

Kostandi, A. B., 1959, Facies maps for the study of the Paleozoic and Mesozoic sedimentary basins of the Egyptian region. *First Arab. Petrol. Congr., Cairo,* **2:**54-62.

Sadek, H., 1959. *The Miocene in the Gulf of Suez region (Egypt).* Geol. Survey Egypt., Cairo, 118 pp.

Said, R., and Shukri, N. M., 1955. Ancient shore-lines of Egypt, part I: The Paleozoic. *Bull. soc. géogr. Égypte,* **28:**41-49.

Schürmann, H. M. E., 1949. The basement rocks of the northern part of the Eastern Desert of Egypt. *Bull. soc. géograph. Égypte,* **23:**35-61.

Shata, A., 1955. Some remarks on the distribution of the Carboniferous formations in Egypt. *Bull. inst. désert Égypt,* **5(1):**241-247.

Swartz, D. H., and Arden, D. D., 1960. Geologic history of Red Sea area. *Bull. Am. Assoc. Petrol. Geologists,* **44:**1621-1637.

Tromp, S. W., 1950. The age and origin of the Red Sea graben. *Geol. Mag.* **87:**385-392.

22

Copyright ©1971 by the Geological Society of London

Reprinted from pp. 247–248, 252, 258, 259–260, 265, and 266-271 of *Geol. Soc. London Jour.* **127**:247–276 (1971)

The structure of the Gulf of Suez (Clysmic) rift, with special reference to the eastern side

DOUGLAS ARTHUR ROBSON

SUMMARY

The stratigraphical sequence includes Palaeozoic (resting unconformably upon a Pre-Cambrian peneplane), Mesozoic and Tertiary rocks. These were deposited uniformly over the entire region although in Egypt the succession thins towards the south. Miocene rocks rest unconformably on earlier strata with which they contrast sharply in showing remarkable facies changes closely related to structural movements. Igneous activity was on a subdued scale and confined to Oligocene times. Dykes generally run parallel to the main Clysmic rift trend.

Rift faulting was initiated in Oligocene and continued into post-Miocene times. In addition to major marginal faults there are many tilted blocks which typify the pre-Miocene faulting. On the west some of these were eroded down to Basement before the deposition of Miocene sediments. In the east many such blocks exhibit only slight erosion and were probably not uplifted until late Oligocene times. The eastern boundary is marked by synthetic faults with downthrows of 1000 metres or more, but in one restricted area there is a downwarp broken by antithetic faults. The Miocene is characterized by anticlinal flexures generally associated with rejuvenation of Oligocene faults. There is no evidence of lateral movement but many of the faults follow ancient Pre-Cambrian trends. On the Sinai peninsula trend-lines show a twofold directional pattern, parallel to the Gulf of

Akada and Gulf of Suez respectively. The angle between them suggests a series of conjugate pairs, originating as transcurrent faults in response to north–south compression. If this was the case, such strike-slip movements ceased before the deposition of the earliest Palaeozoic strata. These are at least as old as Carboniferous.

The area seems to have remained insulated from the great movements of lateral shift which are claimed for the Akaba-Jordan rift and which are considered to be linked with the divergence of the Red Sea and the counter-clockwise rotation of Arabia. However, there is evidence to suggest that, since Eocene times, there has been a growing divergence of the margins of the Clysmic rift towards the south, producing a small counter-clockwise rotation of Sinai.

---

1. Introduction

THE GULF OF SUEZ lies within the northern arm of the Red Sea rift, but because the shoulders of this arm rise far beyond either shore of the gulf itself, Hume (1921) proposed the term 'Clysmic' rather than 'Gulf of Suez' to describe this region. Clysma was the name of the Roman town which stood at the head of the gulf.

The eastern zone of the Clysmic rift is bordered on the west by the Gulf of Suez, and on the east by the high sedimentary plateau of Sinai which, in the south, gives place to the deeply-dissected Pre-Cambrian mountain mass. The geological map (Plate 1) covers most of this eastern zone, and forms the basis of discussion in this paper.

This zone of the rift has been a fruitful source of investigation among geologists for the past 70 years. One of the earliest accounts embracing the whole region was that by Barron (1917), while other accounts by Ball (1916), by Hume, Madgwick, Moon & Sadek (1920A, 1920B) and by Moon & Sadek (1923, 1925) dealt with the geology of selected parts of the region. Later, Busk (1929) described its fault pattern which he considered to be typical for rift valley structures.

The change in the Egyptian law regarding petroleum concessions, together with the discovery of oil at Ras Gharib in 1938, led to an increased interest in both flanks of the rift on the part of a number of oil companies. This interest increased after 1945, but during the past two decades contributions have been made chiefly by geologists of the United Arab Republic. These contributions, up to the year 1961, together with those of earlier workers, have been summarized by Said (1962).

[*Editor's Note:* Material has been omitted at this point.]

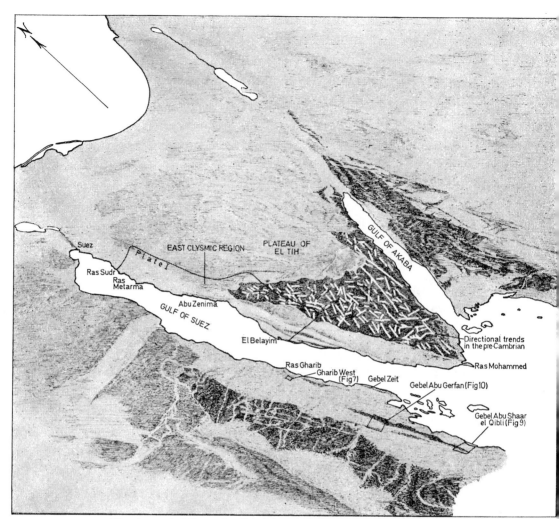

FIG. I. Sketch, copied from the Gemini space photograph, of Sinai, the Gulf of Suez and the Gulf of Akaba.

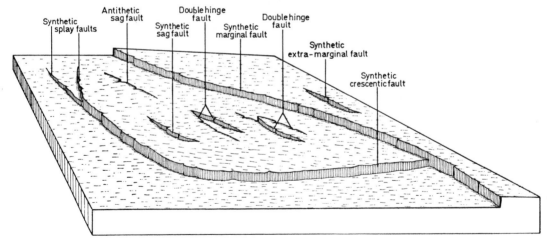

FIG. 6. Perspective block diagram illustrating the general pattern of faulting in the Clysmic rift.

318

(c) DIRECTIONAL TRENDS: THE NATURE OF THE RIFT FAULTS

On the Gemini photograph of the Pre-Cambrian horst which forms the Sinai peninsula, the directional trends can be seen (Fig. 1). These trends have been compiled as a rose diagram in Fig. 2b. Throughout the Pre-Cambrian of Sinai (of some 5000 km²), even in the region nearest to the Clysmic rift, the directional trends seem to have a predominant Akaba, with a more subsidiary Suez orientation. However, in the Palaeozoic and later sediments of the east Clysmic rift, the number of faults having an Akaba trend is low (Fig. 2a), though the Akaba trend must have been present in the Pre-Cambrian beneath those sediments when rifting began.

Therefore, if both Akaba and Suez trends in the Pre-Cambrian are of pre-Palaeozoic origin, it would seem that the stresses that initiated the rift movement in Oligocene times (see below) were so distributed round the Sinai peninsula as to have caused one arm of the rift to select the ancient Suez trend in the Pre-Cambrian (which also must have been dominant throughout the region of the Red Sea itself); while the other arm developed along ancient Akaba trend lines.

Youssef (1968) has observed that the angle between the Suez and Akaba trends corresponds to the angle which would be subtended by a conjugate pair of wrench faults (Anderson 1951). He considers that these faults would have developed in response to a north-south compression, and that this compression would have produced a dextral movement along the Suez gulf and a sinistral movement along that of Akaba (Fig. 11). Youssef goes on to argue, with very little supporting evidence, in favour of a dextral movement of some 60 kilometres along the line of the Gulf of Suez, involving Palaeozoic, Mesozoic and Tertiary sediments.

The explanation for the origin of the Suez-Akaba trends in the Pre-Cambrian in terms of conjugate pairs of faults or joints, may well be correct. It may also be true that, in both the Akaba and Clysmic rifts, the Oligocene rift faulting developed in the Pre-Cambrian rocks along paths established by the earlier transcurrent faults. But the evidence is clear that in the Clysmic region, these movements of rejuvenation were not horizontal but were almost entirely vertical.

Major wrench faults, like those of San Andreas in California (Willis 1938) and the Great Glen in Scotland (Kennedy 1946) are markedly straight fractures, having brecciated zones associated with them, quite unlike the Somar, Gamal and Geba faults of the east Clysmic rift. Apart from the lack of evidence for horizontal movement or fault brecciation, it would be very extraordinary if the movements either of the Somar fault, the Gamal fault or the Geba fault, which bifurcate and pursue curving paths, were other than dominantly vertical.

The trend directions of the synthetic marginal faults of the east Clysmic rift are dominantly those of Suez, though occasionally and for short distances they run

319

parallel with the Gulf of Akaba. If it is admitted, therefore, that the major Clysmic faults developed along established Pre-Cambrian trends, then they selected those having the Suez direction; but at certain points they 'jumped', and the stress system sought relief for short distances along ancient Akaba trends. This may also be true of the synthetic minor faults (Fig. 2a).

Evidence that there was relief of stress across the ancient fracture system of both Suez and Akaba is confirmed by the two dykes which almost intersect at a point south of Gebel Abyad (Pl. 1). The Ras Wata dyke follows a Suez trend for many kilometres; its companion, the Abyad dyke, runs in an Akaba direction for 10 kilometres and then, after a break near Gebel Dahhak, continues far beyond the eastern boundary of the map (but following the same trend) into central Sinai. This evidence lends support to the view that, by Oligocene times at latest, the ancient conjugate fault system was beginning to respond to forces of tension and was becoming ripe for dyke injection. Incidentally, these dykes were intruded in Oligocene times and no subsequent movement has produced in them any significant horizontal shift.

According to van der Ploeg (1953), the Clysmic rift can be divided into a series of tectonic provinces which are delimited by lines or zones crossing the Gulf of Suez with an Akaba trend. Furthermore, v.d. Ploeg considers that the tectonic pattern of each province is distinctive. However, in the analysis of the east Clysmic fault pattern described above, the author finds it difficult to accept the Ploeg hypothesis, at least in regard to the Sinai side of the rift; as the perspective diagrams show (Figs. 3–5), the fault pattern appears to be repetitive.

[*Editor's Note:* Material has been omitted at this point.]

(E) CLYSMIC RIFTING—DEPOSITIONAL AND STRUCTURAL CONSEQUENCES

The constant thickness of the pre-Miocene sedimentary succession on similar latitudes across the rift, from beyond the uplifted shoulders on either side, reflects the stability of the Clysmic region up to the end of the Eocene. Local swells of the Basement floor may have developed between Pre-Cambrian and Oligocene times, but there is no evidence to suggest that such swells heralded the rift faulting, which was not to begin in this region until after the end of Eocene deposition. Isopach contour maps like those of Kostandi (1959) record a great thickness of pre-Miocene within the Clysmic rift, and this evidence could be used to estimate the thickness, before erosion, of the same formations beyond the rift shoulders.

The pattern of the rift itself, not only across the east Clysmic region, but across the Clysmic gulf as a whole, bears out the early picture of rift valleys described by Gregory (1923) and others. Beyond the marginal faults, the strata dip away from the rift at an angle of 8°–10°, as is well displayed for many kilometres on the desert road between the Nile valley and Abu Shaar el Qibli. It marks the remnant of a great elongated arch which foundered along the marginal faults now flanking the Clysmic rift. In this respect, the Clysmic rift affords a more symmetrical example of rift faulting than sometimes occurs in East Africa. Dixey (1965) has drawn

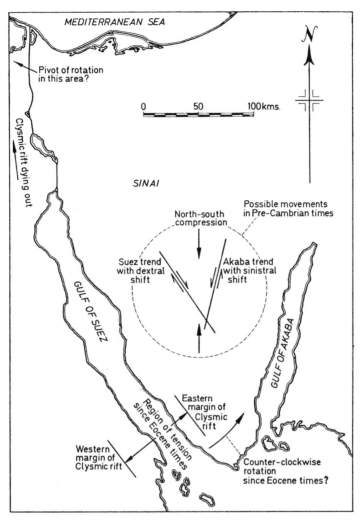

FIG. 11. Possible movements in the Sinai region in pre-Palaeozoic and in post-Eocene times.

attention to the fact that in some parts of the Ethiopian, the Kenyan, the Tanzanian and the Nyasa rifts, marginal faulting on one side may be entirely replaced by downwarping. The same is true of the Dead Sea depression, though the Gulf of Akaba achieves a symmetry comparable to that of the Clysmic rift.

In summary: the Clysmic rift is generally bordered by marginal faults. Of the tilted blocks within the region some, as at Abu Shaar el Qibli, present a Miocene veneer resting on Basement; others, as at Hammam Faraun, have Miocene resting on Eocene, while yet others lie deeply buried beneath a thick succession of Miocene evaporites—for erosion took place to a greater or lesser degree during Oligocene times. Furthermore, the faults which gave birth to the tilted blocks during the Oligocene continued to move during Miocene times, and controlled the thickness of evaporite deposition. The complexities of this Clysmic faulting in relation to the evaporite basins has been summarized by Heybroek (1965).

(F) IGNEOUS ACTIVITY

In his observations on the mechanism of rift faulting, Taber (1927) suggested the analogy of cauldron subsidence, and this was later discussed by Shand (1936). It was thought that just as the extrusion of great volumes of volcanic material from a vent could lead to foundering, so also fissure eruption might bring about rift subsidence. The point has been emphasised by Cloos (1939) and others, including Hills (1965), have observed that the association of rifting and vulcanicity is generally recognized and may be fundamental.

Indeed, many workers have reported volcanic activity along different regions of the African rift system. In Tanzania, Pallister (1965) has cited rift areas both with and without vulcanicity. In the former, the general order seems to have been faulting, succeeded by fissure eruption, followed by repeated faulting and then by extrusion from central volcanoes. Other evidence for the association of volcanic activity with rifting has been recorded by Mohr & Rogers (1965) from Ethiopia, by Baker (1965) from Kenya, by Garson (1965) from Malawi and by de Swardt (1965) from Zambia.

These observations from the African rift system have shown that igneous activity commonly, but not universally, accompanies rift faulting. In the Clysmic rift, the aftermath of this activity, which was confined to the Oligocene period (page 6), is today represented by a few hot springs. Of these, the two which are the most important are Moya Sokhna, beneath the North Qalala plateau, and Hammam Faraun, below the mountain of that name. Both are situated on major synthetic faults.

Within the Oligocene period, evidence of extrusive igneous activity in the Clysmic rift is confined to the two localities from which basalt flows have been recorded (page 5). This paucity of volcanic deposits below the Miocene, even in areas where there was little erosion during Oligocene times, does suggest that igneous activity was only on a reduced scale in the Clysmic rift. Dykes occur in the Gebel Hammam Faraun district and dykes and sills have both been intruded into Cretaceous and Eocene rocks near the Somar fault. Dykes appear to have been introduced only after the initiation of Oligocene faulting, since they are frequently found along rift faults, or along joints with rift trends.

In view of the fact that similar tectonic conditions continued from Oligocene times into and throughout the Miocene, the restriction of igneous activity to the Oligocene is a problem yet to be explained. It may be, if uprise of magma is a pre-requisite for rift foundering (though this may conflict with ideas about convection currents), that the stresses within the Clysmic rift were such as to cause a migration of basalt magma into the nearby region of the Red Sea, where igneous intrusions continued to occur.

(G) THE CLYSMIC RIFT IN ITS WIDER SETTING

The width of the Clysmic rift, between the marginal faults, is under 100 kilometres; that of Akaba, on average, is no more than 25 kilometres, while the Red Sea graben is over 200 kilometres wide between the marginal Pre-Cambrian mountains of Nubia and those of Arabia. The synthetic faults which form the western boundary of the Clysmic rift extend southwards along the Red Sea rift at least as far south as Ras Benas, at latitude 24° north, and probably further. In the Sudan, however, Whiteman (1965) found that the African margin of the Red Sea rift is represented not by faulting but by a nonconformable overlap.

The observations of Wayland (1921) and Bullard (1936) in East Africa, those of Bailey Willis in Palestine (1928) and, later, the comments by Lees (1952), favoured the compression theory for the origin of rift valleys. However, evidence from the field, whether from East Africa (Davies 1951) from the Clysmic rift (including Busk 1929) and elsewhere has unequivocally supported a rifting mechanism which involved extension rather than compression; and, as Holmes (1965) has observed, the ramp hypothesis is now discredited.

In the late fifties, Quennell (1956; 1958) revived an idea which had earlier been promulgated by Dubertret (1932), that the Akaba—Dead Sea depression was not just a simple rift but that it lay along the line of a great sinistral wrench fault which had moved Arabia, with respect to Sinai and Palestine, some 100 kilometres in a north-easterly direction. More recently, Freund (1965) has supplemented the tectonic picture described by Quennell, with certain modifications. On the other hand, Bender (1968), working along the edge of the rift in Jordan, has been unable to subscribe to a great movement of lateral shift.

The work of Drake & Girdler (1964) led to the proposition that a basaltic dyke, or series of basaltic dykes, had been intruded beneath the Red Sea, along its length, as Arabia moved away from Africa by anti-clockwise rotation, and that this process was still continuing. There was no geophysical evidence to show that this intrusive activity extended into the Gulf of Suez. More recently, Gass & Gibson (1969) have cited evidence in favour of the idea that not only was Arabia moving to the north-east, but that the Nubian block was also moving in the same direction, though at a slower rate.

It is important that these movements should be considered in the light of the Clysmic structures. There could have been sinistral movement along the Dead Sea—Akaba rift, with Arabia moving north-eastwards, without any impingement upon the Clysmic structures; Arabia could have been rotating in an anti-clockwise direction without affecting Sinai. But any movement of Africa, relative to Arabia, must undoubtedly have had its impact upon the Clysmic region. Advance of Africa upon Sinai would have produced evidence of compression, whereas north-eastward movement of Sinai, away from Africa, would have been accompanied by extension within the Clysmic rift.

In fact, the horizontal displacement normal to the strike of a single Clysmic fault of 3000 metres' vertical throw, having a dip of the fault plane of 70°, would be somewhat in excess of 1 kilometre. Therefore the total extension across the southern end of the Clysmic rift, where there are at least four synthetic faults of this magnitude, could have amounted to perhaps 5–6 kilometres; possibly more, since the extent of faulting offshore is unknown. In contrast to the faulting at this end of the Clysmic rift, the rift faulting in the Suez area seems to be very much reduced, and at the latitude of the southern shore of the Mediterranean, the vertical movements have probably been negligible. Hence, there must have been an increasing extension of the crust between the Mediterranean and the southern end of the Clysmic rift. This extension might have been caused by an anti-clockwise rotation of Sinai with respect to Africa; on the other hand, the initial arching—and consequent shortening of the crust—which, in early Oligocene times, heralded the development of the Clysmic rift, might have been sufficient to account for extension, by foundering, along the rift faults (Fig. 11).

The fault pattern in the Clysmic rift is very similar to patterns which have been described from other regions of the African rift system, for example by Busk (1939) in Kenya, Hutchinson and Engels (1970) in the Danakil Depression and Bender (1968) in Jordan. Still a region of active earth movement, the fault pattern of the Clysmic rift, as described here, should help towards a fuller understanding of the tectonic events which have been associated with the African rift system.

4. References

ANDERSON, E. M. 1951. *The dynamics of faulting.* 2nd ed. London (Oliver and Boyd).

ANDREW, G. 1937. The late Tertiary igneous rocks of Egypt (Field relations). *Bull. Fac. Sci. Cairo Univ.* **10,** 1–61.

BAKER, B. H. 1965. The rift system in Kenya. In C. L. Drake and I. S. Loupekine (eds), East African rift system. 1. Report on the UMC/UNESCO Seminar on the East African rift system. Univ. Coll. Nairobi. 82–84.

BALL, J. 1916. The geology and geography of west central Sinai. *Egypt. Survey Dept.* Cairo. 1–219.

BARRON, T. 1917. The topography and geology of the peninsula of Sinai (western portion). *Egypt. Survey Dept.* Cairo. 1–241.

BENDER, F. 1968. Über das Alter und die Entstehungs geschichte des Jordangrabens am Beispiel seines Südabschnittes (Wadi Araba) Jordanien. *Geol. Jahrbuch* **86**, 117–196.

BLANKENHORN, M. 1901. Neues zur geologie und paläontologie Agyptens III: Das Miozän *Z. deut. geol. Ges.* **53**, 52–132.

BOWMAN, T. S. 1931. Report on boring for oil in Egypt, Sec. III: Eastern Desert and adjoining islands. *Mines & Quarries Dept.* Cairo. 1–353.

BULLARD, E. C. 1936. Gravity measurements in East Africa. *Phil. Trans. Roy. Soc.* A, **235**, 445–531.

BUSK, H. G. 1929. *Earth Flexures.* Cambridge Univ. Press.

—— 1939. Explanatory note on the block diagram of the Great Rift Valley from Nakuru to Lake Magadi. *Q. Jl geol. Soc. Lond.* **95**, 231–233.

CLOOS, H. 1928. Uber antithetische Bewegungen. *Geol. Rdsch.* **19**, 246–251.

—— 1939. Hebung-Spaltung-Vulkanismus. *Geol. Rdsch.* **30**, 405–527.

DAVIES, K. A. 1951. The Uganda section of the Western Rift. *Geol. Mag.* **88**, 377–385.

DIXEY, F. 1965. Points arising from geological presentations and discussion. In C. L. Drake and I. S. Loupekine (eds), East African rift system. 1. Report on the UMC/UNESCO Seminar on the East African rift system. Univ. Coll. Nairobi. 123, 124.

DRAKE, C. L. & GIRDLER, R. W. 1964. A geophysical study of the Red Sea. *Geophys. Jl Roy. Ast. Soc.* **8**, 473–495.

DUBERTRET, L. 1932. Les formes structurales de la Syrie et de la Palestine; leur origine. *C. R. Acad. Sci. Paris.* **195**, 65–67.

FREUND, R. 1965. A model of the structural development of Israel and adjacent areas since Upper Cretaceous times. *Geol. Mag.* **102**, 189–205.

GARSON, M. S. 1965. Summary of present knowledge of the rift system in Malawi. In C. L. Drake and I. S. Loupekine (eds), East African rift system. 1. Report on the UMC/UNESCO Seminar on the East African rift system. Univ. Coll. Nairobi. 94–104.

GASS, I. G. & GIBSON, I. L. 1969. Structural evolution of the rift system in the Middle East. *Nature, Lond.* **221**, 926–930.

GREGORY, J. W. 1923. Structure of the Rift Valley. *Nature Lond.* **112**, 514–516.

HEYBROEK, F. 1965. The Red Sea Miocene evaporite basin, in Salt basins around Africa. *Inst. Petrol. Lond.* 17–40, (and general discussion).

HILLS, E. S. 1965. *Elements of Structural Geology.* London (Sci. Paperbacks & Methuen.).

HOLMES, A. 1965. *Principles of Physical Geology.* London (Nelson).

HUME, W. F. 1901. Rift Valleys of Eastern Sinai. *Geol. Mag.* **4**, 198–200.

—— 1921. Relations of the northern Red Sea and associated Gulf areas to the "rift" theory. *Proc. geol. Soc. Lond.* **77**, 96–101.

——, MADGWICK, T. G., MOON, F. W. & SADEK, H. 1920A. Preliminary general report on the occurrence of petroleum in Western Sinai. *Petrol. Research Bull.* **2**, Cairo (Govt. Press), 1–15

——, ——, —— & —— 1920B. Preliminary report on the Gebel Tanka area. *Petrol. Research Bull.* **4**, Cairo (Govt. Press) 1–16.

HUTCHINSON, R. W. & ENGELS, G. G. 1970. Tectonic significance of evaporites in the Danakil Depression, Ethiopia. In N. L. Falcon, I. G. Gass, R. W. Girdler and A. S. Laughton (organisers), A discussion on the structure and evolution of the Red Sea and the nature of the Red Sea, Gulf of Aden and Ethiopia rift junction. *Phil. Trans. Roy. Soc.* A **267**, 313–329.

KENNEDY, W. Q. 1946. The Great Glen fault. *Q. Jl geol. Soc.* **102**, 41–76.

KERDANY, M. T. & ABDEL SALAM, H. A. 1970. Bio- and Lith-Stratigraphical Studies of the Pre-Miocene of some offshore exploration wells in the Gulf of Suez. *7th Arab Petrol. Cong.* Kuwait. 1–17.

KHATTAB, H. A. & HADIDI T. A. 1961. A Comparative Study of the Bakr, Kareem and Ras Gharib Oilfields. *3rd Arab Petrol. Cong.* Alexandria. 1–7.

KOSTANDI, A. B. 1959. Facies maps for the study of Palaeozoic and Mesozoic sedimentary basins of the Egyptian region. *1st Arab Petrol. Cong.* Cairo, **2**, 54–62.

LEES, G. M. 1952. Foreland folding. *Q. Jl geol. Soc.* **108**, 1–34.

MOHR, P. A. & ROGERS, A. S. 1965. Status of geological and geophysical studies and resumé of the geology of Ethiopia. In C. L. Drake and I. S. Loupekine (eds), East African rift system. 1.

Report on the UMC/UNESCO Seminar on the East African rift system. Univ. Coll. Nairobi. 47–51.

MOON, F. W. & SADEK, H. 1923. Preliminary geological report on the Wadi Gharandel area. *Petrol. Research Bull.* **10,** Cairo. (Govt. Press) 1–42.

—— & —— 1925. Preliminary geological report on the Gebel Khoshera area (Western Sinai). *Petrol. Research Bull.* **9,** Cairo. (Govt. Press) 1–40.

MORGAN, D. E. & EL BARKOUKY, A. N. 1956. Geophysical history of the Ras Gharib field. *Soc. Explor. Geophys., Geophys. Case Hist.,* **2,** 237–247.

OSMAN, A. 1954. Upper Cretaceous foraminifera of western Sinai. *Bull. Fac. Eng. Cairo Univ.,* **1,** 335–365.

PALLISTER, J. W. 1965. The rift system in Tanzania. In C. L. Drake and I. S. Loupekine (eds), East African rift system. 1. Report on the UMC/UNESCO Seminar on the East African rift system. Univ. Coll. Nairobi. 86–91.

PICARD, L. 1939. Outline of the tectonics of the earth with special emphasis on Africa. *Bull. Geol. Dept. Hebrew Univ. Jerusalem,* **2,** 1–66.

QUENNELL, A. M. 1956. Tectonics of the Dead Sea rift. *Ass. African geol. Surveys. 20th Internat. geol. Cong. Mexico,* 385–405.

—— 1958. The structural and geomorphic evolution of the Dead Sea rift. *Q. Jl geol. Soc. Lond.* **114,** 1–24.

ROBSON, D. A. 1959. The geological structure of the Wadi el Dirba area of Sinai. *Q. Jl geol. Soc. Lond.* **115,** 41–47.

SADEK, H. 1928. The principal structural features of the peninsula of Sinai. *Comp. Rend. 14e. Cong. geol. Internat. Madrid* (1926) fasc. 3. 895–900.

SAID, R. 1962. *The Geology of Egypt.* Amsterdam (Elsevier).

SCHÜRMANN, H. M. E. 1966. *The Pre-Cambrian along the Gulf of Suez and the northern part of the Red Sea.* Leiden (Brill).

SHAND, S. J. 1936. Rift Valley impressions. *Geol. Mag.* **73,** 307–312.

SWARDT, A. M. J. DE 1965. Rift faulting in Zambia. In C. L. Drake and I. S. Loupekine (eds). East African rift system. 1. Report on the UMC/UNESCO Seminar on the East African rift system. Univ. Coll. Nairobi. 105–114.

TABER, S. 1927. Fault troughs. *Jl Geol.* **35,** 577–606.

THIÉBAUD, C. E. & ROBSON, D. A. 1970. Short discussion contribution. In N. L. Falcon, I. G. Gass, R. W. Girdler and A. S. Laughton (organisers), A discussion on the structure and evolution of the Red Sea and the nature of the Red Sea, Gulf of Aden and Ethiopia rift junction. *Phil. Trans. Roy. Soc.* A **267,** 413–415.

VAN DER PLOEG, 1953. Egypt. *In:* V. C. Illing, The World's Oilfields: the Eastern Hemisphere. *The Science of Petroleum.* **6,** pt. 1. 151–157. (Oxford Univ. Press.)

WAYLAND, E. J. 1921. Some account of the geology of the Lake Albert rift valley. *Geogr. Jl* **58,** 344–359.

WILLIS, B. 1928. The Dead Sea problem: rift or ramp valley ?, *Bull. geol. Soc. Am.* **39,** 450–542.

—— 1938. San Andreas rift. *Jl Geol.* **46,** 793–827.

WHITEMAN, A. J. 1965. A summary of present knowledge of the rift valley and associated structures in Sudan. In C. L. Drake and I. S. Loupekine (eds), East African rift system. 1. Report on the UMC/UNESCO Seminar on the East African rift system. Univ. Coll. Nairobi, 34–46.

YOUSSEF, M. I. 1968. Structural pattern of Egypt and its interpretation. *Bull. Am. Ass. Petrol. Geol.* **52,** 601–614.

23

Reprinted from *Royal Soc. London Philos. Trans.* **A267**:413-414 (1970)

A DISCUSSION OF THE STRUCTURE AND EVOLUTION OF THE RED SEA AND THE NATURE OF THE RED SEA, GULF OF ADEN AND ETHIOPIA RIFT JUNCTION

C. E. Thiébaud and D. A. Robson

The outcrops of rock within the bounding faults of the rift valley, both in the Gulf of Suez region and along the northwestern margin of the Red Sea, are numerous and over wide areas they are continuous. The geology of this section of the great rift should therefore be closely considered in any assessment of the tectonic pattern of the Red Sea as a whole.

In our opinion, based on many years of detailed geological mapping, the following facts appear to be of fundamental importance:

1. In the Egyptian region of the great rift valley, Palaeozoic formations represented by the diachronous Nubian Sandstone, lie upon the peneplained surface of the Precambrian. There are many extensive outcrops where this peneplain can be seen in the field. However, as Schürmann has pointed out, and as Bender has observed in Jordan, there are localities where Precambrian has an uneven and irregular base.

2. The Palaeozoic and Mesozoic formations which follow the Precambrian are remarkably constant in thickness over wide areas. There is no significant variation in thickness between sections measured within the confines of the rift and those nearby, but beyond the bounding faults. For example, the succession of Nubian Sandstone, Cenomanian shales, etc. up to and including Eocene on the Plateau of Sinai is closely similar to that, on the same latitude, along the western margin of Sinai, within the rift depression. This same correspondence can be observed when comparing the sections, along similar latitudes, within and outside the rift on the western side of the Gulf of Suez. Therefore, the rift faulting in this region must have started only after Eocene times.

3. In fact, there was a marked lack of disturbance from the end of Precambrian until Oligocene times, except in the north where the folding movements of the Egypto-Syrian arc developed. These movements must have occurred chiefly during Middle Eocene times, as is established by the presence across the Sinai isthmus of Upper Lutetian conglomerates overlying, unconformably, older formations.

4. The rift faulting which began in Oligocene times is normal faulting, with fault planes dipping at about 70°. The rift fault pattern is notable for the enormous vertical throw of the west and east marginal faults, and also for the groups of tilted blocks within the rift valley itself. The tilted blocks described by Hutchinson in the Danakil Depression of Ethiopia, and by Bender in Jordan, can be matched by many similar structures in the Gulf of Suez rift.

5. The absence, at the present time, of rocks of Mesozoic age on some of the tilted blocks— especially on the west side of the Gulf of Suez—is due, not to block-faulting and elevation in

327

Mesozoic times resulting in non-deposition, but rather to continuous deposition throughout Mesozoic times, followed by the birth of block-faulted structures only in post-Eocene times.

6. Faulting in Oligocene times was accompanied by volcanicity, but no igneous activity occurred in the Gulf of Suez region after the end of Oligocene times.

7. Faulting continued throughout Miocene times, and seems to have been an important controlling factor in the production of the enormous thicknesses of evaporite deposits.

8. Further faulting took place in post-Miocene times, and there is abundant evidence to show that rejuvenation of the pre-Miocene faulted blocks produced flexuring and sometimes faulting in the overlying plastic Miocene deposits.

9. The rift faults in the Gulf of Suez, and on the northwestern side of the Red Sea, as seen in plan, are winding, curving structures, with vertical throws of up to 3 000 m. There is no evidence from mapping that there was any substantial amount of horizontal movement along these faults (one of the writers was able to determine horizontal movement in Wadi Dirba, ESE of Suez, of at most 200 m). Furthermore, the well-established feature of a fault which is involved in a large horizontal movement—as, for example, the Great Glen fault in Scotland and the San Andreas fault in California—is that it is notably straight. The faulting along the Gulf of Suez may appear straight on the much-simplified regional map, but on the scale of 1:25 000 or even 1:100 000 the true pattern—that of curving, bifurcating fractures—can be clearly seen. Field evidence would thus oppose the views expressed by Baker, Abdel-Gawad and others, advocating horizontal movements of many tens of kilometres along the length of the Gulf of Suez.

In our view, the field evidence lends support to the following points:

(a) The paucity of dyking and of volcanicity in Oligocene times, and the total absence of any igneous activity from Miocene times onwards, makes it difficult to believe that the Gulf of Suez was undergoing any considerable lateral extension during the great period of post-Eocene rifting. Furthermore, very little lateral extension could be ascribed to the effect of rift faulting, since the dip of the fault planes is invariably high.

(b) The absence of long, straight faults in the region, and along the northwestern side of the Red Sea, suggests that wrench faults, if present at all, played only a minor role.

(c) The Gulf of Suez, together with that region of the Red Sea lying immediately south-southwest of it, was remarkably isolated from the great movements of shear which some workers consider to have taken place along the Akaba–Dead Sea depression. It follows that: (i) If the Red Sea opened up, as postulated from geophysical evidence, this widening must have been due, among other things, to the rotation of Arabia in an anticlockwise direction; any augmenting clockwise rotation of Africa would surely have opened up the Gulf of Suez. (ii) It is difficult to accept the view that any major NNW to SSE horizontal movements can have occurred along the Red Sea rift since Precambrian times; such fractures would undoubtedly have extended northwards to involve the Gulf of Suez also in horizontal movement. Evidence for this is lacking.

Part X

WESTERN ARABIA RIFT SYSTEM

Editor's Comments
on Papers 24 Through 27

GEOLOGY

Paper 24, by Dubertret, contains a concise account of the geology and structure of the Western Arabia Rift System, with emphasis on the northern extension. The latter half of this paper (pp. 9–15) is a discussion of hypotheses of the formation of the zone. Although the author stresses the simplicity of the southern part of the system extending from the Gulf of Aqaba to beyond Lake Houle, there are in fact five contrasting structural forms. The author's account of the structure and geology of the northern extension should be highly regarded as he was the leading contributor to the geology of Lebanon and Syria. The sections are valuable and complement those in Burdon (Paper 25, Figure 3).

For the Gulf of Aqaba, the western flank is described by Hume (1901, 1925) and the morphology on the map by Awad (1952). Said (1962) gives a general description and refers to Hume (1906) and Sadek (1928). For the eastern flank of the gulf, there is the map published by the U.S. Geological Survey (1963). Mitchell (1957) includes a map and gives a detailed description of the eastern flank, including the faulting of the NW Hegaz. Quennell (1958, Figure 1) and Paper 26 (Figure 6) gives submarine topography taken from Admiralty Chart, *Suez to the Brothers*, 1953.

For the Wadi Araba, the Dead Sea, Jordan Valley and Lake Tiberias, Paper 25, excerpted from Burdon, gives a graphic description of the geomorphology, geology, and structure not only of the rift but of adjacent Arabia, which is to be clearly distinguished from Sinai-Palestine on the west. The geomorphic analysis by Quennell (1958), which uses projected profiles and reconstructed stream profiles, complements Burdon's study.

The earliest published accounts of the geology date from the early nineteenth century. Of these, Lynch (1852) includes a bathymetric chart of the Dead Sea; Lartet (1877) discusses the formation of the Dead Sea basin and other salt lakes; while Blanckenhorn (1914) and Krenkel (1957) describe, among other things, the geology and structure of the Arabian rift system.

The first systematic geological mapping of Palestine was by Blake (Blake, 1939; Blake and Goldschmidt, 1947) and Shaw (1947). These papers were followed for the eastern flank by Blake (Blake and Ionides, 1939); Quennell (1951, 1959); Wetzel and Morton (1959); Burdon (Paper 25); and Bender (1968, 1974a, 1974b). The last reference is to a detailed account of geology and structure of the eastern flank of the Wadi Araba. For the western flank, an early general account is by Picard (1943), whereas the geological map (Picard et al., 1965) in two sheets is a standard reference. Bentor (1961) describes the history of the Dead Sea and the Jordan Valley.

For the geology of the Lebanese and Syrian rift zones, Dubertret (Paper 24), provides a concise account of structure and stratigraphy. This is supplemented by earlier maps by Dubertret and Vautrin for Syria and Lebanon. Lebanon is incompletely covered by sheets on scale 1:50,000 with explanatory notes. The work on Lebanon was later consolidated on a map on scale 1:200,000 with explanatory notes (Dubertret, 1955). More recent work in Syria was carried out by a team of Soviet geologists (Ponikarov, 1964), and published on scales 1:100,000, 1:500,000, and 1:200,000. The text (in Russian) has been translated but not published. The field work for these maps resulted in a number of works published only in Russian. For references see Razvalyayev (1972). The Regional Geology of the Earth series includes *The geology of the Syria and Lebanon* by Wolfart (1967). The structure and stratigraphy of the rift zone are described in relation to petroleum prospects by Renouard (1955) and Beydoun (1977).

Paper 27 by Freund et al., although primarily concerned with tectonic hypotheses, contains a valuable account of the stratigraphy of the flanking blocks with columnar sections chiefly of the Upper Cretaceous marine rocks.

It is seen from Papers 24 and 27 and the works referred to, that there is a progressive change in the nature of the geology from the Red

Sea to where the rift system meets the Amanos fault strike-slip zone at the western end of the Border Folds of southern Turkey. The Amanos faults carry the rift system into the Toros-Zagros structural belts by the meeting of the former with the East Anatolian fault.

From the Red Sea (Zone E-D on map, Paper 20B) the Precambrian crystalline complex descends gently northward but without matching of the flanks. On the eastern flank from the south end of the Dead Sea, terrestrial sediments overlie and are in turn overlain by gently folded and faulted marine rocks (Zone D-C). These continue with more pronounced folding (Zone C-B) into Syria (Zone B-A). From there to the Turkish border, a great thickness of Mesozoic and Palaeogene marine sediments probably overlies a floor of oceanic crust (Kozlov et al., 1965; Wolfart, 1967). On the western flank the complex disappears beneath Paleozoic sediments not far north of the Gulf of Aqaba.

Freund and Garfunkel (1981) contains 16 papers on the geology and geophysics of the Gulf of Aqaba and the Dead Sea Rift.

Geophysics

Gravity data, which could illuminate the problem of the nature of the crust of the Dead Sea–Jordan Valley zone, is not readily available but such as is accessible (Bruyn, 1955; Knopoff and Belshe, 1965; Woodside and Bowen, 1970) suggests that the continental crustal margin is sigmoidal in plan with the straight middle arm subparallel to the Dead Sea–Jordan River zone. From the Dead Sea it swings to the southwest and west (the northern margin of the continental part of the African plate). Along the first mentioned zone (C-D on map, Paper 20B) there is a "steep" gradient with a Bouger value difference of 100 mgal, suggesting an abrupt westward change from continental to oceanic crust. Along the upper Litani River (Zone B-C) is a "gentle" gradient zone across which gravity contrast is 100 mgal (Plassard and Stahl, 1957). The zone swings to E—W along parallel 35° and contrast is 60 mgal (Lejay, 1938; Bourgoin, 1950). Between here and the Turkish Border Zone at parallel 36.5° is the 150-km-wide belt of high anomaly.

Seismotectonics of the West Arabian Rift System (Levant fracture zone of Nowroozi, 1972) has not yet made a real contribution. The Dead Sea Rift has enjoyed a historical notoriety for earthquake shocks since Jericho fell to the Israelites. Ambraseys (1971, pers. comm.) indicates historical macroseismal epicenters and discusses (1976) comparison with those instrumentally recorded. Plassard and Kogoj (1968) list historical shocks as far back as 590 *B.C.* Ben Menahem and Aboodi (1981) summarize microseismicity for the Dead Sea (see also

Wu et al., 1973) and macroseismicity for the whole length of the rift zone. On the Jordan-Dead Sea-Wadi Araba zone (C-D on map, Paper 20B); and the Lebanon-Bekaa zone (B-C), there have been more than fifty instrumentally recorded earthquake shocks of magnitude greater than 3.6, three of which had magnitudes greater than 6.0.

For the whole length of the rift system, the seismicity is highest between latitudes 30.8°N and 34.2°N, that is, from 30 km south to 30 km north of the Dead Sea (Arieh, 1967; Shalem, 1952). To the south, along the Aqaba zone (D-E), there is no recorded macroseismic activity, whereas along the Syrian zone (A-B) few shocks are recorded. Along the Amanos fault zone, which the Syrian Rift zone meets north of Antikya, there are many recorded shocks with magnitudes as high as 7.0. All the foregoing have depths to hypocenter of less than 70 kms, all except two being assigned an arbitrary depth of 33 km.

There are no published fault-plane solutions along the rift system although at the south end of the Gulf of Suez there is one solution indicating normal or dip-slip motion (Nowroozi, 1971, 1972; and McKenzie, 1972). The close recording stations, Helwan, Istanbul, and Ksara in Lebanon (Plassard and Kogoj, 1968), have supplied most of the information over a longer period, but all lie to the west of the rift zone. The usefulness of the epicentral data in seismotectonic study is thus reduced and, in any case, coincidence of foci with faults is not to be expected or relied upon (Richter, 1958, pp. 314–315; Ambraseys, 1976). However, repeated shocks from a single location can have significance. The value of seismic study lies in its identification of those zones where crustal conditions favor earthquake shocks. First motion studies directed at fault-plane solutions, more reliably accurate epicentral locations and hypocentral depths would give information essential for further study. However, it must be recognized that the application of seismotectonics to construction of plate tectonic models should not be undertaken as hastily as it may have been in some recent studies (Ambraseys, 1976).

Two papers on paleomagnetism (Van Dongen, 1967; Gregor et al., 1974) record the results of sampling of Jurassic to Recent rocks of the Lebanon mountain range. The latter paper is of special interest because the implications for the plate tectonics of the region are discussed with references to other work. There has also been a 70° counter-clockwise rotation and a possible northward shift relative to Africa since Jurassic times. Van der Voo (1968) also includes results from Turkey and other areas of the Arabian plate. Helsley and Nur (1970) give results from Israel.

TECTONIC HYPOTHESES

As stated, Paper 27 contains details of the geology of the flanks of the Dead Sea Rift. The authors' purpose was to demonstrate the left-lateral shift along the rift. However, their historical review of hypotheses advanced forms a good introduction.

Other reviews can be found in Picard (1931, pp. 90–95) where he recounts the early history of investigation. For the northern extension, Dubertret (Part II of Paper 24) reviews and discusses the different hypotheses and (1970) their regional implications. For development of his own ideas his paper (1971–1972) is the best account. Bender (1974a) also reviews some hypotheses (pp. 122–123) and includes refutation of some of the evidence for the transcurrent hypothesis. The choice of the favored hypothesis tends to reflect the leanings of the author toward one or another of two schools of thought: (1) the vertical (or radial) tectonics school, which is based on the contracting earth concept and which denies any major horizontal translation of continental masses; and (2) the horizontal or tangential tectonics school, which is the framework of continental drift and more recently of plate tectonics.

The Dead Sea Rift was first stated to be the result of normal faulting by von Buch (1841); he was followed by Hull (1889), Blanckenhorn (1914, 1929), and Krenkel (1925, 1957). (However, Willis (1928) advanced the "ramp" or compression hypothesis.) Blake (1937) explained the sigmoidal plan of isopachs as limits of Mesozoic marine transgression. Wetzel and Morton (1959) advocate normal faulting. Picard (1953, 1970) remains the principal advocate, and Vroman (1961) advances experimental evidence. Lees (1952) and Razvaljaev (1972) discuss graben structural development as it relates to "basement." Bender (1974b) also advocates vertical displacement but would accept limited lateral shift.

Lartet (1869) first suggested the lateral shift or wrench faulting hypothesis. He was followed by Huddleston (1882) who related this hypothesis to the formation of the Red Sea. Dubertret (1932) outlined the hypothesis of lateral shift that is at present held but has since (1970) emphasized the difficulties arrising from geological and structural evidence. Wellings (pp. 659–660 in Willis, 1938) supported, and continues to support, Dubertret's views of 1932. Quennell (Paper 26) found support for Dubertret's and Willings's original construction from detailed mapping of the eastern flank. There was also support from Carey (1958). De Sitter (1962) and Holmes (1965) accept lateral shift. Zak and Freund (1966) and Freund et al. (Paper 27) have supported and advanced the hypothesis with additional evidence (see also Freund and Garfunkel, 1981). In two papers Dubertret (1970,

1971-1972) reviews the regional structure and introduces new concepts related to the structure of Arabia (Brown and Coleman, 1972).

In Paper 26, as mentioned above, Quennell advances evidence to support the hypotheses of Lartet, Dubertret and Wellings. He quantified and dated the amount of shift by matching homologous geological and structural features. From the evidence of sedimentation in the Dead Sea "rhombochasm" (Carey, 1958), Quennell demonstrated that movement took place in two unequal phases since the Oligocene. From the circular arc plan of the fault trace features of the Wadi Araba—Dead Sea—Jordan Valley wrench faults (scarplets, springs, displaced fans), he located geometrically the center of the arc, the first published pole of rotation. In Paper 27, Freund et al. produced further evidence, as described above.

The difficulties of reconciling the tectonics of the Dead Sea Rift with the structural pattern beyond where it meets the Litani River, in Lebanon and Syria, still remain, but the seismological, gravity and paleomagnetic evidence has not yet been profitably exploited in finding explanations that no doubt lie in the deep crustal structure.

PLATE TECTONICS OF THE AFRO-ARABIAN RIFT SYSTEM—SUMMARY

There are four possible plates (1) Nubia (Africa); (2) Sinai-Palestine (which may not be independent of Africa); (3) Arabia; and (4) Somalia (which may not be separated from Nubia). Therefore, there may be only two plates: Africa (Nubia, Sinai-Palestine, and Somalia) and Arabia. McKenzie et al. (1970) state the basic situation; Roberts (1969, 1970) argues for the existence of a plate margin between (1) and (4); Searle (1970) for the extension of this possible margin southward as the Eastern African Rift System; Mohr (1970) argues for (1), (3), and (4) as plates. Darracott et al. (1973) accept (1) and (3), but they are uncertain about (1) and (4). Girdler and Darracott (1972) define poles of rotation for (1), (3), and (4). Quennell (Paper 26) describes, in preplate tectonic terms, the motion of (3) relative to (2) about a pole of rotation. Girdler (1978) contains the most recent review. Nowroozi (1972) presents the overall picture presented by seismic activity. Richardson and Harrison (1976) and Garson and Krs (1976) suggest two-stage motions for the separation of Nubia and Arabia.

The tectonics of the Syrian part of the Arabian plate and of the Lebanon-Syrian part of the Sinai-Palestine plate (north of the Emeq-Yizreel depression) including the east Mediterranean, are complex because these are probably the oceanic parts of each plate. For this reason, too, the rift system shows important differences north of where it meets the Litani River. Nowroozi (1972) and McKenzie (1970, 1972) deal with seismicity of the northern boundaries of the region.

REFERENCES

Ambraseys, N. N., 1971, Value of historical records of earthquakes, *Nature* **232:**375-379.

Ambraseys, N. N., 1976, Middle-East—a reappraisal of the seismicity, *Geol. Soc. London Engineering Group,* 48p.

Arieh, E. J., 1967, Seismicity of Israel and adjacent areas, *Israel Geol. Survey Bull.* **43:**1-14.

Awad, H., 1952, Presentation d'une carte morphologique du Sinai au 1:500,000, *Inst. Fuad Ier du Désert Bull.* **2:**132-138.

Ben-Menahem, A., and E. Aboodi, 1981, Micro-macroseismicity of the Dead Sea Rift and off-coast eastern Mediterranean, *Tectonophysics* **80:**199-234.

Bender, F., 1968, *Geologie von Jordanien,* Gebrüder Borntraeger, Berlin, 230p.

Bender, F., 1974a, *Geology of Jordan,* M. K. Khader and D. H. Parker, trans., Gebrüder Borntraeger, Berlin, 196p.

Bender, F., 1974b, Explanatory notes on the geological map of the Wadi Araba, Jordan, *Geol. Jahrb.* **B10:**3-62.

Bentor, Y. K., 1961, Some geochemical aspects of the Dead Sea and the question of its age, *Geochim. et Cosmochim. Acta* **25:**239-260.

Beydoun, Z. R., 1977, Petroleum Prospects of Lebanon—re-evaluation, *Am. Assoc. Petroleum Geologists Bull.* **61:**43-64.

Blake, G. S., 1937, Old Shore Lines of Palestine, *Geol. Mag.* **74:**68-78.

Blake, G. S., 1939, *Geological Map of Palestine on Scale 1:250,000,* Survey of Palestine, Jerusalem.

Blake, G. S., and M. G. Ionides, 1939, *Report on the water resources of Transjordan and their development,* Crown Agents, London, 372p.

Blake, G. S., and M. J. Goldschmidt, 1947, *Geology and water resources of Palestine,* Government Printer, Jerusalem, 413p.

Blanckenhorn, M., 1914, Syrien, Arabien und Mesopotamien, in *Handbuch der Regionalen Geologie,* vol. 5, Carl Winter, Heidelberg, 157p.

Bourgoin, A., 1950, Sur les anomalies de la pésanteur en Syrie et au Liban. Discussion et interpretation géologique des observations par R. P. Lejay en 1936, in *Notes and Memoires, Le Liban, La Syrie et le Loyen-orient,* vol. 4, L. Dubertret, ed., Dir. Gén. Relation Culturelle, Beyrouth, pp. 59-89.

Brown, G. F., and R. G. Coleman, 1972, The tectonic framework of the Arabian Peninsula, *24th Intern. Geol. Congr. Proc.* **3:**300-305.

Bruyn, J. W. de, 1955, Isogam maps of Europe and North Africa, *Geophys. Prosp.* (Netherlands) **3:**1-14.

Buch, L. von, 1841, Letter in Robinson E., *Biblical Researches in Palestine,* vol. 2, appendix, London, pp. 673-675.

Burdon, D. J., 1959, *Handbook of the Geology of Jordan to Accompany and Explain the Three Sheets of the 1:250,000 Geological Map of Jordan East of the Rift by A. M. Quennell,* Government of Jordan, Amman, 82p.

Carey, S. W., 1958, The dilatational origins of rift valleys, in *The Tectonic Approach to Continental Drift,* University of Tasmania, Hobart, pp. 180-190, 252-253.

Darracott, B. W., R. W. Girdler, J. D. Fairhead, and S. A. Hall, 1973, The East African Rift System, in *Implications of Continental Drift to the Earth*

Sciences, D. H. Tarling and S. K. Runcorn, eds., Academie Press, London, pp. 757–766.

DeSitter, L. U., 1962, Structural development of the Arabian shield in Palestine, *Geologie en Mijnbouw* **41:**116–124.

Dubertret, L., 1932, Les formes structurales de la Syrie et de la Palestine; leur origine, *Acad. Sci. Comptes Rendus* **195:**66.

Dubertret, L., 1955, Carte géologique du Liban au 1:200,000, Ministère des Travaux Publics, Beyreuth.

Dubertret, L., 1970, Review of structural geology of the Red Sea and surrounding areas, *Royal Soc. London Philos. Trans.* **A267:**9–20.

Dubertret, L., 1971–1972, Sur la dislocation de l'ancienne plaque sialique Africa-Sinai-Peninsule Arabique, in *Notes et Memoires sur le Moyen-Orient,* vol. 12 Museum National D'Histoire Naturelle, Paris, pp. 227–243.

Eyal, M., Y. Eyal, Y. Bartov, and G. Steinetz, 1981, The tectonic development of the western margin of the Gulf of Elat (Aqaba) rift, *Tectonophysics* **80:**39–66.

Freund, R., and Z. Garfunkel, eds. 1981, The Dead Sea Rift; Selected papers Intem. Symp. on the Dead Sea Rift, 1979 *Tectonophysics* **80:**1–304.

Garfunkel, Z., I. Zak, and R. Freund, 1981, Active faulting in the Dead Sea Rift, *Tectonophysics* **80:**1–25.

Garson, M. S., and M. Krs, 1976, Geophysical and geological evidence of the relationship of Red Sea tectonics to ancient fractures, *Geol. Soc. America Bull.* **87:**169–181.

Girdler, R. W., 1978, Comparison of the East African Rift System and the Permian Oslo Rift, in *Tectonics and Geophysics of Continental Rifts,* I. B. Ramberg and E. R. Neumann, eds., D. Reidel, Dordrecht, pp. 329–345.

Girdler, R. W., and B. W. Darracott, 1972, African poles of rotation, *Comments on Earth Sciences: Geophysics* **2:**131–138.

Gregor, C. B., S. Mertzman, A. E. M. Nairn, and J. Negendan, 1974, The paleomagnetism of some Mesozoic and Cenozoic volcanic rocks from the Lebanon, *Tectonophysics* **21:**375–395.

Helsley, C. E., and A. Nur, 1970, The paleomagnetism of Cretaceous rocks from Israel, *Earth and Planetary Sci. Letters* **8:**403–410.

Holmes, A., 1965, *Principles of Physical Geology,* rev. ed., Thomas Nelson, London, 1288p.

Huddleston, W. H., 1885, *On the Geology of Palestine,* London, pp. 39–73.

Hull, E., 1889, Formation of the Jordan-Arabah Valley, in *Memoir on the Geology and Geography of Arabia Petraea, Palestine Exploration Fund,* London, pp. 104–109.

Hume, W. F., 1901, *The Rift Valleys and Geology of Eastern Sinai,* Dulau and Co., London, pp. 1–21.

Hume, W. F., 1906, *The Topography and Geology of the Peninsula of Sinai (Southern Eastern Portion),* Egyptian Survey Department, Cairo, 280p.

Hume, W. F., 1925, *Geology of Egypt,* vol. 1, Egyptian Survey Department, Cairo, 408p.

Knopoff, L., and J. C. Belshe, 1965, Gravity observations of the Dead Sea Rift, I.U.M.P. Report 9, *Canada Geol. Survey Paper 66–14,* pp. 5–21.

Kozlov, V. V., V. P. Ponikarov, A. V. Razvalyaev, E. D. Sulidi-Kondratier, and V. A. Farajev, 1965, The Syria's Cretaceous deposits, *Moscow Soc. Naturalists Geol. Bull.* **40:**57–68.

Krenkel, E., 1925, *Geologie Afrikas,* vol. 1, Borntraeger, Berlin, 1918p.

Krenkel, E., 1957, Der Graben des Roten Meeres: die Erythreis, in *Geologie Afrikas,* vol. 1, rev. ed., Borntraeger, Berlin, pp. 123-128.

Lartet, L., 1869, Essay on La géologie de la Palestine, *Ann. Sci. Geol., Paris,* **1:**16-18.

Lartet, L., 1877, Formation du Bassin de la Mer Morte, in *Exploration géologique de la Mer Morte, de la Palestine et de l'Idumée,* Bertrand, Paris, pp. 241-268.

Lejay, P., 1938, Exploration gravimetrique des Etats du Levant sous Mandat Français, *Comité National Francais de Géodésie et de Géophysique,* Paris, 54p.

Lees, G. M., 1952, Foreland folding, *Geol. Soc. London Quart. Jour.* **108:**1-34.

Lynch, W. F., 1852, Map accompanying *Official Report of the U. S. Expedition to Explore the Dead Sea and the River Jordan,* Baltimore (Refer to Paper 26, Figure 7).

McKenzie, D. P., 1970, Plate tectonics of the Mediterranean region, *Nature* **226:**239-243.

McKenzie, D. P., 1972, Active tectonics of the Mediterranean region, *Royal Astron. Soc. Geophys. Jour.* **30:**109-185.

McKenzie, D. P., D. Davies, and P. Molnar, 1970, Plate tectonics of the Red Sea and East Africa, *Nature* **226:**243-248.

Mitchell, R. C., 1957, Fault patterns of northwestern Hegaz, Saudi Arabia, *Eclogae Geol. Helvetiae* **50:**257-270.

Mohr, P. A., 1970, The Afar triple junction and sea-floor spreading, *Jour. Geophys. Research* **75:**7340-7352.

Nowroozi, A. A., 1971, Seismo-tectonics of the Persian Plateau, eastern Turkey, caucasus and Hindu-Kush regions, *Seismol. Soc. America Bull.* **61:**317-341.

Nowroozi, A. A., 1972, Focal mechanism of earthquakes in Persia, Turkey, West Pakistan and Afghanistan and plate tectonics of the Middle East, *Seismol. Soc. America Bull.* **62:**823-850.

Picard, L., 1931, *Geological Researches in the Judean Desert,* Hebrew University, Jerusalem, 108p.

Picard, L., 1943, *Structure and Evolution of Palestine; With Comparative Notes on Neighbouring Countries,* Hebrew University, Jerusalem, 134p.

Picard, L., 1953, Disharmonic faulting, a tectonic concept, *Israel Research Council Bull.* **3:**132-135.

Picard, L., 1970, On Afro-Arabian Graben Tectonics, *Geol. Rundschau* **59:**337-381.

Picard, L., Y. K. Bentor, 1965, Geological map. Scale 1:250,000. North sheet: L. Picard and U. Golani; South sheet: Y. K. Bentor, A. Vroman, and I. Zak, *Inst. Petroleum Research Geophys.,* Survey of Israel, Tel Aviv.

Plassard, J., and B. Kogoj, 1968, Catalogue des Seismes Ressentio au Liban, 2nd ed., refondue, *Annales-Memoires de l'Observatoire de Ksara* **4:**1-28.

Plassard, J., and P. Stahl, 1957, Complements sur la gravimetre au Liban, *Annales de l'Observatoire de Ksara* **2:**12-15.

Ponikarov, V. P., 1964, Geological map of Syria, Scale 1:1m (also on scales 1:500,000 and 1:200,000), *Soviet Geologists V/O Technoexport contract 944.*

Quennell, A. M., 1951, The geology and mineral resources of former Trans-Jordan, *Col. Geol. Min. Res.* **2:**85-115.

Quennell, A. M., 1958, The structural and geomorphic evolution of the Dead Sea Rift, *Geol. Soc. London Quart. Jour.* **114:**1-24.

Quennell, A. M., 1959, *Geological Map of Jordan East of the Rift,* Government of Jordan, Amman.

Razvalyaev, A. V., 1972, Some distinctive aspects of the structure and geologic history of the Western Arabia rift system, *Internat. Geology Rev.* **14:**738-747.

Renouard, G., 1955, Oil prospects of Lebanon, *Am. Assoc. Petroleum Geologists Bull.* **39:**2125-2169.

Richardson, E. S., and C. G. A. Harrison, 1976, Opening of the Red Sea with two poles of rotation *Earth and Planetary Sci. Letters* **30:**135-142.

Richter, C. F., 1958, *Elementary Seismology,* Freeman, San Francisco, 768p.

Roberts, D. G., 1969, Structural evolution of the rift zones in the Middle East, *Nature* **223:**55-57.

Roberts, D. G., 1970, A discussion mainly concerning the contributions by Hutchinson and by Baker, *Royal Soc. London Philos. Trans.* **A267:**399-405.

Sadek, H., 1928, The principal structural features of the Peninsula of Sinai, *14th Intern. Geol. Congr. Proc.* **3:**895-900.

Said, R., 1962, *Geology of Egypt,* Elsevier, Amsterdam, 377p.

Searle, R. C., 1970, Lateral extension in the East African rift valleys, *Nature* **277:**267-268.

Shalem, N., 1952, La seismicité au Levant, *Israel Research Council Bull.* **2:**1-16.

Shaw, S. H., 1947, *Southern Palestine, geological map on 1:250,000 with explanatory notes,* Government Printer, Jerusalem.

U.S. Geological Survey, 1963, *Geological Map of the Arabian Peninsula,* Scale 1:2m, Arabian Oil Co., Saudi Arabia, U.S. Dept. of State.

Van der Voo, R., 1968, Jurassic, Cretaceous and Eocene pole positions from northeastern Turkey, *Tectonophysics* **6:**251-269.

Van Dongen, P. G., R. Van der Voo, and Th. Raven, 1967, Paleomagnetic research in the Central Lebanon Mountains and in the Tartoies Area (Syria), *Tectonophysics* **4:**35-53.

Vroman, A. J., 1961, On the Red Sea rift problem, *Israel Research Council Bull.* **10G:**321-338.

Willis, B., 1928, Dead sea problem: rift valley or ramp valley, *Geol. Soc. America Bull.* **39:**490-542.

Willis, B., 1938, Wellings's observations of Dead Sea structure (with discussion by Bailey Willis), *Geol. Soc. America Bull.* **49:**659-668.

Wetzel, R., and M. Morton, 1959, Contribution à la Géologie de la Transjordanie, in *Notes et Mémoires sur le Moyer-Orient,* vol. 7, Museum National d'Histoire Naturelle, Paris, pp. 95-191.

Wolfart, R., 1967, *Geologie von Syrien und dem Libanon,* Gebrüder Borntraeger, Berlin, 326p.

Woodside, J., and C. Bowen, 1970, Gravity anomalies and inferred crustal structure in the eastern Mediterranean Sea, *Geol. Soc. America Bull.* **81:**1107-1125.

Wu, F. T., I. Karcz, E. J. Arieh, U. Kafri, and U. Peled, 1973, Micro-earthquakes along the Dead Sea Rift, *Geology,* **1:**159-161.

Zak, I., and R. Freund, 1966, Recent strike slip movements along the Dead Sea Rift, *Israel Jour. Earth Sci.* **15:**33-37.

24

Copyright ©1967 by Masson et Cie, Paris

Reprinted from *Rev. Géographie Phys. et Géologie Dynam.* **9**:3-16 (1967)

REMARQUES SUR LE FOSSÉ DE LA MER MORTE
ET SES PROLONGEMENTS AU NORD JUSQU'AU TAURUS

par L. DUBERTRET *

RESUME. — Une dérive de la péninsule Arabique vers le N, le long du sillon de la mer Morte, semble en faveur actuellement comme explication de la tectonique et de la morphologie des régions voisines. Tandis que certains auteurs la font commencer au début du Miocène et admettent qu'elle se poursuit de nos jours, les données de terrain indiquent qu'elle pourrait avoir débuté à la limite du Précambrien et du Cambrien et s'être poursuivie tout au plus jusqu'à la fin du Jurassique. — Les grandes failles méridiennes du sillon de la mer Morte et leurs prolongements au N du Taurus seraient des traits anciens ayant déterminé l'ordonnancement de la chaine de massifs trapus et de dépressions longeant les rives orientales de la Méditerranée. Leur genèse est cependant expliquée par la subsidence d'une aire aujourd'hui noyée sous la Méditerranée.

ABSTRACT. — While large scale northward drift of the Arabian peninsula relative to Sinaï and Palestine in early Miocene and in Pleistocene times has been favoured recently to explain the tectonics and the morphology of the areas surrounding the Dead Sea rift, field evidence points to wrench faulting going back at least as far as Upper Jurassic time and originating perhaps even in the late Precambrian. The ancient fault system along the Dead Sea rift and its extension to the North as far as to the Taurus ranges gave its pattern to the chain of large massives and depressions which border the eastern shore of the Mediterranean Sea; but these are considered due to the subsidence of a landstrip lying now beneath the Mediterranean Sea.

Voici bien des années je faisais part à mon regretté Maître Léon Lutaud de mon étonnement devant l'interprétation communément acceptée de la chaîne des massifs et dépressions formant la bordure orientale de la Méditerranée. Sa réponse fut : « Mais décrivez-la donc ». Une description structurale complète implique une analyse de la stratigraphie, puisque celle-ci permet de reconstituer la genèse de l'état actuel. En attendant de pouvoir répondre dans ce sens à l'avis de mon Maître, je vais m'efforcer de dégager les traits majeurs de la région intéressée, de rappeler comment elle a été interprétée; je présenterai enfin quelques suggestions.

Ce mémoire s'appuie sur les cartes géologiques et les études stratigraphiques précises, qui existent désormais pour l'ensemble de la bordure orientale de la Méditerranée. Ces documents représentent pour nous un grand avantage sur les géologues de jadis, qui ne pouvaient que parcourir des itinéraires et rapporter leurs observations à des cartes imparfaites, sur lesquelles l'ordonnance structurale apparaissait mal.

L'alignement des cassures sensiblement méridiennes qui s'étend de l'entrée du golfe d'Akaba jusqu'au pied du Taurus, à Maraş, sur 1 150 km de longueur, et les massifs et dépressions qui les jalonnent constituent l'un des traits marquants de la péninsule Arabique. Par leur dimension, leur style structural et leur histoire, les massifs appartiennent à une même famille et leurs caractères diffèrent de ceux des autres structures de la péninsule Arabique.

Ce grand ensemble demande une description précise, car la richesse de ses manifestations ne peut se réduire à des formules simples, comme horsts et fossés, ou anticlinaux et synclinaux; il ne peut être figuré par deux grandes failles parallèles comme celà se fait parfois.

De plus une explication peut être tentée; les hypothèses ont été nombreuses et n'ont pas toujours tenu compte des faits.

* C.N.R.S. Centre de Recherches sur le Moyen-Orient. Institut de Paléontologie, Muséum national d'Histoire naturelle, 8, rue Buffon, Paris (5ᵉ).

I. — APERÇU STRUCTURAL

A) LE SILLON DE LA MER MORTE

La *partie méridionale* du grand ensemble (fig. 1) frappe par sa simplicité : large trait linéaire, légèrement concave vers l'W, un sillon contenant en son fond la *mer Morte*, la dépression la plus basse du globe, avec plan d'eau à —392 m, et en outre profonde de 400 m. De part et d'autre, au S et au N, s'étendent des plaines de 5 à 25 km de largeur : la mer Morte est séparée du golfe d'Aqaba par le *Ghor* et l'*Wadi Arabah*; la partie septentrionale du sillon sert de vallée au Jourdain : il jaillit du pied de l'Hermon et traverse successivement la plaine du Houlé, cote + 70 m, aujourd'hui drainée. et le *lac de Tibériade*, cote —212 m.

Par *sillon de la mer Morte* sera désigné ici tout l'ensemble depuis le golfe d'Aqaba jusqu'au pied de l'Hermon. Il sépare le plateau de la Transjordanie et du Haouran (Syrie) de la presqu'île du Sinaï et de la Palestine : les deux régions juxtaposées sont fort dissemblables.

Le *plateau transjordanien* et le *Haouran* se rattachent à la plateforme Arabique, ils en constituent la marge. Du fait d'un très léger pendage vers le NE et le N, le socle précambrien. qui apparaît en surface au-dessus du golfe d'Aqaba, s'enfonce vers le N sous le sédimentaire; il reste visible le long du pied du plateau transjordanien jusqu'à la pointe S de la mer Morte. Il est recouvert par des grès rouges : au-dessus d'Aqaba ceux-ci sont cambriens et ordoviciens, puis crétacés; les grès crétacés sont coiffés par des calcaires cénomaniens-turoniens, des formations crayeuses, siliceuses et phosphatées du Sénonien; la surface du plateau est au voisinage de la cote 1 400 m. Vers le N le complexe gréseux se modifie, la lacune entre l'Ordovicien et le Crétacé se réduit, et successivement apparaissent des indentations calcaires : calcaires cambriens moyens à une cinquantaine de kilomètres au S de la mer Morte, calcaires triasiques vers la pointe N, et calcaires jurassiques un peu plus au N. Les calcaires crétacés s'enfoncent au N du Jebel Adjloun sous des craies sénoniennes et éocènes, et celles-ci, à leur tour disparaissent au N du Nahr Yarmouk sous les nappes basaltiques du Haouran. La surface du Haouran se tient entre 500 et 1 000 m.

Sur toute sa longueur, le bord du plateau transjordanien est haut et net : incontestablement une grande faille maîtresse ou bien une suite de grandes failles se relayant longent son pied. Ces failles sont directement visibles de loin en loin, mais sur de

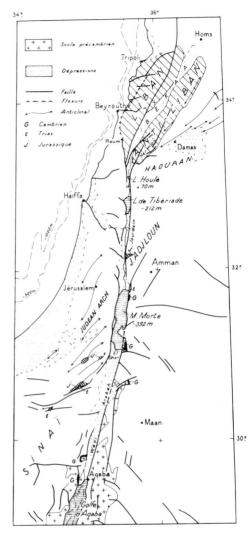

FIGURE 1

Le sillon de l'Wadi Arabah, de la mer Morte et du Jourdain, et ses prolongements à travers le Liban, la Békaa et l'Anti-Liban; esquisse structurale.
Echelle 1/4 000 000

Sensiblement méridien, le sillon est légèrement concave vers l'W. Remarquer le décrochement des affleurements les plus méridionaux des calcaires cambriens (G), triasiques (t) et jurassiques (j) de part et d'autre du sillon.

D'après la feuille 16 de la carte tectonique d'Europe; données fournies par F. Bender pour la Jordanie et par Y. K. Bentor pour Israël.

longs tronçons leur trace est ensevelie sous le colmatage néogène et pléistocène du fond du sillon. Dans la région dominant Aqaba, le socle et sa couverture de grès paléozoïques sont affectés par des failles parallèles aux côtes de la mer Rouge et aussi par des tronçons de failles parallèles au sillon de la mer Morte. Plus au N, de petites failles de diverses directions pénètrent dans le bord du plateau, comme des déchirures.

La région *Sinaï - Palestine* est très différente : elle est *basse*, comme l'avait fait remarquer M. Blanckenhorn; elle est faillée, morcelée, plissée selon des axes SSW-NNE et son dispositif structural est tout différent, ce n'est pas un plateau.

Le socle précambrien n'y est visible que jusqu'à une trentaine de kilomètres au N du golfe d'Aqaba; le calcaire cambrien moyen affleure au golfe; le Trias et le Jurassique s'avancent bien au S de la pointe S de la mer Morte. Il y a comme un *décrochement*, de l'ordre de 150 km, dans la distribution des indentations calcaires cambriens moyens, triasiques et jurassiques de part et d'autre du sillon de la mer Morte (F. E. Wellings, 1928).

De la côte vers l'intérieur, il y a montée des couches, culmination, puis descente vers le sillon de la mer Morte. Ce mouvement d'ensemble constitue en particulier le « Judean arch », à noyau calcaire cénomanien-turonien et à flancs sénoniens et éocènes. Jérusalem édifiée sur la partie culminante de la voûte, est à la cote 900 m. Ainsi le Sénonien et l'Eocène descendent dans l'Wadi Araba et dans la vallée du Jourdain et s'enfoncent sous le colmatage du fond du sillon, sans paraître limités par des failles.

Il existe cependant des failles sur la retombée vers le sillon de la mer Morte. La plus marquée est celle qui longe la rive W de la mer Morte et se prolonge de part et d'autre, atteignant ainsi une longueur d'une centaine de kilomètres : elle est juxtaposée au creux le plus profond du sillon. — La dépression du Houlé est encadrée à l'W comme à l'E par des failles (fig. 4, coupe aa'); il est vrai que de part et d'autres s'étendent des plateaux : plateau du Liban sud à l'W, plateau du Haouran à l'E.

L'examen de la carte géologique d'Israël (L. Picard, 1959) suggère très fortement l'existence *d'une* grande faille maîtresse, à grand rejet, longeant le pied du plateau transjordanien, et le long de laquelle la région Sinaï-Palestine serait juxtaposée à la région Transjordanie - Haouran. La faille de la rive W de la mer Morte et les autres failles méridiennes décelables au bas de la retombée des couches de la région Sinaï - Palestine, de rejets

moindres, représenteraient un cortège de failles subsidiaires.

L. Lartet, dès 1877 (pp. 259-262), était arrivé à cette conclusion et cherchait la faille unique, maîtresse, dans l'axe de la mer Morte.

B) LES PROLONGEMENTS DES STRUCTURES DU SILLON DE LA MER MORTE VERS LE N, JUSQU'AU TAURUS

La prédominance d'une faille maîtresse et ses possibles subdivisions en un faisceau de failles subsidiaires de plus ou moins grand rejet sont confirmées par l'ordonnancement des prolongements des structures du sillon de la mer Morte vers le N jusqu'au Taurus.

Liban, Békaa et Anti-Liban.

La faille occidentale de la dépression du Houlé forme un faisceau dont le prolongement principal est la *faille de Yammouneh* (fig. 2), de direction SSW-NNE tout le long de la retombée orientale du Liban, et se poursuivant au-delà en direction méridienne. La *faille du Roum*, deuxième branche passe à une ligne de flexures enveloppant le haut-massif du Liban à l'W. Enfin la *faille de Hasbaya*, qui s'incurve vers l'Anti-Liban, ne joue qu'un rôle subordonné. La faille orientale de la dépression d'Houlé se prolonge par la *faille de Rachaya* et la *faille de Serrhaya*; cette dernière, parallèle à la faille de Yammouneh, coupe l'Anti-Liban en diagonale et joue un rôle dominant.

Le rôle de ces diverses failles apparaît sur les deux coupes transversales passant l'une par le S du système Liban, Békaa, Anti-Liban (la faille de Hasbaya n'est pas figurée) et l'autre par le N (fig. 4, coupes bb' et cc'). Le schéma structural apparaît dans les coupes, fig. 5, où n'est montrée que la surface structurale du Jurassique.

La partie axiale du massif du *Liban* se dégage bien en « touche de piano [1] » entre la faille de Yammouneh et la faille de Roum et les flexures qui prolongent celle-ci. L'*Anti-Liban*, qui dans l'ensemble se présente aussi comme une « touche de piano », est cependant plus complexe. L'Hermon est une voûte à axe SW-NE, cassée sur sa retombée vers le SE. En schématisant à l'extrême, on peut considérer que depuis la faille de Yammouneh jusqu'au sommet de l'Hermon, il y a montée monoclinale, et au delà retombée par cassures. La Békaa, dépression encaissée entre le Liban et l'Anti-Liban, occupe dans sa partie S le bas du monoclinal de l'Hermon, et dans ses parties centre et N le fond

1. Terme suggéré par M. Lugeon à la vue de la première carte géologique Liban-Syrie au 1/1 000 000°.

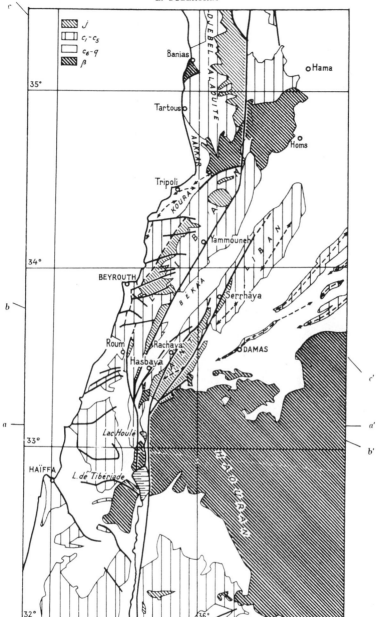

FIGURE 2
**Liban et Anti-Liban, entre la vallée supérieure du Jourdain
et la montagne Alaouite; esquisse structurale.**
Echelle 1/2 000 000
Les deux failles méridiennes encadrant la dépression du Houlé éclatent
au N en un faisceau de failles SSW-NNE; seule la faille de Yammounech
se poursuit dans le massif Alaouite; elle prend la direction méridienne
dès la pointe N du Liban.

343

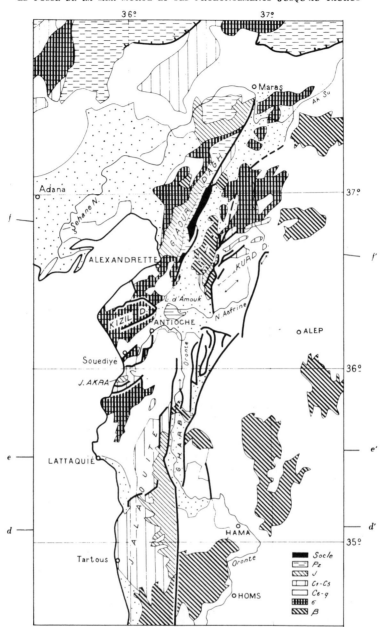

FIGURE 3

Massif Alaouite, Gharb et Jebel Zawiyé, Amanus (Kizil Dagh et Giaour Dagh) et Kurd Dagh; esquisse structurale.
Echelle 1/2 000 000

Dispersion des failles au N du Gharb et sur le pourtour de l'Amouk; mais une faille maîtresse continue le long du pied du Giaour Dagh, jusqu'à Maraş, où elle s'efface devant le front des chevauchements du Taurus.

du compartiment affaissé entre la faille de Yammouneh et la faille de Serrhaya et plié en synclinal.

Les failles méridiennes qui encadrent la plaine
du Houlé passent ainsi vers le N à un faisceau de
failles dans l'ensemble couchées dans la direction
SSW-NNE. Deux d'entre elles découpent les grandes
unités structurales du Liban, il y a harmonie entre
le tracé de ces failles et la structure du massif. Les
autres recoupent *obliquement* l'Anti-Liban et en
causent la complexité. Les deux failles de Yammouneh et de Serrhaya sont parallèles, elles donnent
toutes deux des rejets équivalents. Le compartiment intercepté comprend, dans le S, le massif de
l'Hermon (2 814 m), où le Jurassique culmine; à
hauteur de Serrhaya, il est ployé en synclinal dont
le fond est occupé par la plaine alluviale de la
Békaa proprement dite; plus loin vers le N, la
faille de Serrhaya s'estompe en une flexure, le fond
synclinal est occupé par des poudingues néogènes.

Massif Alaouite, Gharb et Jebel Zawiyé.

De ce faisceau de failles, seule la faille de Yammouneh se poursuit vers le N en *pays Alaouite*.
Bientôt elle se dédouble, puis se disperse en éventail vers le NE (fig. 3 et fig. 4, coupes dd' et ee').
Dans sa partie méridionale, une coupe transversale
du massif Alaouite montre ainsi une simple juxtaposition par faille d'un plateau très doucement
redressé, jurassique dans sa partie haute, et d'un
plateau bas, cénomanien-turonien (coupe dd'). A
noter le broyage des calcaires dans la zone de la
faille. — La partie septentrionale du massif est d'un
relief plus vigoureux, sa structure se rapproche
davantage d'un horst N-S. Au pied des crêtes jurassiques, portées à 1 562 m, s'étend la plaine marécageuse du *Gharb*, cote 170 m; une faille sépare
celle-ci à l'E du *Jebel Zawiyé*, crétacé, 751 m
(coupe ee') : le Gharb est effectivement un fossé
encadré de failles, mais les massifs encaissants ne
sont pas équivalents, l'un étant tectoniquement
beaucoup plus haut que l'autre : à l'W, la surface
structurale du Jurassique se situe à 1 560 m, à l'E
elle est au-dessous du niveau du Gharb, c'est-à-dire
environ 1 500 m plus basse.

Giaour Dagh et Kurd Dagh.

A la *hauteur d'Antioche*, la structure se com-
plique (fig. 3) ; la plaine marécageuse de l'Amouk,
cote 80 m, occupe le centre d'un dispositif de failles
en étoile. La basse vallée de l'Oronte, entre Antioche et Souédiyé, est encadrée de failles d'orientation générale SW-NE. L'éventail de failles subméridiennes de la région de Harim s'arrête sur
la faille W-E de Yeni Sehir. Les grandes failles

méridiennes du S trouvent leur prolongement dans
une grande faille SSW-NNE qui longe le pied du
Giaour Dagh.

Ce massif, long d'une centaine de kilomètres
culmine à 2 224 m; il présente une morphologie générale qui rappelle de très près les massifs libano-
syriens; sa coupe transversale (fig. 4, coupe ff') a
la même allure générale. Il offre cependant une ori-
ginalité due à sa proximité du Taurus : un substratum précambrien, plissé dans la direction WSW-
ENE, pointe à sa base; la plus grande partie du
massif est constituée de quartzites, calcaires et
schistes cambriens et ordoviciens, reposant horizon-
talement sur le Précambrien et s'incurvant doucement dans la partie W du massif, de sorte que leur
profil rappelle les profils transversaux des massifs
libano-syriens; des calcaires mésozoïques recouvrent
la retombée à l'W; au bas du versant ils sont sur-
montés de péridotites et serpentines.

En face du Giaour Dagh, à l'E, s'étend le *Kurd
Dagh*, pays bas, crétacé, plissé dans la direction SW-
NE. Une faille de moindre rejet, grossièrement
parallèle à celle du pied du Giaour Dagh, la limite
à l'W. Et entre les deux failles est encaissé le *fossé
du Kara Su*. Au milieu de celui-ci émergent des
collines de péridotites; des nappes basaltiques plé-
istocènes en couvrent la majeure partie.

Comme dans le cas de la partie septentrionale
du massif Alaouite, les massifs encadrant le fossé
du Kara Su ne sont pas équivalents, l'un, à l'W,
étant tectoniquement beaucoup plus haut que l'autre, à l'E : le rejet de la faille du Giaour Dagh est
de l'ordre de 2 500 - 3 000 m; celui de la faille du
Kurd Dagh n'est que de quelques centaines de
mètres.

Les grandes failles subméridiennes s'estompent à
Maras, devant le front WSW-ENE des chevauchements du Taurus.

Cette brève description confirme le contraste
entre le sillon de la mer Morte et ce qui a été consi-
déré comme son prolongement au N de la dépression du Houlé : plus grande complexité, plus grande
variété des structures au N. Cependant le principe
des structures (fig. 5) reste le même depuis Aqaba
jusqu'à Maras. Il peut y avoir simple juxtaposition
de deux régions le long d'une faille unique, comme
dans certaines parties du sillon de la mer Morte ou
dans le S du massif Alaouite. La faille peut aussi se
diviser en deux ou trois ou en un faisceau de failles
parallèles, éclatement favorable à l'affaissement ou
à la montée de longs compartiments, le compartiment le plus élevé faisant face normalement au
compartiment le plus affaissé.

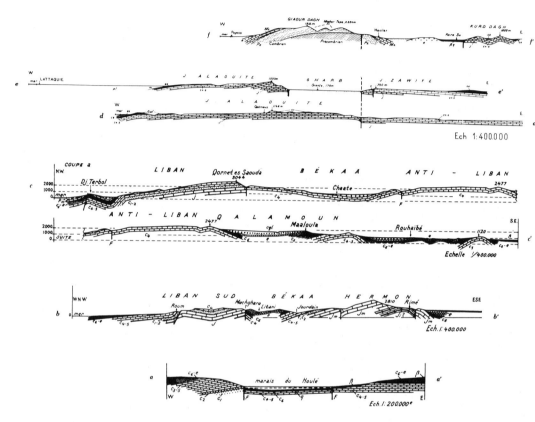

FIGURE 4

Houlé, Liban et Anti-Liban; massif Alaouite, Gharb el Jebel Zawiyé; Giaour Dagh et Kurd Dagh : coupes situées fig. 2 et 3.

ff' : Giaour Dagh, fossé du Kara Su et Kurd Dagh
ee' : N du massif Alaouite, Gharb et Jebel Zawiyé
dd' : S du massif Alaouite et plateau de Hama
cc' : N du Liban, de la Békaa et de l'Anti-Liban
bb' : S du Liban, Hermon
aa' : Dépression du Lac Houlé

Terrains figurés et signes conventionnels : socle précambrien, Paléozoïque inférieur (Pz), Mésozoïque (Mz), Jurassique (j) ; Grès de base du Crétacé (C₁), Aptien (C₂), Albien (C₃), Cénomanien (C₄), Turonien (C₅), Sénonien (C₆), Paléogène (e), Miocène (m), Pliocène (pl.), Quaternaire (q) ; Roches vertes maestrichtiennes (ε), basaltes néogènes et quaternaires (β).
Coupes à l'échelle de 1/400 000, sauf la coupe aa' qui est au 1/200 000.

346

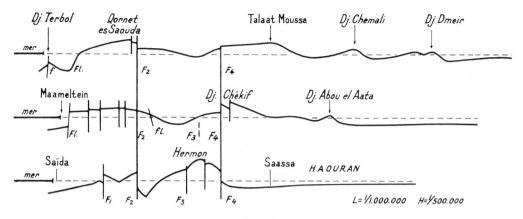

FIGURE 5
Liban, Békaa, Anti-Liban et Damascène, coupes transversales par le N, le centre et le S, montrant l'allure de la surface structurale du Jurassique.
D'un style cassant à l'W, il y a passage à un style plissé vers l'E.

Il est caractéristique que *toujours l'une des failles reste nettement prédominante* par l'importance de son rejet : faille de Yammouneh au Liban, faille occidentale du Gharb, faille au pied du Giaour Dagh ; de sorte que les régions ou massifs encadrant les compartiments affaissés ne sont pas la réplique l'un de l'autre : l'un est d'un relief considérable, l'autre n'est que peu marqué. La fig. 6 résume ce style ; c'est une coupe transversale de la partie N du massif Alaouite, du Gharb et du Jebel Zawiyé, sur laquelle n'est figurée que la surface structurale du Jurassique. Un fossé s'est formé, occupé par les marais du Gharb ; la surface structurale du Jurassique peut y être à la cote —1 000 m. A l'W, dans le massif Alaouite, la surface structurale du Jurassique est à 1 550 m ; à l'E, sous le Jebel Zawiyé, elle doit se situer au voisinage de la cote 0.

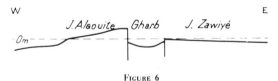

FIGURE 6
N du massif Alaouite, fossé du Gharb et Jebel Zawiyé, coupe W-E montrant l'allure de la surface structurale du Jurassique.
Cette coupe caractérise le style structural de la chaîne des massifs et dépressions longeant la bordure orientale de la Méditerranée. L'une des failles s'affirme comme maîtresse par un rejet nettement prédominant ; les massifs de part et d'autre du fossé sont inégalement exhaussés. Ce n'est pas le schéma d'une voûte à clef effondrée.

Sont caractéristiques également les dimensions et proportions des massifs qui jalonnent ces failles ; ce sont des massifs trapus, de 20-22 km de large et d'environ 150 km de long, très différents des plis qui de la Damascène se dirigent vers le NE.

II. — ESSAIS D'EXPLICATION

Le sillon de la mer Morte et ses prolongements jusqu'au Taurus ont été communément considérés comme les manifestations les plus septentrionales des grands accidents de l'Est africain. Trait remarquable de l'écorce terrestre, il a depuis longtemps suscité l'intérêt des explorateurs et leurs explications.

Cependant les vues exprimées ont concerné plus spécialement le seul sillon de la mer Morte : les problèmes posés par ses prolongements ont été parfois évoqués, mais succinctement. Le grand ensemble n'est pas sans rapports avec la mer Rouge. Seule une large étude régionale pourra être concluante.

Toute tentative d'explication devrait comporter la définition des phénomènes tectoniques qui sont intervenus et des précisions sur l'époque à laquelle ils ont joué. Les mêmes données peuvent en effet s'interpréter différemment selon leur place dans l'histoire tectonique.

Les recherches ont été orientées dans trois directions différentes.

1) Effondrement d'une clef de voûte

E. Suess, d'après O. Fraas, décrit un *effondre-ment* du sillon de la mer Morte entre failles paral-lèles (I, p. 476) : d'après Fraas, « il existe... des failles parallèles en gradins... mais tandis qu'à l'E l'effondrement s'est produit selon une grande faille, il s'est formé à l'W plusieurs fentes parallèles, de telle sorte que la région occidentale s'est affaissée non pas en bloc, mais par gradins successifs, et qu'un fossé dissymétrique a pris naissance.. (p. 477) une faille simple peut former à la surface du sol un gradin, mais non une vallée... il faut que des bandes de terrain se soient affaissées suivant des cassures parallèles, sur une grande longueur et à une pro-fondeur inégale... (p. 478) la mer Rouge elle aussi est un fossé d'effondrement ».

L'effondrement de la mer Morte serait, d'après Fraas (1867, p. 79), antérieur au Tertiaire.

M. Blanckenhorn (1914, pp. 75-78) décrit une série de coupes transversales au sillon de la mer Morte, en insistant sur sa dissymétrie et sa com-plexité; il situe les dislocations à la limite du Plio-cène et du Pléistocène.

L. Picard (1943, p. 4), également partisan de la théorie de l'effondrement, écrit au sujet de la mer Morte, de la vallée du Jourdain et de la dépression du Houlé : " They... originated in two considerable border faults and can therefore be designated as *graben* or *rift valleys*. These ... depressions origi-nated for the most part at the end of the Tertiary or at the beginning of the Quaternary. Neverthe-less the origin of certain important depressions go back as far as Miocène ".

2) La " ramp valley " de B. Willis

B. Willis (1928) cherche une autre explication en posant la question : " Dead Sea, rift valley or ramp valley ? ". Il se résume ainsi (p. 492) : " The Pa-lestine and Transjordan plateau present arched surfaces or swells caused by vertical protrusion of solid rock mass in consequence of deapseated hori-zontal shortening ".

" The cliffs that wall the Dead Sea trough are due to upthrust faults or ramps produced by deep seated shearing. They are conceived to have origi-nated beneath the Palestine and Transjordan pla-teau and to curve upward toward a vertical atti-tude. The plateau blocks have risen on them ".

" The Dead Sea block, the mass beneath the trough and spreading beyond it, has been pushed down by overloading and pressure on the ramps ".

Il situe le phénomène entre la fin du Tertiaire et le Pléistocène.

3) Glissement le long des failles, dérive

Les deux points de vue précédents représentaient l'alternative entre une genèse du sillon de la mer Morte par distension ou par compression. La troi-sième solution possible, un glissement horizontal relatif des deux régions opposées de part et d'autre de la faille maîtresse du sillon a été suggérée par L. Lartet et développée par moi. J'y suis arrivé in-directement.

Dès 1869 (note infrapaginale, p. 13). L. Lartet avait noté : « L'explication de la formation de la mer Rouge par' une fracture de l'écorce terrestre, origine que Dolomieu avait pressentie, est mise en évidence par les contours des deux rivages opposés de ce golfe qui se correspondent obliquement, ainsi que les deux lèvres d'une déchirure. En effet, si par la pensée on rapproche le rivage égyptien de la mer Rouge du rivage arabique, en faisant subir au premier un très léger déplacement vers le Nord, on voit qu'ils s'appliqueraient exactement l'un sur l'au-tre, sauf toutefois vers l'entrée du golfe où des accidents volcaniques d'une grande importance sont venus introduire de profondes modifications dans le relief du sol ».

Frappé par ce même fait, j'avais cherché à re-constituer le bloc unique que l'Afrique, l'Arabie et le Sinaï avaient dû former avant leur disloca-tion, et à apprécier quels mouvements relatifs avaient conduit à la situation présente (fig. 7). Mes conclusions furent les suivantes (1932, p. 66) :

« L'hypothèse d'une dérive des socles étant admise et les déplacements rapportés au socle de l'Arabie supposé fixe, la dislocation du continent primitif peut se ramener à : 1° un glissement du bloc du Sinaï de 160 km vers le Sud, le long de sa bordure orientale; 2° une rotation du socle africain de 6°4, dans le sens des aiguilles d'une montre, autour d'un centre situé dans la mer Ionienne ».

Ces vues, proposées comme hypothèse de travail, ne plurent pas à H. Douville ! Elles comportaient une faute : l'hypothèse d'une crevasse égale à toute la largeur de la mer Rouge, et une omission : je ne précisais pas l'époque de ces mouvements relatifs.

Mais je crus mes vues confirmées lorsque F. E. Wellings, au retour d'une mission dans l'Wadi Arabah, me dit avoir constaté un décrochement de 150 km dans les rivages [2] du Cambrien moyen, du Trias et du Jurassique de part et d'autre du sillon de la mer Morte. Ultérieurement il précisa

2. Il ne s'agit pas exactement de rivages, mais des limites d'extension des indentations calcaires.

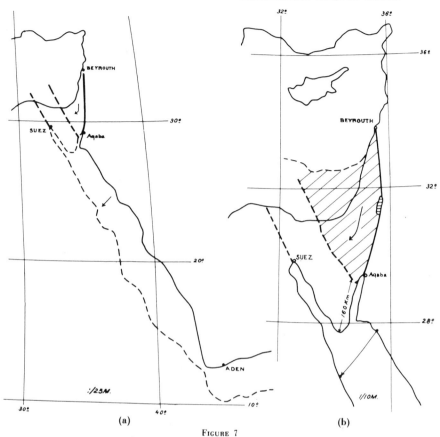

(a)   (b)

FIGURE 7

Schéma de la dislocation du continent Afrique - Sinaï - Arabie, telle qu'elle était conçue par L. Dubertret (1932)

(a) ensemble au 1/25 millions, (b) détail au 1/10 millions.

A l'origine n'auraient existé qu'une grande cassure de la mer Rouge, NNW-SSE et une autre grande cassure du golfe d'Akaba du sillon de la mer Morte, se prolongeant en direction méridienne vers Beyrouth. La péninsule Arabique étant supposée fixe, l'Afrique aurait dérivé vers le SW et le bloc du Sinaï aurait glissé vers le S en accomplissant une légère rotation dans le sens des aiguilles d'une montre. — L'esprit de l'hypothèse reste valable, mais l'origine des mouvements relatifs doit être recherchée dans un passé plus lointain qu'il n'avait été envisagé et l'ampleur du mouvement doit être considérablement réduite.

(1938, p. 660) : "The fault movements can thus be dated at least as far back as the end of the Jurassic ".

En effet, dans notre pensée, le mouvement de glissement vers le S s'était accompli, au N de la dépression de Houlé, *le long de la faille de Roum*, prolongement rectiligne du sillon du Jourdain, et non le long de la faille de Yammouneh, qui est couchée dans la direction SSW-NNE. Or, au N de Roum, le Crétacé passe d'une lèvre à l'autre de la faille sans montrer le moindre signe de dérangement.

B. Willis, en présentant les observations de F. E. Wellings (1938, p. 666), qui semblaient confirmer l'hypothèse de dislocation d'une plus large unité continentale, note : " The seeming reality of this continuity is a mirage which will disappear with the intellectual atmosphere in which it is reflected ".

A. M. Quennell (1956), après un séjour prolongé en Jordanie, pendant lequel il avait dressé une carte géologique au 1/250 000ᵉ du pays, reprit l'idée d'un glissement relatif du bloc Sinaï - Palestine

349

par rapport à la péninsule Arabique, mais en rapportant les mouvements au bloc Sinaï - Palestine supposé fixe. D'après lui, la dérive de la péninsule Arabique vers le N, le long de la faille maîtresse du sillon du Jourdain et *de la faille de Yammouneh* se serait accompli en deux étapes, l'une de 62 km au Miocène inférieur, l'autre de 45 km au Pléistocène; cette dernière se poursuivrait de nos jours. Ses « restaurations » de la situation antérieure à la dérive l'amènent à conclure à la continuité originelle de certains éléments structuraux et morphologiques de part et d'autre du sillon.

Mais il rencontre une difficulté dans l'obliquité de la faille de Yammouneh (1956, p. 394) : " The Lebanon section of the rift, the Yammoûné fault, is oblique to the general direction. Restauration left a gap 25 km wide which is believed to have been formed by crustal shortening due to the Palmyra folding, the upthrusting of the Lebanon and Syrian horst, and by their displacement to the northeast as wedges of the crust between converging wrench faults. The Haouran vulcanicity is believed to be related ". Le contraste est surprenant entre la minutie avec laquelle Quennell retrouve et situe tous les éléments structuraux et morphologiques du sillon de la mer Morte, et l'aisance avec laquelle il fait disparaître une bande d'écorce terrestre de Syrie large de 25 km.

A. J. Vroman (1961) aborde le problème de la mer Rouge par la méthode expérimentale; ses critiques sont pertinentes (p. 324) : " As the adherents of the drift theory date the birth of the Red Sea floor from early Cretaceous time, sediments- either continental or marine -older than Cretaceous must be absent from the rift floor ".

" Older rocks are exposed in one major region in the rift floor : the triangle of the Danakil Hills (W. Ethiopia...). Here several small cores of the old Precambrian basement are surrounded by a large mantle of Jurassic and Triassic sediments. On the basis of the drift theory the Red Sea must therefore be of pre-Triassic, if not of Precambrian age "...

Puis (p. 329) : " ...nowhere along the entire length of the Dead Sea-Jordan rift are there any indications of large-scale longitudinal displacements, and certainly not a slip of 100 kilometers ".

Cette dernière constatation est valable aussi pour la région libano-syrienne. Il a été mentionné déjà que le Crétacé passe par dessus le prolongement de la faille de Roum sans présenter la moindre dislocation; d'autre part le long de la faille de Yammouneh existent des lambeaux qui font pont d'un côté à l'autre de la faille; ces faits sont incompatibles avec des déplacements horizontaux.

Malgré ces avertissements de A. J. Vroman, R. Freund (1965) revient à l'idée " that a post Turonian sinistral normal fault movement of about 70-80 km occured along the Dead Sea fault ".

A. S. Laughton (1966) s'intéresse du point de vue océanographique à une dérive de la péninsule Arabique vers le N et précise qu'un déplacement de 100 km *a été mesuré*; mais il conclut : " The whole story seams plausible enough, but much of it is speculation ". Dans le golfe d'Aden, la dérive de la péninsule Arabique vers le N atteindrait d'après lui 2 cm par an.

III. — DISCUSSION

Une dérive de la péninsule Arabique vers le N le long du sillon de la mer Morte, ou au contraire un glissement de la presqu'île du Sinaï avec la Palestine vers le S (peu importe la parcelle continentale prise comme référence) ne peut être évoquée sans que l'on songe également à un mouvement relatif de la presqu'île du Sinaï et de l'Afrique. Finalement c'est bien de la dislocation d'une plus large unité continentale en trois parcelles dont il est question : rechercher à quel moment elle a eu lieu, quels mouvements elle a impliqués, me paraît une voie de recherche valable et ne pas être plus critiquable que les cascades d'hypothèses de certains auteurs.

C'est l'ensemble de la mer Rouge et du sillon de la mer Morte qu'il faut considérer. L'amplitude d'une dérive ou de mouvements relatifs pourra être appréciée le jour où la structure du fond de la mer Rouge sera connue et où l'on pourra chiffrer la largeur de la « crevasse » supposée exister dans sa partie axiale. Elle ne représente peut-être que le quart ou le dizième de la largeur de la mer Rouge et il est presque certain que l'amplitude du mouvement relatif de la péninsule Arabique et de la presqu'île du Sinaï sera finalement reconnue être bien inférieure aux chiffres avancés jusqu'ici.

La rareté des témoins stratigraphiques rend l'étude de la mer Rouge difficile; par contre le golfe de Suez, le sillon de la mer Morte et ses prolongements sont aujourd'hui bien connus, ce qui permet un certain nombre de *constatations*.

Le sillon de la mer Morte sépare la presqu'île du Sinaï et la Palestine de la péninsule Arabique. A. M. Quennell et R. Freund ont admis comme ligne de séparation des deux blocs, au N de la dépression du Houlé, la faille de Yammouneh, orientée SSW-NNE. Ils se sont ainsi trouvés obligés de faire « disparaître » une bande d'écorce terrestre large de 25 km et longue de 170 km. Je pense que la ligne de séparation devrait être cherchée dans la faille de Roum et son prolongement au N en direction de Beyrouth, sous le Crétacé : cette faille est le prolongement direct, vers le N, de la faille ou des failles du sillon de la mer Morte. La faille de Yammouneh et son cortège de failles subordonnées par contre recoupent la marge de la plateforme Arabique, c'est-à-dire se situent à l'intérieur de celle-ci.

Des témoignages indépendants convergent à constater *l'absence de toutes traces visibles sur le terrain de mouvements horizontaux de grande envergure.* — En ce qui concerne le golfe de Suez, F. Heybroek (1965, p. 19) affirme : " The faults observed in the Clysmic Gulf [3] are all normal faults; there is no simple observations which might point to compressional or *lateral movements*, and I am therefore convinced that the Gulf of Suez and the Red Sea Grabens have been formed by tension ". — A. J. Vroman de son côté note (1961, p. 329) : " But nowhere along the entire length of the Dead Sea-Jordan rift are there any indications of large-scale longitudinal displacements, and certainly not a slip of 100 kilometers ". — Je peux conclure en termes identiques pour les prolongements de la faille du sillon de la mer Morte au N de la dépression du Houlé, que je connais bien pour les avoir suivis au marteau et levés jusqu'à proximité du Taurus.

La conclusion s'impose : *les mouvements de glissement ne peuvent être qu'anté-crétacés.*

M. Blanckenhorn (1914) situait l'origine des grandes failles méridiennes et le début des effondrements du sillon de la mer Morte à la fin du Pliocène. Nous avons appris petit à petit à vieillir ces accidents. Il est d'abord apparu qu'au Liban de nombreuses failles commandant la morphologie des massifs jurassiques étaient injectées par des basaltes jurassiques supérieurs, donc qu'elles remontent au Jurassique supérieur. De cette constatation il n'y avait qu'un pas pour conclure que la faille de Yammouneh et les failles maîtresses qui l'accompagnaient étaient anciennes et avaient rejoué à travers la couverture crétacée. Des vérifications ont été possibles.

Une carte des isopaches du " Judea limestone ", c'est-à-dire des calcaires cénomaniens à santoniens, publiée par R. Wetzel et M. Morton (1959, p. 182, fig. 25), montre qu'au Crétacé la Palestine d'une part et la Transjordanie de l'autre étaient déjà nettement différenciées, qu'entre ces deux régions

3. Golfe de Suez.

existait déjà un contraste de relief semblable à celui que nous observons aujourd'hui. — Tenant compte enfin des décrochements des « rivages » du Cambrien moyen, du Trias et du Jurassique moyen de part et d'autre du sillon de la mer Morte (voir G. S. Blake, 1937, puis R. Wetzel et M. Morton, 1959) on pouvait être incité à conclure que le glissement le long du sillon de la mer Morte et de la faille de Roum s'était accompli à la fin du Jurassique; ainsi s'expliquerait le fait que le Crétacé et le Cénozoïque ne montrent aucune trace de déplacements horizontaux.

Cependant les « rivages » du Cambrien moyen, du Trias et du Jurassique moyen pourraient très bien ne pas avoir été décrochés et leur disposition résulter simplement d'une orographie semblable à celle que suggère la carte des isopaches des calcaires cénomaniens à santoniens : un bord de plateau transjordanien abrupt juxtaposé à une Palestine basse : le mouvement de dérive serait alors anté-Cambrien.

La géologie des environs immédiats d'Aqaba, l'extraordinaire essaim des dykes, basiques et acides, qui y recoupent le granite du socle et les conglomérats précambriens incrustés dans leur surface, enfin certains détails précis, indiquent qu'un événement tectonique majeur s'est accompli à la limite du Précambrien et du Cambrien : ce serait précisément l'apparition des grandes failles du sillon de la mer Morte.

C'est avec intérêt que j'ai lu le témoignage suivant de Y. K. Bentor (1961, p. 245) : " Until the end of Mesozoic times the region of the present Jordan-Aqaba graben was part of the mountain area of Palestine, but separated from the Transjordan block in the east by the *very old geosuture*, now forming the eastern border of the rift ".

Enfin, il est difficile de penser que l'apparition de cette géosuture n'ait pas été précédée ou accompagnée par celle des cassures qui ont donné naissance à la mer Rouge.

La dislocation de l'ancienne grande unité continentale réunissant l'Afrique, l'Arabie et le Sinaï aurait commencé alors dans le Précambrien, et la dérive des blocs formés pourrait s'être amorcée dès cette époque et s'être poursuivie jusqu'à la fin du Jurassique. — Le cortège d'accidents secondaires se serait manifesté ultérieurement à travers la couverture sédimentaire crétacée et cénozoïque, sans qu'une dérive importante n'ait eu lieu depuis la fin du Jurassique.

La genèse de la *chaîne des massifs et dépressions de la bordure orientale de la Méditerranée* s'est

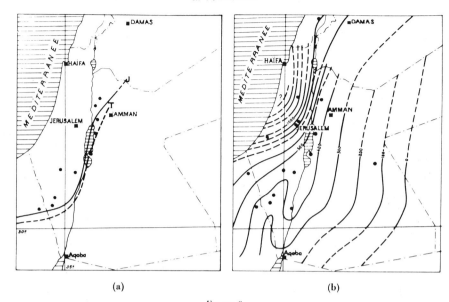

(a)   (b)

FIGURE 8
Israël et Jordanie :
(a) lignes de « rivage » du Trias et du Jurassique, (b) isopaches du « Judea limestone », c'est-à-dire des calcaires cénomaniens à santoniens; d'après R. Wetzel et M. Morton (1959, p. 182, fig. 25).
Le décrochement des lignes de rivage le long du sillon de la mer Morte est l'une des principales données qui aient suggéré une dérive du Sinaï vers le S par rapport à la péninsule Arabique (ou une dérive relative de celle-ci vers le N). La carte des isopaches conduit cependant à une autre explication : la préexistence d'un vigoureux contraste dans le relief de part et d'autre du sillon de la mer Morte, dont l'origine semble devoir être recherchée à la fin du Précambrien.

amorcée à la fin du Cénomanien. Aucune translation horizontale ne lui paraît liée. Ces massifs et dépressions accompagnent les grandes failles méridiennes et subméridiennes qui prolongent vers le N les failles du sillon de la mer Morte. Par leur disposition d'ensemble, par leurs proportions, leur style structural, ils forment, le long des rives orientales de la Méditerranée, un ensemble très particulier, qui se différencie des plissements du Sinaï ou de la Damascène et la Palmyrène.

Dans le double système montagneux du Liban et de l'Anti-Liban, L. Kober (1915) voyait un pays plissé dû à la surrection des chaînes tauriques et jouant par rapport à celles-ci le même rôle que le Jura par rapport aux Alpes. L'image ne me paraît pas pertinente.

En créant la notion « d'arc syrien », E. Krenkel (1924) associait des éléments hétérogènes, sans rien expliquer.

La voie d'une explication me paraît se trouver plutôt dans un schéma de horst donné par B. Wil

lis (1928, p. 500) et accompagné du commentaire suivant : "...a horst — that is, a mass which is elevated by vertical forces and is bounded by normal faults that slope down toward the adjacent basins. A common occurence outside of zones of superficial, horizontal pressure and presumably due to deep seated compression ".

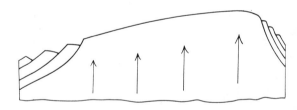

FIGURE 9
Coupe d'un horst d'après B. Willis (1928, p. 500, fig. 4)
Ce schéma rappelle la coupe transversale du Liban : les massifs de la bordure orientale de la Méditerranée semblent s'expliquent par des poussées verticales plutôt que par une compression horizontale.

La coupe du horst donnée par B. Willis (fig. 9) est à peu près celle du massif du Liban : région comprise entre une grande faille verticale à l'E et une flexure à l'W.

Sous l'effet de poussées horizontales, un massif limité par des faces divergeant en profondeur, devrait être refoulé en profondeur, m'avait fait remarquer, voici bien des années, le Professeur B. G. Escher (Leyden). Mais précisément je pensais que des massifs comme le Liban et l'Anti-Liban s'étaient mis en relief sous l'effet de poussées *verticales* causées par un écoulement, de la région méditerranéenne vers l'intérieur des terres, du matériel plastique sous-jacent à l'écorce terrestre solide, comme conséquence d'une subsidence d'une zone située aujourd'hui en mer (1947, pp. 17-19) : le problème se situerait (?) au niveau du manteau supérieur. Les grandes failles méridiennes ont joué un rôle déterminant, en permettant la mobilité des compartiments adjacents. Une interdépendance subsiste cependant entre les compartiments voisins, l'un ne pouvant s'enfoncer sans refouler latéralement le matériel plastique sous-jacent ou l'autre ne pouvant s'élever en créant un vide au-dessous de lui.

Une coupe transversale type se rapprocherait de celle passant par le N du massif Alaouite, le fossé du Ghab et le Jebel Zawiyé (fig. 6). La dissymétrie de part et d'autre du passage d'une faille maîtresse est caractéristique. Lorsque celle-ci éclate en un faisceau de failles parallèles, un compartiment étroit, situé du côté de la lèvre haute, se trouve mis encore davantage en relief, tandis qu'en face, de l'autre côté de la faille maîtresse, un autre compartiment étroit se trouve entraîné en profondeur et forme un fossé. Ce qui montre la complexité de l'effet de l'écoulement du matériel plastique qui soulève un massif, au-delà de la faille maîtresse qui lui a donné sa mobilité. Le phénomène est assez caractéristique pour mériter une étude expérimentale sur modèles réduits. Des structures de ce type sont tellement différentes du dispositif en voûte avec clef effondrée qu'elles ne devraient pas être appelées graben ou rift, mais recevoir un nom nouveau.

Dans le cadre de ces vues, la mer Morte ne correspondrait pas à une sorte de trou dans l'écorce solide, ainsi que le suggèrent les schémas de A. Quennell, mais résulterait de la subsidence d'une bande de terrain entre le bloc de la Palestine et celui de la Transjordanie.

Les vues exposées sur la tectonique du sillon de la mer Morte et de la chaîne des massifs et dépressions longeant la rive orientale de la Méditerranée, sont celles d'un géologue de terrain; la solution du problème nécessite une large confrontation de données diverses, de l'océanographie et de la géophysique en particulier.

Il est curieux que chacun des auteurs qui s'est penché de près sur le problème semble avoir détenu une parcelle de vérité. Nous ne sommes peut-être pas loin d'une synthèse susceptible de rallier le plus grand nombre d'entre eux.

Note. — L'auteur de cet article sera reconnaissant à ceux qui voudront bien lui faire part de leurs remarques sur le problème discuté.

BIBLIOGRAPHIE SOMMAIRE

BENDER, F. (1963). — Stratigraphie der « Nubischen Sandsteine » in süd-Jordanien. *Geol. Jahrb.*, 81, 237-276, II fig., 6 pl., 1 carte au 1/100 000.

BENTOR, Y. K. (1961). — Some aspects of the Dead Sea and the question of its age. *Geochimica et Cosmochimica Acta*, 25, 239-260, 1 pl.

BLAKE, G.S. (1937). — Old shorelines of Palestine. *Geol. Mag.* LXXIV. 68-78.

BLANKENHORN, M. (1914). — Syrien, Arabien, Mesopotamien. *Handbuch d. reg. Geologie*, V, 4.

DUBERTRET, L. (1932). — Les formes structurales de la Syrie et de la Palestine. *C.R. Acad. Sci.*, 195, p. 66.

DUBERTRET, L. (1946). — Problèmes de la Géologie du Levant. *Bull. Soc. Géol. Fr.*, (5), **17**, 3-31.

DUBERTRET, L. (1962). — Carte géologique Liban, Syrie et bordure des pays voisins au 1/1 000 000 *in Lexique Stratigraphique International*, III, 10 C1.

FRAAS, O. (1867). — Aus dem Orient, Geologische Beobachtungen am Nil, auf der Sinai-Halbinsel und in Syrien, p. 76.

FREUND, R. (1965). — Structural development of Israël and adjacent areas since Upper Cretaceous times. *Geolog. Mag.*, 102, (3), 189-205, 5 fig.

HEYBROEK, F. (1965). — The Red Sea Miocene evaporite basin. *The Institute of Petroleum*, London, 17-40, 13 fig.

KOBER L. (1915). — Geologische Forschungen in Vorderasien A) Das Taurugebirge, B) Zur Tektonik des Libanon. *Denkschr. math. naturw. Kl. Kais. Akad. Wissenschaften Wien*, 91, 379-427, 3 pl.

KRENKEL, F. (1924). — Der syrische Bogen. *Centralbl.* 274-281 et 301-313.

LARTET, L. (1869). — Essai sur la géologie de la Palestine et des contrées avoisinantes, telles que l'Egypte et l'Arabie. 292 p., 1 pl.

LARTET, L. (1877). — Exploration géologique de la mer Morte, de la Palestine et de l'Idumée. 326 p., 12 pl. — Formation du bassin de la mer Morte, 241-268.

LAUGHTON, A.S. (1966). — The birth of an Ocean. *New Scientist*, 27.1.66, 218-220.

PICARD, L. (1943). — Structure and evolution of Palestine. *Bull. Geol. Dep. Hebrew Univ.*, Jérusalem, IV, 2-4, 134 p., 18 fig., carte géolog. au 1/500 000ᵉ.

PICARD, L. (1959). — Geological map, Israël, 1/500 000 *in* Lexique Stratigraphique International, III, 10 C2, Israël.

PICARD, L. (1963). — The Quaternary in the northern Jordan valley. The Israël Acad. of Sciences and Humanities, 1, 4, 34 p., 1 carte géol. en couleurs.

PONIKAROV, V.P. (1964). — Geologic map of Syria, 1/1 000 000 .

QUENNELL, A.M. (1956). — Geologic map of Jordan, 1/250 000°; Amman, Karak, Maan.

QUENNELL, A.M. (1958). — The structural and geomorphic evolution of the Dead Sea rift. *Quart. Journ. Geol. Soc. London*, CXIV, 1-24, 2 pl. (note présentée le 12.XII.56).

QUENNELL, A.M. (1959). — Tectonics of the Dead Sea rift. *XXth Session Intern. Geol. Cong.*, Mexico, 1956.

SCOTT, C.G. (1966). — Dérive ou permanence des continents. *Atomes*, n° 233, juin, p. 319-328, 15 fig.

SUESS, E. — Traduction par E. de Margerie : La face de la Terre. I (1897) (pp. 471-478, 542-543, 820), II (1902) (pp. 724-729).

VROMAN, A.J. (1961). — On the Red Sea rift problem. *Bull. Research Council of Israël*, 10 G, 1-2, 321-338, 10 fig.

WELLINGS, F.E. (1938). — Welling's observations of Dead Sea Structure (with discussion by Bailey WILLIS. *Bull. Geol. Soc. Am.*, 49, 659-668.

WETZEL, R. & MORTON, M. (1959). — Contribution à la Géologie de la Transjordanie. *Notes et Mém. sur le Moyen-Orient*, VII, 95-191, 25 fig. (v. fig. 25).

WILLIS, B. (1928). — Dead Sea problem: rift valley or ramp valley ? *Bull. Geol. Soc. Am.*, 39, 490-542, pl. 14-18.

Manuscrit déposé le 16 août 1966.

25

PHYSIOGRAPHY

David J. Burdon

[*Editor's Note:* In the original, material precedes this excerpt.]

A. REGIONAL AND RELIEF

JORDAN EAST OF THE RIFT is a land lying on the north-western edge of the Arabian sub-continent. Its position is more readily visualised if it is imagined that a rise of some 250 metres in the level of the seas has permitted the Gulf of Aqaba to rush up the Wadi Araba and fill the depression of the Jordan Rift Valley with a new gulf extending to the north of Lake Houle. While such an imaginary gulf would communicate with the Mediterranean through the Plain of Esdraelon, its main effect would be to emphasise the continuity between East Jordan and the rest of the Arabian Peninsula. Thus the three lakes of Houle, Tiberias and the Dead Sea, which now mark the line of the Jordan Rift Valley, may be considered as indications of this postulated gulf and show how the Valley forms a natural western boundary to the areas described in this Handbook.

The Jordan Rift Valley (see Fig. 4) extends for 375 kilometres in a general north–south direction, from the southern shores of Lake Tiberias to the top of the Gulf of Aqaba. Lake Tiberias is at −212 metres below sea-level (1937 survey), and the Dead Sea is at −392 metres below sea-level (1937 survey). The Jordan River drains Lake Tiberias into the Dead Sea. Ninety-six kilometres south of the Dead Sea, there is a divide in the Wadi Araba at Ghor el Ajram, where the floor of the Rift rises to 240 metres above sea-level. Thence it falls for some seventy-five kilometres to the level of the open sea at the head of the Gulf of Aqaba.

The 375 kilometres of the Jordan Rift Valley is, of course, but a small fraction of the Rift System, which is one of the great tectonic features of East Africa and Asia Minor (Dixey, 1957). This system of rift tectonics may be considered to extend over some 65° of latitude, from the Mozambique Channel separating Madagascar from the mainland to the Ægean Sea, separating Europe from Asia Minor. The Gulf of Aqaba and the Jordan Rift Valley is but a branch from this great system, but it does continue north of Lake Tiberias through Lake Houle (seventy-two metres above sea-level) and so up to the Bekka of Lebanon, which rises to 1100 metres elevation near Baalbek, yet remains a marked depression separating the great ranges of the Lebanon (crest at 3088 metres) from those of the Anti-Lebanon, much of which is above 2000 metres, but which does not reach 3000 metres in height. Rift tectonics continue northwards, being expressed in the Boukaia depression, by the faults near Massiaf, and in the Ghab Depression and the Antioch Plain of the Lower Orontes. Beyond Antioch, rift tectonics are smothered in the fold tectonics of the forefront of the Taurus Mountains.

The northern limit of the area under discussion here, however, may be considered as the southern edges of the great Tertiary basalt flows of the Hauran Plain and of the Djebel Druse mountains. From Deraa in Syria to Adasiya in the Jordan Valley, this limit is marked by the Yarmouk River, deeply incised into the rocks of the Belaq Series. On the north, and locally in the south, the chalk cliffs are capped by basalts which are the aquifers bringing groundwater from the north to the great springs of the Mzeirib and Tell Shihab areas in Syria. East of Deraa (elevation around 500 metres), the Jordan–Syrian frontier lies on the basalts, swinging to the south-east and rising higher to pass across the southern slopes of the Djebel Druse mountains. While the highest peak of these mountains reaches just 1800 metres in Syria, the frontier on their southern slopes does not rise above 1250 metres.

The open country between the deeply incised Yarmouk River and the wild, rugged, well-nigh impassable mountains of the Djebel Druse forms a gap which has always been of major importance. As may be seen from the 200 mm. isohyete of Fig. 7, it is a 'threshold' across which one steps from the pastures of the steppe, narrowing up from the south-east, on to the wheat-growing plains of the settled southern Hauran. In the past the gap was guarded by the Greek–Roman–Byzantine city of Adara (now Derra) and by the Nabataean–Ghassanid city now called Busra-eski-Cham, lying on the lower slopes of the Djebel Druse mountains at an elevation of about 850 metres and distant some thirty-seven kilometres from Deraa. The Byzantines were defending this gap when they were defeated by the Muslim invaders at the battle of the Yarmouk in A.D. 636, and by this victory the Arabs burst out of the Arabian steppe into the fertile areas of Syria. Today the Hijaz railway, as well as the main Damascus–Amman road, pass through this same Deraa gap.

The basalts of the Djebel Druse narrow southwards towards Azraq, which lies at the head of the great Wadi Sirhan depressions. A northern extension of this depression may be traced up the Wadi er Ratam and along the mud-flats bordering the basalts as far as Qasr Rammam es Sarkh (E.279 : N.166; Elev. 600). Along this line the flow southwards of the basalts may have been halted by the necessity of infilling a sunken area. The basalts continue to the east of Azraq, and extend south into Saudi Arabia, cutting off the sedimentary outcrops of the 'corridor' leading to Iraq from those of the rest of East Jordan.

The main pools and the marshes of Azraq Shishan lie just above the 500 metres contour. In addition to the surface drainage of the lightly incised desert wadis, the depression is watered by springs, some from basaltic aquifers, others possibly from deep-seated aquifers of the Ajlun Series. It is a true oasis on the steppe, and as an area for potential development it is now under intensive examination. The great Wadi Sirhan stretches away to the south-south-east; the actual depression lies with Saudi Arabia, but its western rim forms the boundary between Jordan and that Kingdom for a distance of almost 200 kilometres. The Wadi Sirhan is a major depression, the area below 500 metres elevation being long and narrow, suggesting a down-faulted or down-warped graben. Its tectonic significance and possible rift associations are discussed in Chapter IV, B; 5 (b) and at the end of Chapter IV, D; 2.

Much of the East Jordan plateau slopes eastwards towards the Wadi Sirhan. In general, this slope is so gentle as to be attributed to the regional tilting of the block, but towards the edge of the great Wadi, the slope increases, This general regional slope brings surface run-off into the wadi. But in it there are also numerous springs, some fed by Eocene limestone aquifers, and others which may be supplied by water rising under hydrostatic pressure from the deeply buried limestones of the Ajlun Series. Some 200 kilometres from Azraq, the Wadi Sirhan narrows and turns away to the south-east, while the Jordanian–Saudi Arabian frontier turns due south along

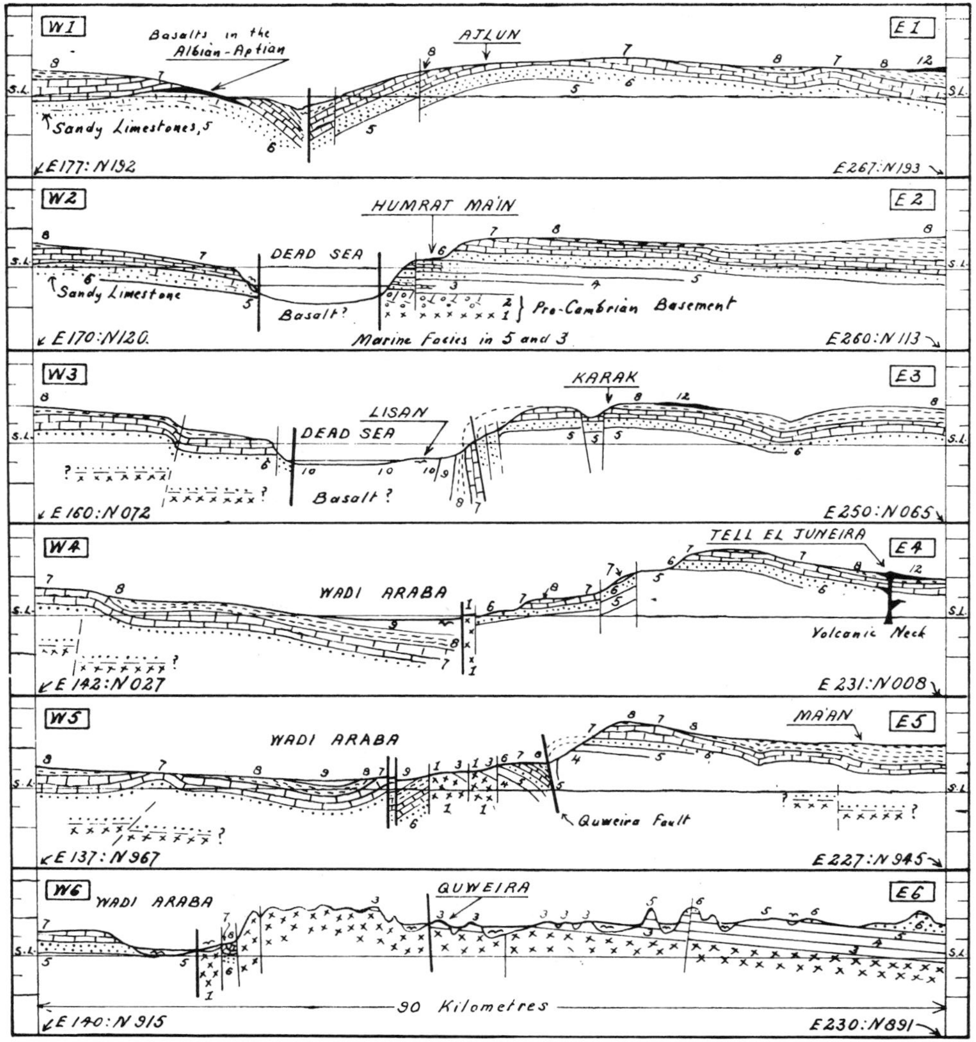

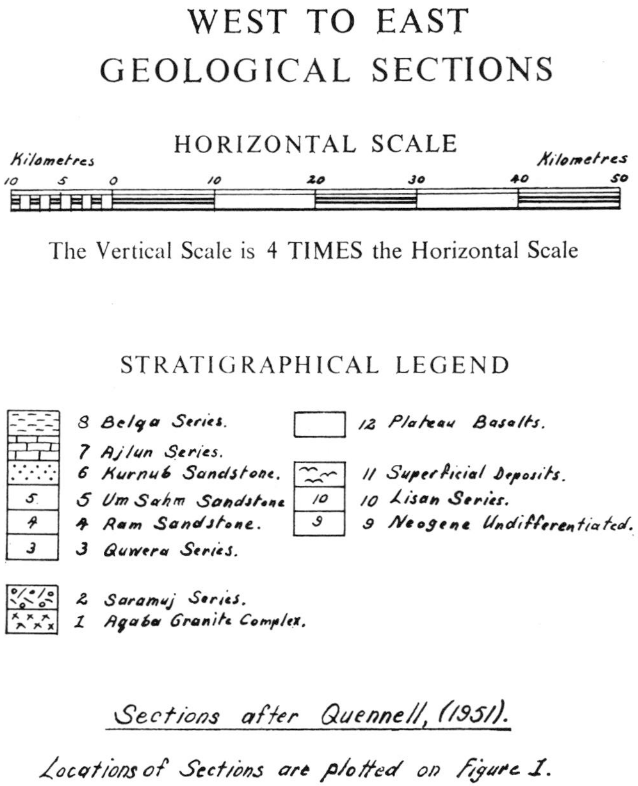

WEST TO EAST GEOLOGICAL SECTIONS

HORIZONTAL SCALE

The Vertical Scale is 4 TIMES the Horizontal Scale

STRATIGRAPHICAL LEGEND

8 Belga Series.
7 Ajlun Series.
6 Kurnub Sandstone.
5 Um Sahm Sandstone.
4 Ram Sandstone.
3 Quweira Series.
2 Saramuj Series.
1 Aqaba Granite Complex.

12 Plateau Basalts.
11 Superficial Deposits.
10 Lisan Series.
9 Neogene Undifferentiated.

Sections after Quennell, (1951).

Locations of Sections are plotted on Figure 1.

Fig. 3. Six West to East Vertical Geological Sections, with a vertical scale which is four times the horizontal scale. The locations of the sections are shown on Fig. 1. The sections are based on those published by Quennell in 1951.

357

the 38° of longitude. After some fifty kilometres in this direction, the frontier turns west-south-west towards El Mudawwara Station on the old Hijaz Railway. This station may be seen in the extreme south-east corner of the mapped area on Sheet 3. Thence the frontier runs towards Aqaba, passing through the 'Southern Desert', described in more detail under B, 4, of this Chapter.

The foregoing description covers the physical limits of East Jordan, with a few excursions farther afield as required by the geological implications. The internal physiography is dealt with in some detail in the succeeding section of this chapter. However, a synopsis of the internal topography may be gained from Figs. 3 and 4. The geological sections, of course, are designed primarily to depict the geology in that third dimension which cannot be supplied by maps; but they also illustrate the relationships between the floor of the Jordan Rift Valley, the escarpment leading up to the plateau, the zone of rejuvenated drainage and the general slope of the plateau towards Wadi Sirhan and the East. Though the vertical scale in Fig. 4 is still more exaggerated, it does emphasise the relationships between the Jordan Rift Valley and the country which borders it on the east. Viewed from the Rift Valley, this bordering country appears as a long range of mountains, dissected at intervals by canyons eroded by the great wadis; viewed from the East Jordan Highlands, the same country appears but as the fretted-away edge of the Plateau itself.

In studying the geology and structure of the Jordan Rift Valley, frequent reference will be made to the country to the west of the Valley. Since the physiography of this area has been covered by the publications of Blake, Shaw, Picard and others, it will not be described here, more so since such an undertaking would involve this Handbook in a description of the large physiographical unit lying between the Jordan Rift Valley and the Mediterranean Sea.

B. EVOLUTION OF THE TOPOGRAPHY

East Jordan is readily subdivided into four topographical sub-units, each of which can be referred with certainty to different periods and conditions in the geological history and evolution of the country. These subdivisions may be listed as follows:

1. The Eastern Plateau
2. The Rejuvenated Drainage to the Jordan Rift Valley
3. The Floor of the Jordan Rift Valley
4. The Southern Desert

The Eastern Plateau is the oldest topographical unit, and may be considered as a relic of the peneplain developed during the Oligocene; it is named the 'Arabia Surface (900 m.)' by Quennell (1956b). The Rejuvenated Drainage into the Rift commenced in the Miocene and has cut back farther and farther to the east, and so into the old peneplain, as the base-level has been lowered with each renewed downward movement of the Rift itself. The Floor of the Rift Valley commenced to form at the same time as the zone of Rejuvenated Drainage; but whereas the passage of time has permitted the latter to reveal older and older rocks, the former has concealed its past under an ever-increasing thickness of sediments. The Southern Desert shows a more mature topography, and since this topography has been sculptured from the Eastern Plateau, it suggests that a rejuvenated drainage into the Gulf of Aqaba arose much earlier than that into the Jordan Rift Valley.

1. The Eastern Plateau. The Eastern Plateau comprises all the country lying to the east of the Hijaz Railway from Mafraq to Ma'an. The precise boundary lies along the watershed dividing the inland drainage from the Jordan Rift Valley drainage. It is flat open country, with lightly incised wadis draining inland, towards the Jaffr depression, the Wadi Sirhan with the Azraq depression at its northern head, and the areas around Qasr el Burqu to the east of the basalt flows. It has not been possible to reconstruct the drainage which formed the peneplain, and the effect of uplift and regional tilting to the east have been masked by the development of the Wadi Sirhan on the east, and the outpourings of the basaltic mountains of the Djebel Druse in the north. According to Quennell (1956b), the amount of tilting is not uniform, and has been correlated by him with the Rift tectonics.

The plateau has been carved from the marine chalks, chalks-with-flint and marls, with subsidiary limestones, of the Belqa Series, embracing the Senonian and the Eocene. These beds are not folded and lie sensibly horizontal, or with slight dips. Gravels, sands, silts and muds of terrestrial, and sometimes lacustrine, origin infill the depressions, ranging in age from possible Miocene through Pliocene to certain Quaternary; to the latter must be assigned the moving gravels and silts covering the wide floors of the wadi depressions. In numerous areas, the higher ground is covered with flint lag-gravels, residuals after the aeolian erosion of the chalks which once enclosed these flints. Loess has recently been identified by Dr. Moormann.

Rare hills rise up from this generally open topography and form clear landmarks in the steppe-desert. Thus the flinty limestone of the Lower Eocene forms two prominent peaks at Thlatawat, north of Bayer, which are known as 'The Light-houses'. In general, these inselbergs, cuestas and mesas which break the monotony of the topography are of limestone, harder chalks or flinty chalks, able to resist the erosional forces which carved the plateau. But the volcanic cones and hills are, of course, of another origin, representing either the necks and craters of volcanoes, or hills and mountains built up by successive flows of plateau-type basalts.

2. The Rejuvenated Drainage to the Jordan Rift Valley. Each successive opening and downward movement of the Jordan Rift Valley has lowered the base-level and rejuvenated the streams draining into this tectonic depression, thus permitting them to cut farther and farther eastwards into the plateau and capturing additional areas for the drainage basin of the Rift. Quennell's reconstruction (1956b) of the old stream profiles combined with a study of old terraces, platforms and surfaces of deposition indicate that there are seven low-level surfaces (at 650–500 m., 300 m., 180 m., 100 m., −20 m., −100 m., and −290 m.) formed in connection with the first phase of movement, plus the present −392 m. base, established by the second phase of movement. Repeated lowerings of base-level were more important than uplift of the plateau.

It is of interest to note that maximum eastwards extension of the rejuvenated drainage has taken place opposite the Dead Sea, where base-level is now lowest. Here the canyons of Wadi Mojib and Wadi Hasa have cut through some 1750 metres of strata ranging in age from Uppermost Cretaceous down into the Pre-Cambrian and produced spectacular scenery reminiscent of the Grand Canyon of Colorado (which is 6250 feet deep and exposes strata ranging from a 1000 feet of Pre-Cambrian to Permian). South of the Dead Sea, the eastern extension is less and has encroached but little on plateau drainage through Ma'an to Jaffr. This may be

359

COMBINED PROFILES OF JORDANIAN HIGHLANDS AND JORDAN RIFT VALLEY

FROM LAKE TIBERIAS IN THE NORTH TO THE GULF OF AQABA IN THE SOUTH

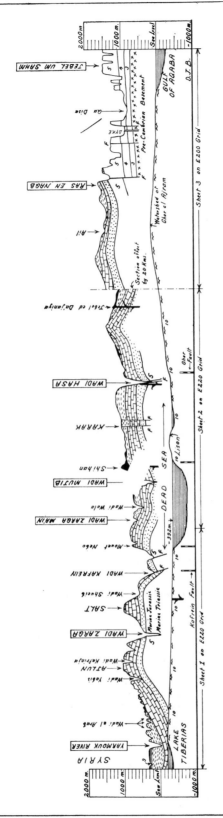

STRATIGRAPHICAL LEGEND

MENOGENE and QUATERNARY

RECENT .... 11 Superficial Deposits.
PLEISTOCENE .... 10 Lisan Series.
NEOGENE .... 9 Neogene Undifferentiated.

UPPER CRETACEOUS to EOCENE .... 8 Belqa Series.
MIDDLE CRETACEOUS .... 7 Ajlun Series.
UPPER JURASSIC to LOWER CRETACEOUS ... 6 Kurnub Sandstone.
TRIASSIC to MIDDLE JURASSIC .... 5 Um Sahm Sandstone.
ORDOVICIAN to PERMIAN .... 4 Ram Sandstone.
CAMBRIAN .... 3 Quweira Series.

PRE-CAMBRIAN .... 2 Saramuj Series.
                                              1 Aqaba Granite Complex.

12 Plateau Basalts.

HORIZONTAL SCALE

Km. 10   0   10   20   30   40   50   60   70   80   90 100 110 120 130 140 150 160 170 180 190 200 Km.

The Vertical Scale is 20 TIMES the Horizontal Scale

Fig. 4. North–South Section, showing the topographical and geological relationships between the Jordan Rift Valley from Lake Tiberias to the Gulf of Aqaba and the Highlands which border it on the east.

360

due to lesser rainfall in this area, to a higher base-level in the Wadi Araba, or to a stronger eastern drainage induced by greater tilting of the plateau to the east. South of Naqb Ishtar the area of rejuvenated drainage to the Rift coalesces and overlaps with the older drainage to the Gulf of Aqaba which has given rise to the southern desert.

In the north, the watershed between the Jordan Rift Valley and the inland drainage has been influenced and moved eastwards by the eruption of the basaltic mountains of the Djebel Druse. In the Amman–Ajlun–Yarmouk area, the high rainfall, the general folded nature of the strata and the presence of faulting, have combined to form a hill region of diversified scenery and innumerable outcrops of limestones, sandstones, chalks and flints. Here erosion has removed almost all the Belqa Series, and it is possible that the Ajlun dome itself may have emerged from beneath the sea in Eocene times before the general emergence, so that Eocene strata were never deposited in this area.

In studying the belt of rejuvenated drainage, it is well to remember that all the material eroded to form these innumerable incisions and valleys in the old plateau has been deposited to form a partial infilling of the Rift Valley. That equilibrium between erosion in the belt of rejuvenated drainage and deposition in the Rift Valley has not been reached is proved by the existence of solid rock outcrops in the floors not only of the Wadi Hasa, Wadi Karack, Wadi Moujib and Wadi Zarqa Main, draining into the Dead Sea, but also in the Yarmouk River, as revealed by the recent drillings near Adasiya and Wadi Khalid. In some of the other wadis, however, base has been attained in the lower reaches, though the headwaters are still actively cutting back into the plateau.

3. The Floor of the Jordan Rift Valley. By contrast with the bordering mountainous country, the Floor of the Rift Valley, whether of land or water, presents a tranquil if barren and arid aspect. In the north, water is, or was, dominant in Lake Tiberias, in the lacustrine Lisan Series and in the Dead Sea, with the Lisan Series extending to form its southern shore. But southwards through the Wadi Araba to Aqaba, terrestrial deposits predominate, and talus fans, gravel outwash plains and sand-dunes testify to, and combine with, the lack of water in the past and in the present, so as to produce a scenery which begins to resemble that of certain portions of the Southern Desert.

Between Tiberias and the Dead Sea, the Jordan River meanders through its flood-plain, known as the Zhor, incised to a depth of some fifty metres below the Ghor, as the main plain of the valley is known (see Plate IX). Tiberias is at 212 metres, and the Dead Sea at about 392 metres below sea-level, so that the Jordan has a fall of 180 metres over a distance of some 104 kilometres. The meanderings show that base-level has been attained and that no marked changes have occurred in the recent past between the relative levels of Lake Tiberias and the Dead Sea; they suggest that any small changes have been in the nature of a lowering of Lake Tiberias or raising of the Dead Sea level. The Zhor is flanked by bad-lands, formed by erosion of the soft Lisan marls; these latter underlie the Ghor, where they are generally concealed by a shallow cover of recent deposits.

The Dead Sea has a maximum depth of some 401 metres, at a point near its eastern coast midway between the mouths of Wadi Hasa and Wadi Zarqa Ma'in; here the bottom is at some 793 metres below sea-level. Beneath much of this northern area the sea is more than 300 metres deep; the shore is formed of precipitous cliffs, through which the Wadi Zarqa Ma'in and Wadi Moujib have cut narrow gorges to discharge direct into the sea without any intervening delta. The absence of such deltas is a most extraordinary fact, when the amount of material eroded by the Wadi Zarqa Ma'in, and above all by the huge Wadi Moujib and its major tributary the Wadi Wala, is taken into consideration. Where has all this material gone? A possible solution is given in Chapter IV, where Quennell's theory is advanced that the present deep portion of the Dead Sea is of recent origin, and the eastern shore has moved northwards for long distances. For south of the Lisan Peninsula, the Dead Sea is but eight to ten metres deep, and the floor of this portion is considered to be older and infilled prior to the opening and deepening of the northern portion.

The Wadi Araba is that portion of the Rift Valley which leads from the Dead Sea to the Gulf of Aqaba. Commencing at −392 metres at the Dead Sea, it rises for a distance of ninety-six kilometres southwards to +240 metres at Ghorel 'Ajram; from this divide it extends a further seventy-five kilometres south to the sea at Aqaba. The edges of the depression are far from clear-cut, and it varies in width from some twenty kilometres in the north to but some ten kilometres south of the divide. While this portion of the Rift may have opened in Miocene time, there has never been a northwards extension of the sea from the present head of the Gulf of Aqaba, for Neogene or Quaternary beds are unknown in the Wadi Araba; and while ancestors of the present Dead Sea may have extended south of its present limits, it is certain that the Dead Sea has never joined with the Gulf of Aqaba through the Wadi Araba, though a river may have flowed from the first Dead Sea to the Gulf. As noted above, the Wadi Araba is distinguished by the presence of all types of arid weathering, whose products cannot be removed by the infrequent floods.

4. The Southern Desert. The Southern Desert is that area lying south of the escarpment which extends east-south-east from Naqb Ishtar; it is therefore, for all practical purposes, the area of outcrops of sandstones (Quweira, Ram, Um Sahm and Kurnub) and granite in the south of Sheet 3 (Ma'an). Here erosion of the original plateau has advanced much farther than along the edge of the Jordan Rift Valley, suggesting that erosion has proceeded for much longer into the Gulf of Aqaba than into the Rift Valley. The Southern Desert has reached the stage of maturity in the erosion cycle, and the low grade of the wadis and flats are reminiscent of the Eastern Plateau.

But the dominant topographical feature of the Southern Desert are the mountains of sandstone, often standing on a plinth of granite, rising in steep slopes and even sheer cliffs from the flat base-level of erosion, which is often covered by drifting sand or emphasised by extensive mud-flats. The classical desert forms are all to be seen. Arid erosion agencies have played the dominant part in carving these wide valleys, but there has been sufficient water to carry off the weathering products, most of which must have passed through the Wadi Ytum into the Rift. Even now, boulders as large as a cottage may be seen in Wadi Ytum, waiting for the next flood to move them in its thick muddy water a little farther towards the sea.

MAJOR STRUCTURAL FEATURES
SCALE

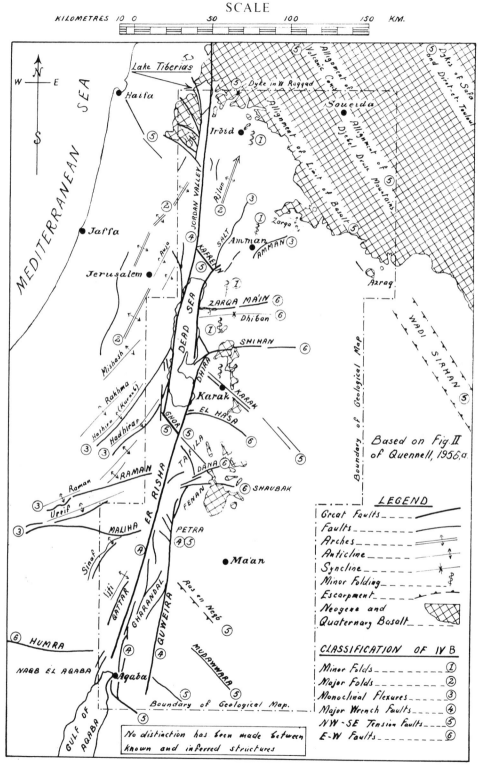

Fig. 5. Map showing the Major Structural Features of the Region; based on fig. II of Quennell (1956a).

[*Editor's Note:* In the original, material follows this excerpt. Only references cited in this excerpt are reprinted here.]

REFERENCES

Dixey, F. (1956b). The East African Rift System. *Colon. Geol. min. Resour.,* Supplement Series No. 1. H.M. Stationery Office, London.
Quennell, A. M. (1956b). The Structural and Geomorphic Evolution of the Dead Sea Rift. Read before the Geological Society, London, on 12/12/1956.

ERRATUM

Page 9, line 22 should read: "Minor (Dixey, 1956b). . . ."

26

Reprinted from *20th Intern. Geol. Congr. Proc., Assoc. Afr. Geol. Surv.,*
1959, pp. 385-403

TECTONICS OF THE DEAD SEA RIFT

A. M. QUENNELL *

ABSTRACT

Structures associated with the Dead Sea Rift are classified and described. Movements begun in the Mesozoic were resumed in the Neogene. The structural units comprise: thrust faults reflected as monoclines trending north-east-south-wset; longitudinal wrench faults of the Rift; transverse faults in the margins of the blocks; and normal faults trending north-west - south-east. As the Jordan Valley and the Wadi Araba wrench faults are "en echelon", sinistral horizontal movement of appreciable amount was possible only after failure of the intervening zone by tension. Progressive widening of this transverse fracture, by movement on the wrench faults, totalled 62 kilometres by the end of the first phase. The Dead Sea then extended from the Ghor Safi to El Lisan. Further movement, now totalling nearly 45 kilometres but probably still continuing, began probably in the Pleistocene, and resulted in the formation of the deep northern half of the Dead Sea in which sedimentation has hardly begun. The present relative positions of many geological features which can be matched on the two sides of the Rift, provide evidence for this interpretation. The traces of the main faults are seen in the valley-floors. Warping of segments of the margins of the blocks has resulted from the sinuous plan of the faults. Igneous activity of two distinct types, has been recorded.

INTRODUCTION

This hypothesis for the formation of the Dead Sea Rift derives from structural information obtained during a geological survey of former Trans-Jordan (Quennell, 1951). Geological maps by the writer on 1:250,000, with explanatory notes by Dr. D. J. Burdon, are to be published shortly. The structure of the eastern margin is now known in as great detail as is that of the west.

The hypothesis is based on a suggestion by Lartet (1869), elaborated in part by Dubertret (1932) and by Wellings (1938). In its present form, it involves horizontal displacement in a sinistral sense on wrench faults which form the Rift. Geological and structural evidence is reviewed, and an account is given of the structural history. The present paper is concerned primarily with the problem as it is seen in plan. A detailed analysis of vertical movements and their relation to the horizontal displacements is required to complete the study.

---

* *Geological Survey Department, Dodoma (Tanganyika).*

I. HISTORICAL AND GENERAL

Study of the Rift problem in general has been hampered by the tacit assumption that the mechanics of formation of all features described as "Rifts" must be the same. Such profound structural features which must penetrate the crust at least as deep as the base of the continental or sialic layer, must be tectonically fundamentally related. However, as further observations are recorded and differences in structure are revealed, it becomes increasingly apparent that the mechanics of formation of, for example, such widely differing features as the Rhine graben, the Dead Sea Rift, the Western Rifts, and the Eastern Rift System of Africa, cannot all be the same. A given hypothesis may appear to explain one or oven two of these quoted cases, but not the others. For example, crustal contraction involving compression and downthrusting, cannot satisfactorily account for the formation of the Red Sea. Similarly, neither tension nor simple vertical stress can explain upthrust blocks such as Ruwenzori or Mbeya Mountains.

In their simplest forms the possible mechanisms can be classified under four heads: (a) radial forces (principal stress axis vertical) causing relative upward and downward movement of adjacent blocks; (b) tangential tension (principal axis horizontal, least axis vertical) causing normal faulting perpendicular to the direction of stress; (c) tangential compression (principal axis horizontal, least axis vertical) causing thrust ("ramp") faulting perpendicular to the principal stress direction; and (d) tangential compression (principal axis horizontal, least axis horizontal) causing wrench faulting at the appropriate angle (45° ±) to the principal stress direction. These hypotheses are not necessarily mutually exclusive, but combinations of two or even three, could have operated.

At any one time and over a given region, one mechanism will have dominated. It is believed that the last mentioned, tangential compress with failure on wrench faults, dominated in the formation of the Dead Sea Rift features, and that the Rift faults are the major wrench faults. Wrench faults of such order of magnitude are already known. The San Andreas fault of California, the Great Glen fault of Scotland, and the Alpine fault of New Zealand, among others, have all been shown to have had horizontal displacement of continental or sub-continental dimension.

The different hypotheses already advanced for the Dead Sea Rift cannot be discussed in detail. Gregory (1921) and Cloos (1939) mis-applied to the Dead Sea Rift, explanations which they had evolved for the African and Rhine Rifts. The value of Krenkel's views (1925) is reduced because they were based on interpretations of the observations of others. The observations of Blanckenhorn (1914), and Picard (1931, 1943, etc.), Blake (1928, etc.), Shaw (1947), and others, on geology and structure, have lead to better understanding of the problem.

Blanckenhorn (1914), Gregory (1921), Picard (1931, 1943) and others, have tended to support some form of the tension hypothesis. On the other hand, Bailey Willis (1928) has advocated compression, within movement on "ramp" faults .However, none of these authors has attempted to apply one particular hypotheses to all the differing structures such as the apparent syncline of the Jordan valley; the opposing fault scarps of the Dead Sea; the overturned monocline of Dhira; and the uplift blocks at Naqb el Aqaba.

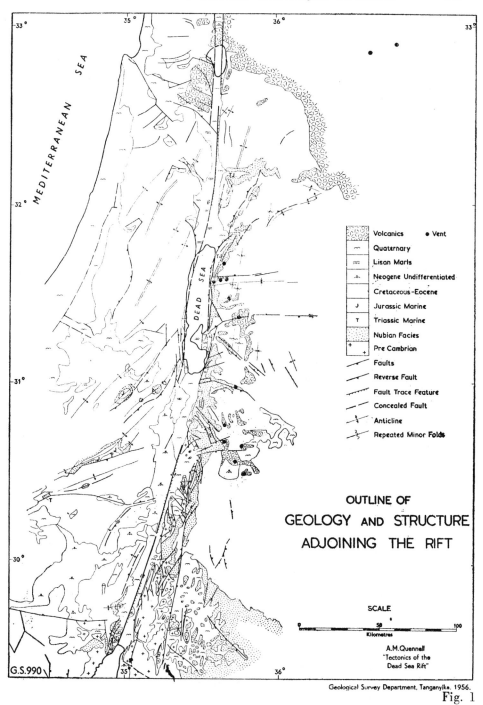

OUTLINE OF

GEOLOGY AND STRUCTURE

ADJOINING THE RIFT

SCALE

Volcanics	• Vent
Quaternary	
Lisan Marls	
Neogene Undifferentiated	
Cretaceous-Eocene	
Jurassic Marine	
Triassic Marine	
Nubian Facies	
Pre Cambrian	
Faults	
Reverse Fault	
Fault Trace Feature	
Concealed Fault	
Anticline	
Repeated Minor Folds	

A.M.Quennell
"Tectonics of the
Dead Sea Rift"

Geological Survey Department, Tanganyika. 1956.

Fig. 1

Any hypothesis, to be acceptable, must explain, not only some, but all features.

Dubertret (1932) suggested the fragmentation of the African continent with the formation of thres *socles*. He postulated horizontal movement of 160 kilometres in a sinistral sense between the Sinai-Palestine and the Arabian blocks, with relative clockwise rotation of Africa.

Wellings in a letter to Willis (1938) referred to Dubertrets' views as being confirmed by his own observations. He also attributes the original suggestion to Lartet (1869). Wellings records evidence for horizontal movement from the structural pattern, and the lack of correspondence of facies of the same age on the two sides of the rift. He states that the evidence for compression is overwhelming on both sides of the Dead Sea trough.

Willis welcomed the evidence for compression and made comparisons with the San Andreas Rift. He did not, however, accept the displacement of the amount suggested by Wellings because of his reluctance thus to lend support to the concept of continental drift.

II. OUTLINE OF THE GEOLOGY

The Precambrian crystalline basement consists of granites, gneisses and schists, with intrusive dolerite and porphyry. The surface was planed at the end of the Precambrian.

Conglomerates and sandstones make up a thick succession of Palaeozoic and Mesozoic continental sediments. The land mass remained relatively undeformed and relative movement of land and sea levels was either epeirogenic or eustatic in nature. Occasional short-lived marine transgressions occurred during the Cambrian, the Carboniferous, the Triassic, the Jurassic, and finally during Upper Cretaceous and Eocene. The deposition of terrestrial sediments continued during the Neogene but their character was determined by the relief following folding and faulting.

Volcanic activity accompanied the tectonic deformation. Pyroclastics and lavas were erupted from isolated vents, and in the north-east there was outpouring of the Hauran flood basalts.

III. STRUCTURAL FEATURES

PRECAMBRIAN

No major structures which can be related to the Rift Valleys existed in Precambrian times. Dolerite dykes are grouped into three sets, the relative age of which can be determined from displacement of the older by the younger. The earliest has a strike of approximately 80°, the second 45°, but the third set is irregular in strike, parallelism and occurrence. West of Quweira, the members of the latter are sub-parallel to the present Rift direction. Close to the Rift, however, they are not found. In the Saramuj rocks at the south-east corner of the Dead Sea, the dykes strike at 20° to the Rift direction. In the south, minor shears with northerly strike have sinistral displacement.

PALAEOZOIC

In the Quweira and Ram sandstones of the eastern block are joints which do not penetrate upwards into the overlying Um Sham sandstone, and do not appear to have a relation to earlier or later structures.

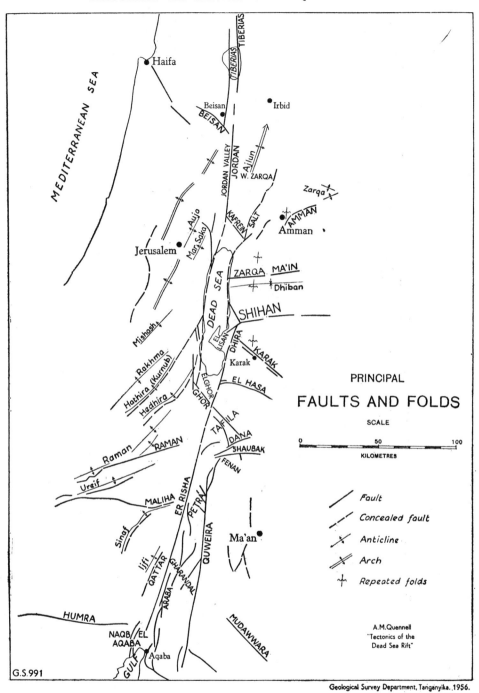

PRINCIPAL

FAULTS AND FOLDS

SCALE

0   50   100

KILOMETRES

Fault
Concealed fault
Anticline
Arch
Repeated folds

A.M.Quennell
"Tectonics of the
Dead Sea Rift"

G.S.991

Geological Survey Department, Tanganyika. 1956.

Fig. 2

MESOZOIC AND PALAEOGENE STRUCTURES

The next structures of known age are the minor close folds in the alternating chert and chalk formation of the Belqa Series (Senonian) seen in a number of restricted areas on the Arabia block north of Wadi Mujib. The strike is 15° to 18°, the wavelength of the order of 100 metres, and amplitude five to then metres. The axial planes hade to the west. They are believed to be post-contemporaneous to the laying down of the beds.

Some of the structures described below under Neogene may have originated in earlier times, but evidence is inconclusive. In particular, the major folds, certain of the monoclines, and certain of the transverse faults may have been initiated in late Cretaceous or Palaeogene times, but if so, the stresses responsible ceased and were reactivated in early Neogene times.

NEOGENE STRUCTURES

Major Folds (Figures 1 & 2)

The complex *Judean arch* is the dominant structure west of the Dead Sea, and is built of a number of individual structures arranged *en echelon* (Picard, 1943). In the Arabian blocks is the *Ajlun* anticline, which is a broad fold with its west flank descending to the Jordan valley.

Monoclines trending north-east

The structures discussed here could perhaps be described as asymmetric anticlines; but Blake (1928, 1937) referred to them as "monoclines" to the definition of which they conform. In plan they are convex to the steep flank; the amplitude decreases towards the ends; the folds generally are unbroken; where exposed by erosion .the folding is seen to be similar; the axial planes have a dip of approximately 45°. In Sinai-Palestine the steep flank faces south-east, and in Arabia it faces north-west.

They are the reflection of thrust faults in the rigid basement. The dip of the thrust plane is regarded as being generally 45° or more because the steeper limb is rarely overturned or broken.

On the west (Ball and Ball, 1953) the most important are the *Hathira (Kurnub)* monoclines, with closure 800 metres; and the *Hedhira* monocline, closure 300 metres; the *Raman* monocline faulted close to the axial plane; the *Ureif* monoclines; and the *Sinaf* which is "squeezed" along its crest, the drag of the folds revealing complex horizontal movement.

On the east are the *Salt* monocline, which continues the east Dead Sea fault to the north-east and has amplitude of 800 metres; the *Amman* monocline with a length of 20 kilometres, and amplitude 500 metres; and others associated with these.

In the *Dhira* structure, to the east of the Dead Sea, the fault is exposed at its southern end, by the dissection of the anticlinal bend. South of the road the monocline is overturned, but northward bifurcates into two gentler structures.

Faults longitudinal to the Rift

These are wrench faults with sinistral displacement. They either make the topographic feature which characterise the Rift, or they are buried

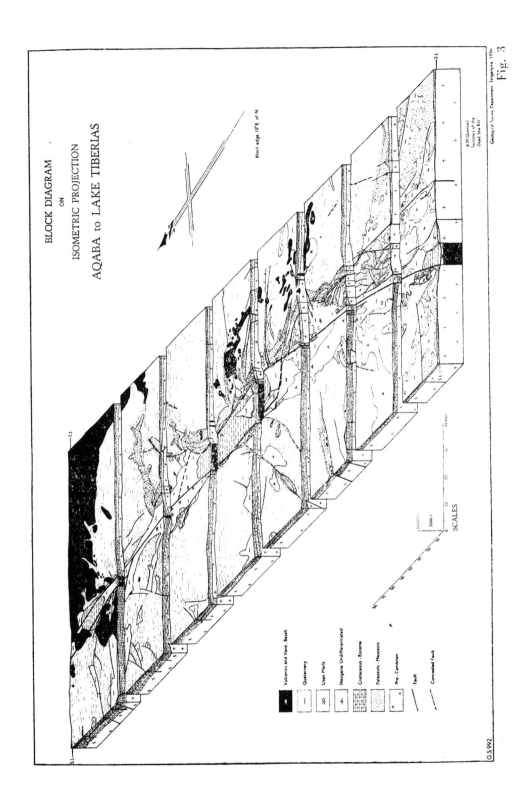

BLOCK DIAGRAM
ON
ISOMETRIC PROJECTION
AQABA to LAKE TIBERIAS

Block edge 10° E. of N

A.M.Quennell
Tectonics of the
Dead Sea Rift

Geological Survey Department, Tanganyika 1956

Fig. 3

SCALES

Volcanic and Vent. Basalt
Quaternary
Lisan Marls
Neogene Undifferentiated
Cretaceous - Eocene
Paleozoic - Mesozoic
Pre - Cambrian
Fault
Concealed Fault

G.S.992

371

beneath the valley floors, and only fault trace features indicate their presence. Of the former, the *Araba faults* (see Fig. 2) bound the Arabia block northward from Aqaba. The indicated stratigraphic throw is 2,200 metres, but the total vertical component may be even greater. The *Dhira fault* exposes a section 1,500 metres thick. The *East Dead Sea fault*, throw 1,500 metres, curves north-eastward into the Salt and Amman monoclines. Lake *Tiberias* has a fault on the east.

The *West Dead Sea fault* curves to the north-west and dies away, and to the sout-west is continued by the Hathira and Hedhira monoclines.

Of the faults sub-parallel and located close to the Rift, the *Quweira fault*, with total length of 120 kilometres, is the monst important. Its apparent throw is greatest in the north but this decreases and it becomes a double hinge fault. There is some horizontal movement. Faults at *Naqb el Aqaba* have their throw away from the Rift. Minor faults near *Gharandal* have variable throw and some sinistral horizontal movement.

On the west the *Sinaf*, the *Ijifi* and the *Qatter* monoclines are probably the reflection of faults of this group. Some minor faults lie between the Hedhira and Hathira monoclines and the Dead Sea.

In the Wadi Araba fault-trace features, such as minor scarplets and shutter ridges, have marked alignment and reflect the concealed *Er Risha wrench fault*. Throw is in general to the east, but may be to the west. Recent sinistral horizontal displacement of approximately 10 metres has been detected in photographs.

In the Jordan Valley is another series of aligned fault-trace features, and on the same alignment steeply-dipping young limestones. This is the trace of the *Jordan* fault.

Faults transverse to the Rift

In the margin of the Arabia block are transverse faults. The *Shaubak fault* is a thrust with minor dextral horizontal movement an dthrow of 800 metres. The *Wadi Hasa fault* is mainly a thrust, but drag of folds indicates dome dextral horizontal movement. The *Shihan fault*, vertical displacement a few hundred metres, and with a sinistral horizontal component, is continued by a "squeezed" and faulted minor fold. The *Dhiban fault* follows a "squeezed" syncline. The *Zarqa Ma'in fault* has 300 metres throw, and eastwards becomes a monocline.

In the Sinai-Palestine block the *Humra fault* lies west of the Gulf of Aqaba.

Faults trending north-west to south-east

The *Mudawwara fault* has a throw to the south-west of less than 100 metres. The *Karak graben* one kilometre wide, has high-angle normal faults which are exposed. The displacement is seen to be small. The faults of the *Petra* group lie across the Quweira "splinter", and have considerable throw. The *Tafila* group have small throw, but are well-defined.

In the Palestine-Sinai block faults with this strike are important north of the Dead Sea, and have greatly influenced the topography. The Wadis Farid, Tabor, the Emeq, and other elongated tectonic depressions, have been formed

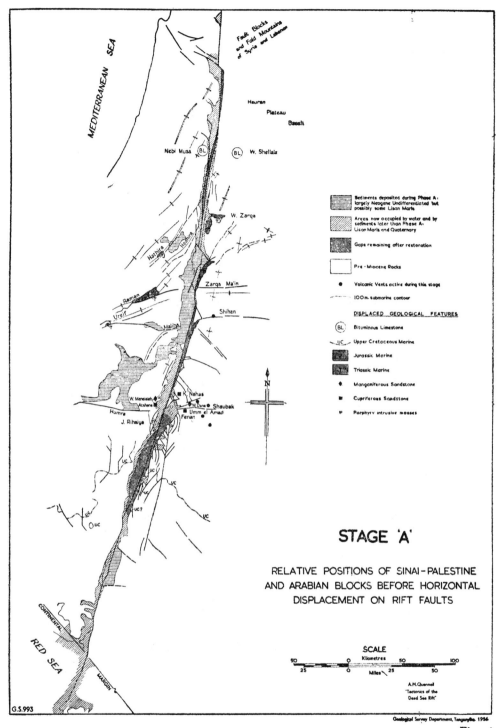

MEDITERRANEAN SEA

Fault Blocks
and Fold Mountains
of Syria and Lebanon

Hauran
Plateau
Basalt

Nebi Musa (BL)   (BL) W. Shellale

W. Zarqa

Hathira

Zarqa Ma'in

Raman

Ureif   Malih   Shihen

K. Nahas
W. Menelsieh   W. Dana ● Shaubak
Atshane   ■ Umm el Amad
Humra   Fenan
J. Rihaiya

UC
UC
UC
UC?   UC
○ UC

CONTINENTAL

RED SEA

MARGIN

G.S.993

Sediments deposited during Phase A:
largely Neogene Undifferentiated but
possibly some Lisan Marls

Areas now occupied by water and by
sediments later than Phase A:
Lisan Marls and Quaternary

Gaps remaining after restoration

Pre-Miocene Rocks

● Volcanic Vents active during this stage

----- 100 m. submarine contour

DISPLACED GEOLOGICAL FEATURES

(BL) Bituminous Limestone

UC Upper Cretaceous Marine

Jurassic Marine

Triassic Marine

◆ Manganiferous Sandstone

■ Cupriferous Sandstone

P Porphyry intrusive masses

N

STAGE 'A'

RELATIVE POSITIONS OF SINAI-PALESTINE
AND ARABIAN BLOCKS BEFORE HORIZONTAL
DISPLACEMENT ON RIFT FAULTS

SCALE
Kilometres
50   0   50   100
25   0   25   50
Miles

A.M.Quennell
"Tectonics of the
Dead Sea Rift"

Geological Survey Department, Tanganyika. 1956

Fig. 4

by faults of this group. The Wadi Sirhan which is believed to be an infilled graben (F. R. S. Henson and D. J. Burdon, personal communications) has this direction, and, with the Red Sea, is believed to be genetically related to this group.

IV. STRUCTURAL PATTERN

The *folds* are broad, symmetrical anticlines and synclines. Their axes have a consistent north-north-east strike, in both the Arabia and the Sinai-Palestine blocks.

The elements of the *fault* pattern are: (a) Monoclines, reflecting deep-seated thrust faults with north-east to south-west strike; (b) Wrench faults, longitudinal to the Rift; (c) Complex faults, either thrust or normal, with horizontal movement, transverse to the Rift; (d) Normal tension faults, striking north-west to south-east.

This structure pattern is consistent with a stress system with the principal axis of compressive stress horizontal and directed west-north-west to east-south-east for the folds, and north-west to south-east for the faults. Major horizontal movement on the Rift faults has resulted from the application of this system of stresses. The evidence for such movement is reviewed.

V. PRE-NEOGENE RESTORATION

For the purpose of testing the hypothesis, maps drawn on the Palestine transverse mercator projection with standard meridian of 35°, were available. The map was cut along the Araba and Jordan faults, and the margins of the Dead Sea and of the Gulf of Aqaba. The Arabia block was moved in a dextral sense in relation to the Sinai-Palestine block for 107 kilometres along the faults. It was found (Fig. 4) that, in addition to the alignement of the Precambrian margins of the north-east of the Gulf of Suez and of the Red Sea. (a) there was matching of the Precambrian margins of the Gulf of Aqaba; (b) there was a gap between the Arabia block and the western margin of the northern part of the Gulf; (c) along the Er Risha fault in the Wadi Araba there was no separation; (d) there was a gap at the northern end of the Wadi Araba on the east of the fault; (e) the bulge at the northern end of the Dead Sea occupied an area of Recent sediments; (f) there was correspondence of the scarp of El Ghor, south of the Dead Sea with the margin of Ghor Kafrein at the north end; (g) there was a gap along the Jordan fault; (h) the Rift valley north of the Beisan fault, infilled with young sediments and volcanics, was eliminated.

The discrepancies or "gaps" found on restoration are interpreted as being the result of faulting or warping of the block margins where they projected across the alignment along which the horizontal movement later took place.

The Lebanon section of the Rift, the Yammoûné fault, is oblique to the general direction. Restoration left a gap 25 kilometres wide which is believed to have been formed by crustal shortening due to the Palmyra folding, the upthrusting of the Lebanon and Syrian horsts, and by their displacement to the north-east as wedges of the crust between converging wrench faults. The Hauran vulcanicity is believed to be related.

The features which were found to match when the restoration (Fig. 4) are now discussed.

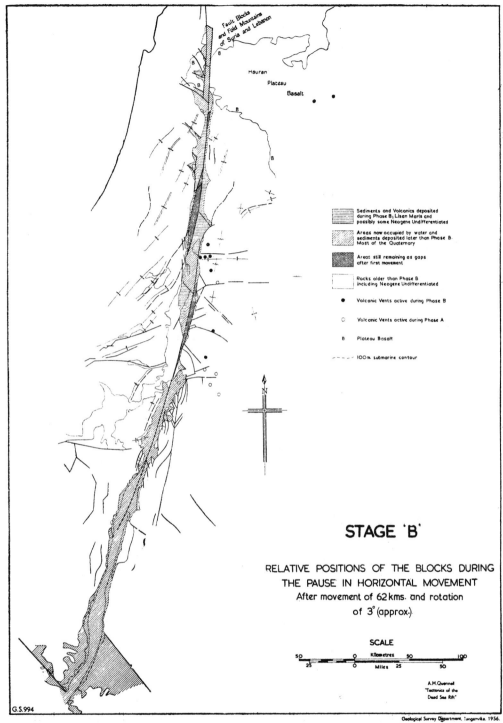

Fault Blocks
and Fold Mountains
of Syria and Lebanon

Hauran
Plateau
Basalt

Sediments and Volcanics deposited
during Phase B, Lisan Marls and
possibly some Neogene Undifferentiated

Areas now occupied by water and
sediments deposited later than Phase B.
Most of the Quaternary

Areas still remaining as gaps
after first movement

Rocks older than Phase B
including Neogene Undifferentiated

●   Volcanic Vents active during Phase B

○   Volcanic Vents active during Phase A

B   Plateau Basalt

‒ ‒ ‒ ‒ ‒   100m submarine contour

N

STAGE 'B'

RELATIVE POSITIONS OF THE BLOCKS DURING
THE PAUSE IN HORIZONTAL MOVEMENT
After movement of 62 kms. and rotation
of 3° (approx.)

SCALE

50   Kilometres   50   100
25   O   Miles   25   50

A.M.Quennel
"Tectonics of the
Dead Sea Rift"

G.S.994

Geological Survey Department, Tanganyika. 1956.

Fig. 5

(a) Three *pairs of faults*, the Humra and Shaubuk faults; the Ureif-Maliha and Shihan faults; and the Raman and Zarqa-Ma'in faults, are found to coincide.

(b) Blake (1936) described from near Fenan a "huge *porphyry dyke*" and "a more basic variety". He also described from the Naqb el Aqaba locality, "a dark igneous mass (? boss) of intrusive rock — appears to be a complex porphyry and a more basic variety such as occurs at Fenan." These masses of rock are in similar structural settings.

(c) Wellings (1938) records the presence of *black Cambrian limestones* in the Wadi Munei'iya section. In the Wadi Dana, are limestones which closely resemble those of Wadi Munei'iya. There are none further south. The actual age of these two occurrences is not established (Shaw, 1947).

(d) Glueck (1940) describes *cupriferous sandstone* in the area of Fenan. This is probably the Quweira sandstone. Copper-bearing sandstones are found also in the Wadi Munei'iya of probably the same age.

(e) In the Wadi Dana is *manganiferous sandstone*. Blake (in Ionides, 1939) compares it with manganiferous sandstones in the Wadi Munei'iya Atshana.

(f) The most southerly occurrence in the east of *marine Triassic* is in the Wadi Zarqa Ma'in. In the west the most southerly occurrence is in the erosion cirque of the Raman monocline (Shaw, 1947).

(g) The most southerly occurrence in the east of *Jurassic marine rocks* is in the Wadi Barud, south of the Wadi Zarqa. In the west the most southerly occurrence is in the Raman monocline.

(h) On the east of the Rift no *Lower Cretaceous marine* beds are exposed, but on the west, Lower Cretaceous marine rocks of Albian age, thick limestones with marls, occur as far south as Jerusalem.

(i) The *maximum marine transgression* was in Senonian and Lower Eocene times. In Sinai these rocks are found 100 kilometres further south than on the east.

(j) In the Wadi Shellala, east of Lake Tiberias, Maestrichian *bituminous chalk* is found 100 metres thick, but elsewhere only in thin beds (Blake, 1939). On the west similar rocks with the same thickness are found only at Nebi Musa near Jerusalem.

When the restoration is made as described, these geological features in every case lie in juxtaposition across the Rift. The Tethys shore line is seen to have had the generally accepted east-west direction during Cambrian (?), Triassic, Jurassic, Lower Cretaceous, and Cretaceous times. This is in contrast with the plan suggested by Blake (1937). Several of the lines of evidence listed here, however, are not related to marine transgressions.

The foregoing comprises the geological and structural evidence that this was the relationship of the blocks in pre-Neogene times. It is here called Stage A (see Fig. 4), and is believed to have ended not earlier than the close of the Oligocene. Confirmation for this restoration is found in the history of sedimentation in the Dead Sea Basin, described below.

VI. STRUCTURAL HISTORY

Separation of the original Arabia-Sinai-Palestine continental mass from Africa, with the formation of the Red Sea and the Gulf of Suez, could have

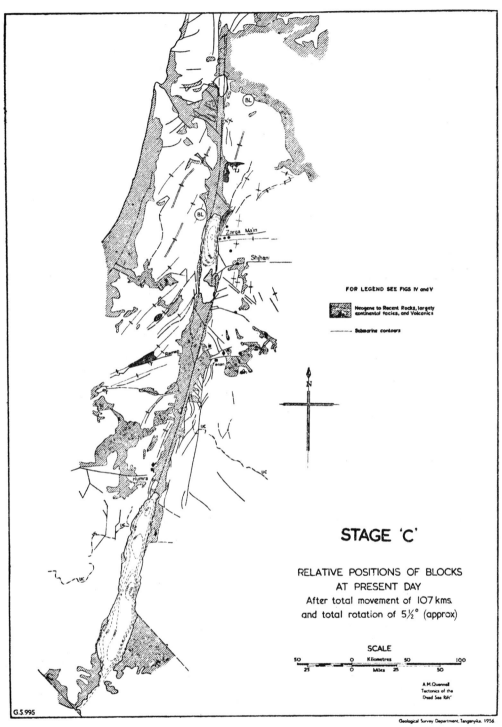

FOR LEGEND SEE FIGS IV and V

Neogene to Recent Rocks, largely
continental facies, and Volcanics

Submarine contours

STAGE 'C'

RELATIVE POSITIONS OF BLOCKS
AT PRESENT DAY
After total movement of 107 kms.
and total rotation of 5½° (approx)

SCALE

50   0   Kilometres   50   100
    25   0   Miles   25   50

A.M.Quennell
"Tectonics of the
Dead Sea Rift"

G.S.995

Geological Survey Department, Tanganyika. 1956

Fig. 6

377

taken place as early as the Jurassic. Until the beginning of the Neogene this northerly movement is believed, for reasons given below, to have taken place without rotation relative to Africa.

The structural history has been governed by a stress system in which the principal axis of compressive stress has veered in direction from 105° to 135°. There have been two phases of stress and movement.

The earliest folding movements which produced the Judean Arch and related structures, and monoclines, and the formation of the three pairs of transverse faults may be as old as Jurassic. Blake (1937) suggested that the absence of Eocene sediments on the Judean Highlands was evidence for their emergence at that time. The first episode, however, whose age is reasonably certain is the post-depositional close folding of the Senonian chert-chalk formation with strike 15°.

The second phase, the Lower Miocene diastrophism, followed the same lines as the earlier faulting and folding pattern but on a grander scale, causing the distortion and disruption of the Oligocene erosion surface, and the local deposition of terrestrial sediments.

Neogene movements, then, began probably with resumption of folding of the Judean Arch and related structures. The general direction of the axes of this folding is 20° to 25°. It possibly reflects the symmetrical deformation of the crystalline basement with axis of maximum compressive stress at 110° to 115°. The basement was either fractured or deformed (Lees, 1952), according to whether the material behaved elastically or plastically (Nadai, 1931). This would depend on depth of burial and the rate of application of the stresses. With a thick cover of sedimentary rocks, folding rather than fracturing could extend down into the basement and would be more likely to produce symmetrical folds in the overlying sediments. The axis of least stress was vertical.

With sligh re-orientation of stress, thrust faulting (or its resumption) in the basement, produced the Hathira, the Hedhira, and the Salt monoclines as reflections. It is suggested that the basement in this case behaved elastically because it was there not so deeply buried beneath a load of sediments. The avis of least stress was vertical, and the principal axis normal to the tangent to the crescentic folds.

With change in the direction of principal stress to about 160°, the Raman and Amman structures were formed, or their formation resumed. There would then tend to be a sinistral horizontal component on the earlier formed monoclines, the thrusts becoming oblique-slip faults.

Sustained application of these stresses culminated in shearing along the Rift zone. Relief of stress ultimately became less easy by folding and thrust faulting than by horizontal shear on vertical planes oblique to the stress direction, which formed along one zone, the Rift, which may habe been a pre-determined line of weakness, possibly by thinning of the sial.

Locally, where the axis of least compressive stress was reversed and became tension, and was also horizontal, structures such as Karak graben and the normal faults were formed, normal to the least stress axis.

The amount of movement along the newly-formed *Er Risha* and *Jordan* wrench faults would at first be limited to the amount of horizontal movement on the monocline thrust faults. The Hedhira and Salt monoclines were then opposite each other, separated by a small zone. (See Fig. 4). Failure of this zone by tension would allow large scale horizontal movement on the Rift wrench faults. (Stage A. Figs. 4 and 8).

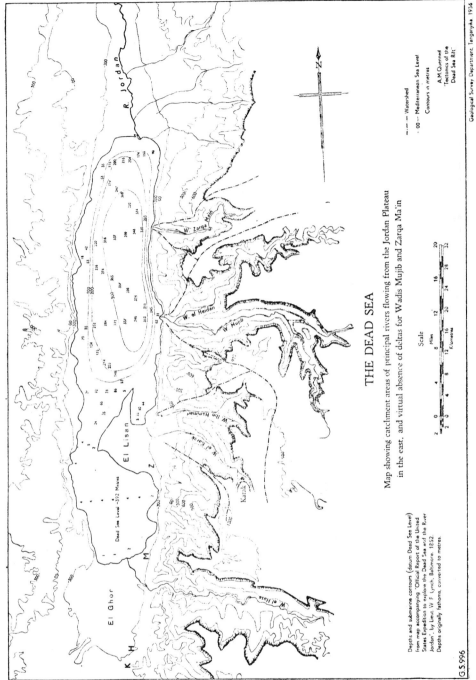

THE DEAD SEA

Map showing catchment areas of principal rivers flowing from the Jordan Plateau
in the east, and virtual absence of deltas for Wadis Mujib and Zarqa Ma'in

Scale

--- -- Watershed

- 00 -- Mediterranean Sea Level

Contours in metres

A.M.Quennel
'Tectonics of the
Dead Sea Rift'

Geological Survey Department, Tanganyika 1956

Fig. 7

Depths and submarine contours (datum Dead Sea Level)
from map accompanying "Official Report of the United
States Expedition to explore the Dead Sea and the River
Jordan", by Lieut. W. F. Lynch, Baltimore, 1852.
Depths originally fathoms, converted to metres.

G.S.996

Wrench faults must follow either a great circle on the earth's surface or a small circle, corresponding, for limited areas, to either a straight line or the arc of a circle. Sinistral movement on a system of parallel rectilinear yrench faults, arranged *en echelon* so that each succeeding fault is beyond and to the left, can take place when the zones between the faults fail under tension. For groups of arcuate faults with same centre but different radii, analogous conditions apply. If the *en echelon* fault pattern has each succeeding fault beyond and to the right, movement in a sinistral sense can take place only when the intervening zones are folded, thrust or wrench faulted, corresponding to failure under compression.

The conclusion is that irregularities and complexities in a wrench fault system, such as *en echelon* patterns, *bulges* and *gaps*, tend to be eliminated to give a single continuous wrench fault, rectilinear or arcuate in plan. The corollary appears to be that all long rectilinear or arcuate faults are most probably wrench faults of appreciable displacement.

Areas W and X in Stage A, Figure 8, are *bulges* where the margins of the blocks originally projected beyond the general line of shear and which with sinistral movement were down-warped first against A' and A, respectively, and then against each other, thus producing the present Jordan fault, and the *apparent* synclinal structure of the Jordan Valley which actually consists of opposing down-warps. The area Y, a bulge acting against *shoulder* B, was downthrust along the Dhira fault and monocline. The area Z appears to have been down-faulted along the Qatter fault agains the shoulder C, the blocks now lying below the Wadi Araba. The shoulder C, on the other hand was upthrust at the same time, to give the Naqb el Aqaba upthrust blocks.

The Dead Sea is a gap. The terminating cross faults, Ghor, and Kafrein, (Fig. 2), are actually the walls of a tension fissure oriented north-west to south-east, which has widened with each succesive horizontal movement on the Rift wrench faults. The Lake Tiberias gap was formed concurrently and in the same manner, but has been largely infilled by late sediments and by basalt. The Gulf of Aqaba is, in part, a *gap*, but without terminating cross faults, the opposite sides moving apart obliquely.

The *bulges* and *gaps* became apparent when the restoration was made. The *bulges* appeared as areas of discrepancy from which sections of the blocks have been eliminated by faulting, folding or warping, while the *gaps* are now occupied by bodies of water, or by sediments and lava.

The wrench faults of the Rift follow circular arcs having a common centre at the approximate position, latitude 33° north, longitude east 24°. Movements on the arcs, means in effect, the rotation of the Arabian block about this centre, the Sinai-Palestine block being regarded as remaining stationary.

The first phase of movement, between Stages A and B, was 62 kilometres and the rotation 3°. The second phase of movement was 45 kilometres with a rotation of $2\frac{1}{2}$°.

The total rotation is $5\frac{1}{2}$°. Dubertret (1932) stated that the angle through which the African pedestal had rotated in relation to Arabia was 6° 04'. This is the amount of angular divergence of the coast of the Red Sea, and agrees clossely with the angle subtended by the length of movement on the arc, 107 kilometres at radius 1,100 kilometres. The coasts of the Gulf of Suez, on the other hand, are essentially parallel. The original Sinai-Palestine-Arabia

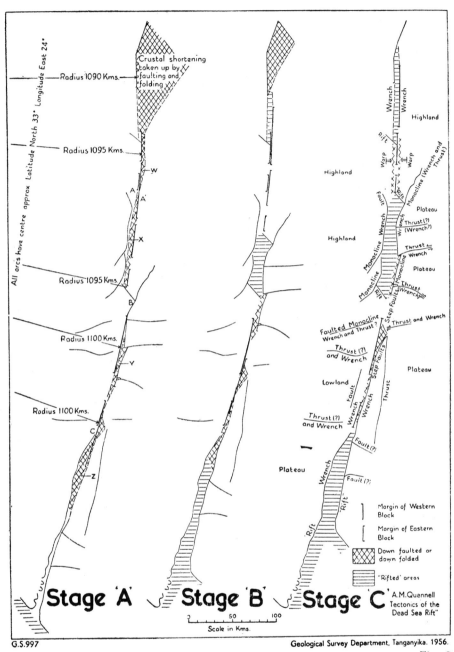

All arcs have centre approx Latitude North 33° Longitude East 24°

Radius 1090 Kms.

Crustal shortening taken up by faulting and folding

Radius 1095 Kms.

W

A A'

X

Radius 1095 Kms.

B

Radius 1100 Kms.

Y

Radius 1100 Kms.

C

Z

Stage 'A'

Stage 'B'

Stage 'C'

Wrench Wrench

Highland

Rift

Warp Warp

Highland

Monocline (Wrench and Thrust)

Plateau

Thrust (?) (Wrench?)

Thrust = Wrench

Highland

Plateau

Thrust (Wrench?)

Faulted Monocline Wrench and Thrust ?

Thrust and Wrench

Thrust (?) and Wrench

Lowland

Steep faults

Plateau

Thrust (?) and Wrench

Fault (?)

Plateau

Fault (?)

Rift

Wrench

Rift

Margin of Western Block

Margin of Eastern Block

Down faulted or down folded

"Rifted" areas

A.M.Quennell Tectonics of the Dead Sea Rift"

Scale in Kms.
0   50   100

G.S.997

Geological Survey Department, Tanganyika. 1956.

Fig. 8

block appears to have moved initially without rotation relative to Africa, the Sinai-Palestine block today retaining the position reached when the Arabia block was separated by the Rift faulting. It continued moving and rotated, as described, to reach its present position.

As the continental or sialic block was torn along the Rift, with widening of the tension fissures, there was rise of the basaltic material of the underlying *sima*. The movement was intermittent, metres at a time. It is suggested that the temperature pressure conditions at the sial-sima interface favoured selective melting of the basaltic material on release of pressure by the opening of the tension fissures. This would have resulted in the injection of successive parallel adjacent dykes. The material in the dykes would probably have solidified before it had risen sufficiently to give hydrostatic equilibrium. This would result in negative anomalies, such as are recorded on the Gulf of Aqaba (de Bruyn, 1955). There are no published values for gravity over the Dead Sea Rift, but surveys have indicated the presence of a salt dome under the Lisan Peninsula and of 10,000 feet or more of light sediments at the south end of the Dead Sea. (Ball and Ball, 1953).

VII. HISTORY OF SEDIMENTATION

Study of the nature and disposition of the sediments deposited in the tectonic basins furnishes confirmatory evidence for the hypothesis here advanced.

The present Dead Sea has a shallow southern half, a few metres deep, of which El Ghor to the south, a few metres above sea level, is an extension; and a deep northern half, with a recorded depth of more than 400 metres. (Fig. 7).

Before the actual break took place on the Ghor-Kafrein tension fissure and movement began on the wrench faults of the Rift, an elongated basin existed between the Salt and Hedhira monoclines. The sediments belong to the Neogene Undifferentiated, and in particular, to the Hasb Series. Possibly the first of the Lisan Marl Series were also then laid down. The basin and lake probably extended from Beisan to near the present divide. (Fig. 4).

During the first phase of movement and the pause, Stage B, the Usdum Series and some of the Neogene Undifferentiated were deposited, infilling the Dead Sea of that time.

During Stage B (Fig. 5), the mouths of the wadis Zarqa Ma'in, Mujib, Ibm Hammad and El Karak were at the positions Z, M, H and K (Fig. 7). The El Lisan peninsula was then the delta of the River Jordan, and the mouth was west of El Lisan.

The volume of excavation of the Wadi Zarqa Ma'in is 22.5 cubic km; of the Wadi El Heidan and Mujib, 115 cubic km; and of the Wadis Karak and Ibm Hammad, 60 cubic km.

These wadis thus have a total measured excavated volume of 197.5 cubic km. The material which this represents, together with that from smaller tributaries, from the erosion of the slopes, and from wadis on the west, infilled the Dead Sea during Stage B. The volume of the infilled part of the basin is believed to have been of the order of 200 cubic kilometres. This is argued in the following manner. The phase of movement after Stage B, of 45 kilometres, produced the present basin which has a volume of 135 cubic kilometres after some sedimentation. The movement before Stage B was 62 kilometres, and the volume produced was in proportion, that is, of the order of 200 cubic kilometres.

After Stage B, movement of the eastern block, 45 kilometres (Fig. 8, Stage C) produced the present deep northern half whose bottom configuration has hardly yet been affected by new sedimentation. The last phase of movement must thus have been very late, and of short duration.

The present volume, 135 cubic kilometres. is contrasted with that of the gorges of the principal wadis, Zarqa Ma'in, El Heidan and Mujib with a total of 137.5 cubic kilometres. The deposition of this quantity of material in the present Dead Sea (Fig. 7.) could not have taken place. Submarine cones and deltas are virtually absent, and the material could not have been so distributed as to give the present form.

BIBLIOGRAPHY

BALL, M. A. AND BALL, D. 1953. Oil Prospects of Israel. *Bull. Amer. Assoc. Petroleum Geologist*, Vol. 37, (1) :1.

BLAKE, G. S. 1928. *Geology and Water resources of Palestine.* Jerusalem.

— 1937. Old Shore Lines of Palestine. *Geol. Mag., London*, 74:68-78.

— (IN IONIDES, M. G.) 1939. *Report on the water resources of Transjordan and their development.* (Incorporating a report on geology, soils and minerals and hydro-geological correlation).*London*, 1 vol.

BLANCKENHORN, M. 1914. Syrien, Arabien und Mesopotamien. *Handb. Reg. Geol.*, 5(4).

BRUYN DE, J. W. 1955. Isogam maps of Europe and North Africa. *Geoph. Prospecting*, III, (1).

CLOOS, H. 1939. Hebung-Paltung-Vulkanismus. *Géologisch Rundschau.* 30

DUBERTRET, L. 1932. Les formes structurales de la Syrie et de la Palestine; leur origine. *C. R. Acad. Sci. Colon., Paris*, Vol. 195.

GLUECK. N. 1940. The other side of the Jordan. *Amer. Sch. Orient. Res.*, U.S.A.

GREGORY, J. W. 1921. *The Rift Valleys and Geology of East Africa.* London. 1 vol.

KRENKEL, E. 1925. *Geology Afrikas,* Erster Teil, Berlin. 1 vol.

LARTET, L. 1869. La géologie de la Palestine. *Ann. Sci. Geol., Paris.* (1).

LEES, G. M. 1952. Foreland Folding. *Quart. J. Geol. Soc. London.* CVIII, (1).

NADAI, A. 1931. *Plasticity. A Mechanics of the Plastic State of Matter.* New York, 1 vol.

PICARD, L. 1931. *Geological Researches in the Judean Desert.* Jerusalem.

— 1943. *Structure and Evolution of Palestine.* (With comparative notes on neighbouring countries). Jerusalem.

QUENNELL, A. M. 1951. The Geology and Mineral Resources of (former) Trans-Jordan. *Col. Geol & Min. Res.*, 2, (2).

SHAW, S. H. 1947. *Southern Palestine, geological map with explanatory notes.* Jerusalem.

WELLINGS, F. E. 1938. in "Wellings observations of Dead Sea Structure" by Bailey Willis. *Bull. Geol. Soc. America*, 49:659-668.

WILLIS, BAILEY. 1928. Dead Sea Problem: Rift Valley or Ramp Valley. *Bull. Geol. Soc. America*, 39:490-512.

— 1938. Wellings' Observations of Dead Sea Structure. (With discussion by Bailey Willis). *Bull. Geol. Soc. America,* 49:659-668.

27

Reprinted from *Royal Soc. London Philos. Trans.* **A267**:107–130 (1970)

The shear along the Dead Sea rift

By R. FREUND, Z. GARFUNKEL, I. ZAK

Department of Geology, The Hebrew University, Jerusalem

M. GOLDBERG, T. WEISSBROD

Geological Survey of Israel, Jerusalem

AND B. DERIN

Israel Petroleum Institute, Tel Aviv

Recent surface and subsurface geological investigations in Israel and Jordan provide new data for the re-examination of Dubertret's (1932) hypothesis of the left-hand shear along the Dead Sea rift. It is found that while none of the pre-Tertiary sedimentary or igneous rock units extend right across the rift, all of them resume a reasonable palaeographical configuration once the east side of the rift is placed 105 km south of its present position. It is therefore concluded that the 105 km post-Cretaceous, left-hand shear along the Dead Sea rift is well established.

The 40 to 45 km offset of Miocene rocks and smaller offsets of younger features indicate an average shear movement rate of 0.4 to 0.6 cm a⁻¹ during the last 7 to 10 Ma. Unfortunately, the 60 km pre-Miocene movement cannot be dated yet.

Along the Arava and Gulf of Aqaba and in Lebanon the shear is divided over a wide fault zone within and outside the rift.

HISTORICAL REVIEW

The verse '...and the Mount of Olives shall cleave in the midst thereof towards the east and towards the west,...and half of the mountain shall remove towards the north, and half of it towards the south' (Zechariah, xiv. 4), can probably be regarded as a description of left-hand shear by an ingenious observer who lacked the professional terms. One century ago Lartet (1869) noticed that Arabia and Africa might have drifted apart in an oblique left-hand direction to open up the Red Sea.

About 60 years later Dubertret (1932), following Lartet's idea, advocated also by Bogolepov (1930) and von Seidlitz (1931), suggested that a 160 km left-hand shear along the Dead Sea rift, associated with a 6° rotation between Arabia and Africa, might explain several structural relations in the Levant. Wellings (in Willis 1938) realized that this hypothesis corresponds to the offset of the marine Cambrian and Jurassic beds across the rift south of the Dead Sea. Willis, however, rejected this hypothesis, and during the following 20 years it was completely neglected in papers about the geology of the Levant (e.g. Picard 1943; Dubertret 1947). It was not mentioned even in papers which discussed other strike slip faults in this region (Bentor & Vroman 1954; Renouard 1955; Vroman 1957).

Quennell (1958, 1959) revived Dubertret's hypothesis, though to his opinion the shear amounts to 107 km as indicated by the offsets across the rift of a Precambrian porphyry body, of Cambrian limestone, copper and manganese sandstone, of marine Triassic and Jurassic beds, of the southern extent of the marine Albian and the Upper Cretaceous transgressions, of Senonian bituminous chalk, and of three pairs of transverse faults. Quennell suggested that a Pleistocene movement of 45 km explains several geomorphic features, the most prominent of which is the shape of the deep depression of the Dead Sea. The remaining 62 km shear was in his opinion of Miocene age.

During the following decade the discussion developed according to two main general considerations: palaeographical and structural. Burdon (1959) reviewed Quennell's arguments, supplementing them by his observation of recent horizontal displacements of about 10 m in the rift, but he found difficulties in tracing either the continuation or the structural effects of the termination of the shear movement southwards and northwards. Consequently he preferred an alternative explanation of the facies offsets discussed by Quennell, namely that they reflect an original bend of the ancient coast lines along the place occupied now by the rift. Henson (1956) and Wetzel & Morton (1959) shared this opinion.

Bentor & Vroman (1960) carried this explanation one step further, suggesting that the bend of the coast was controlled by the east fault of the rift, which raised the east side higher than the west side already during the Jurassic, so that the transgression of the Jurassic sea from the northwest reached farther south on the west side than on the east side. In this way they explained the 600 m thick sequence of marine Jurassic in Massada-1 test well on the west margin of the Dead Sea, right opposite the complete absence of the Jurassic on the east side.

A detailed study of the Lower Turonian beds (Freund 1961) showed that various lithofacies and biofacies are offset across the rift by about 100 km left hand, in accordance with the offsets mentioned by Quennell. De Sitter (1962) accepted the Jurassic and Turonian data as evidence supporting the shear hypothesis.

Swartz & Arden (1960) discussed the rotational opening of the Red Sea and the Gulf of Aden (for which they used the term 'Pa'ar' proposed by Shalem 1954), but they did not refer to the shear movement along the Dead Sea rift. Vroman (1961) maintained, on the other hand, that the occurrence of Precambrian to Jurassic rocks in the Danakil Alps within the Red Sea depression, contradicts the opening of the Red Sea, and thereby also the shear along the Dead Sea rift. Later oceanographic studies (see other papers in this volume) provided, however, arguments in favour of the crustal separation in the Red Sea, making thus the southward continuation of the shear less of a problem.

As to the northward continuation of the shear, Freund (1965) suggested that it extends northwards along the rift, up to southern Turkey, and that the high structures of Lebanon are due to the excess of material formed by the northward shear movement along the N 30° E trending Yamuneh segment of the rift. This model seemed to forbid a shear movement exceeding 70–80 km, which was thought to be still compatible with the recorded offsets. To Dubertret (1967), however, the structures in Lebanon are far too gentle to accommodate this shortening. He was, therefore, reluctant to accept the shear hypothesis, maintaining that the shear movement could have continued northwards only along another fault (Roum Fault) which cuts northwards to the Mediterranean Sea near Beirut. Since Roum Fault disappears beneath Cenomanian beds, the shear movement could have been, according to Dubertret, only of pre-Cenomanian age.

Freund (1965) regarded the folding and the normal faulting of the Levant as secondary structures of the shear. Deducing from their age he concluded that the shear occurred in several phases since the Upper Cretaceous. The relation between the shear and the fold structures was criticized by Garfunkel (1966), but Vroman (1963, 1967) also considered the Levant folds as secondary to shear movements, admitting thereby a shear movement on a minor scale along the Dead Sea rift.

Bender (1965, 1968a) commented that detailed mapping in Jordan invalidated several of Quennell's arguments for the shear hypothesis. The 100 km offset of the Cambrian limestone, copper and manganese sandstone, and Precambrian porphyry across Wadi Arava should not be regarded as evidence in favour of the shear movement, because the same rocks occur in several intermediate locations, forming thus an almost continuous belt across the rift. Nevertheless, he accepted a 25–35 km shear movement according to the offset of a certain pegmatite hornblendite from a position opposite Timna on the east side to the vicinity of Elat on the west side.

Zak & Freund (1966) recorded horizontal displacements of about 150 m younger than the Lisan Marl (23 ka b.p.). Freund, Zak & Garfunkel (1968) showed that Miocene rock bodies are offset by 40 to 45 km across the rift, and suggested therefore that the average rate of the shear movement during the last 10 Ma is about 0.4 cm a^{-1}.

In conclusion, no general agreement about the shear movement along the Dead Sea rift has been reached yet, thus Neev & Emery (1967), discussing the geology of the Dead Sea, accepted it, whereas Picard (1966, 1968) did not.

Method

There is no intention to present any new hypothesis in this paper, but rather to examine whether Dubertret's and Wellings's shear hypothesis is, after some 40 years of extensive geological investigations, still conjectural or if it is well established. It should perhaps be recalled that hypotheses in natural sciences cannot be proven right or wrong beyond their correspondence to all known facts on one hand, and their success in predicitng new facts on the other. Quennell argued that the score of features that are offset by about 100 km across the rift establish the shear hypothesis; if this hypothesis is correct, then the prediction that all the other geological features are likewise offset should never fail. The results of the geological investigations along the Dead Sea rift during the last decade show that this prediction is indeed correct, though sometimes in a somewhat complex manner.

The description of the majority of the rock units along the Red Sea rift from Precambrian to Miocene is presented in a series of illustrations, each of them accompanied by a brief explanation of the main relevant features. It has been preferred to present the data in stratigraphic columnar sections rather than in isopach and lithofacies maps, because the interpretative nature of such maps might have cast some doubts about the validity of the data. Anyway, the results of the two methods of representation are equal, as the trend of the lithofacies lines is usually between ENE and NNE (see, for example, Aharoni 1964).

A second set of illustrations shows the same details of the various rock units from Precambrian to Upper Cretaceous on maps where the eastern side has been shifted 105 km southwards relative to the western side with a clockwise rotation of 6° of the east side, similar to the model proposed by Freund (1965).

Several rock units are not described because they lack significant facies which can be compared on the two sides of the rift (e.g. Upper Turonian, Maastricht-Paleocene). Others are not included because they occur only in very small isolated patches (Upper Eocene–Oligocene) or they have been insufficiently studied (Pliocene–Pleistocene).

The Dead Sea and Jordan River

Along the median part of the Dead Sea rift shear movement is most probably confined within the rift, as there are no faults running parallel to the rift outside it. The entire stratigraphic column is known on the two sides of the Dead Sea; it is exposed on the east side, and is drilled through on the west side.

110

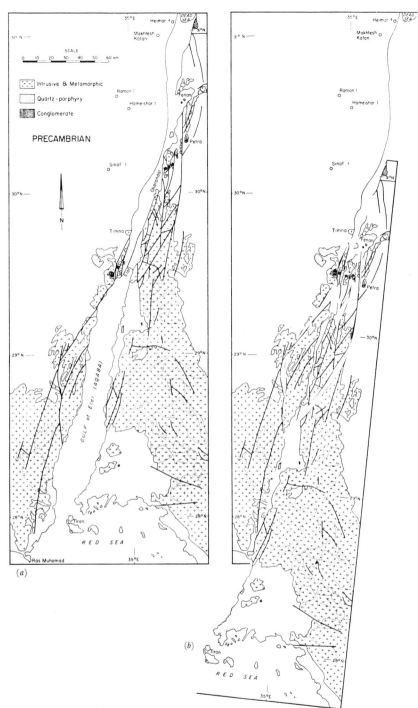

FIGURE 1. (a) Precambrian: quartz prophyry associated with conglomerates occurs in the vicinity of Elat in southern Israel, and from Gharandal to Fenan in Transjordan. A very thick sequence of arkose which underlies the Cambrian in Sinaf 1 and the Permo-Carboniferous in the other bore holes farther north is most probably of Precambrian or Infracambrian age. This arkose is identical to the arkosic cement of the Saramuj Conglomerate exposed on the southeast side of the Dead Sea beneath the Cambrian. Note the numerous anastomosing arcuate faults from Fenan to Aqaba on the east side and from Timna southwards on the west side. (References: Transjordan: Bender 1968a; Saudi Arabia: Bramkamp *et al.* 1963; Sinai: Geological Map of Egypt 1928; Southern Negev: Bentor & Vroman 1955; Z. Garfunkel (unpublished).)

(b) Schematic reconstruction of the shear. The two sides are shifted back 105 km, and the blocks between them are shifted and rotated by various amounts along the anastomosing faults. Thereby the quartz porphyry and conglomerate are brought in line.

387

Cambrian sediments are exposed on the east side of the Dead Sea in Wadi Hesi where their thickness exceeds 300 m, and in Zerka Main (figure 2a). On the west side of the Dead Sea the Cambrian is probably absent, and the Upper Palaeozoic sediments are overlying unconformably the Precambrian (or Infracambrian) arkose. Upper Palaeozoic (Upper Carboniferous–Permian) sedimentary and igneous sequence of 650 to 750 m has been encountered in all the deep test wells on the southwest side of the Dead Sea, from Zohar-8 and Lot-1 to Hameishar-1 (figure 3). Upper Palaeozoic rocks are absent on the east coast of the Dead Sea, and have recently been reported (Bender 1968b) from Safra-1 test well east of Amman.

Triassic section of 550 to 750 m occurs above Upper Palaeozoic sequence southwest of the Dead Sea, and is absent on the east side (figure 4a). The Triassic appears in Transjordan north of Wadi Mujib (Wetzel & Morton 1959; Bender 1968b). On both sides the section becomes thinner southwards both by attenuation of individual parts of the sequence, and by a truncation of the Triassic by the Jurassic and Lower Cretaceous unconformities. The gypsiferous facies of Mohila Formation (the uppermost one) extends on the west side from Ramon through Makhtesh Katan to Lot-1. This facies disappears both southeast and northwest from this belt. On the east side of the rift this gypsum occurs in Wadi Zerka.

The Jurassic sequence (figure 5a) on the west side of the rift is sandy from Ramon in the south to Lot-1 in the north. A few sand beds occur in Massada-1 and Halhul-1, but northwards the sandstone disappears. On the east side of the Dead Sea the Jurassic sequence is absent. Its southernmost occurrence is in Jordan Valley-1 well, and in Wadi Zerka the facies is still sandy. In Zohar wells, in Massada-1, Halhul-1 and Ramalla-1 occurs the Kidod Shale (Callovian–Oxfordian) at the bottom of which occur numerous small ammonites. This shale disappears northwards on the west side, but occurs on the east side of the rift on the southeastern slope of Mount Hermon.

The Lower Cretaceous Hathira formation (figure 6a) is completely sandy on the east side of the Dead Sea, measuring about 200 to 250 m, and the southernmost extent of thin beds of marine Albian limestones occur only as north as Wadi Zerka. In Massada-1 well the Hathira formation is about 400 m thick and half of it consists of marine limestone and shale. In Jerusalem (Motsa-1) and Wadi Malih the Lower Cretaceous sequence is about 700 m thick, more than two-thirds of which are marine limestones and shales. From Wadi Faria northwards on the west side the lower part of this section consists of volcanic rocks; whereas these volcanic rocks appear on the east side only on the southern slopes of Mount Hermon. The base conglomerate of the Lower Cretaceous occur only in Wadi Zerka east of the rift, and in and around Ramon on the west side.

The so-called Cenomanian part of Judea Limestone (figure 7a), the lower part of which is of Albian age, is about 650 m thick in Jerusalem, but only 300 m in Wadi Zerka and about 200 m in Zerka Main and Ed Dhira. In the southern Negev on the west side, it is 100 to 200 m thick, whereas in Naqb Ishtar it is less than 50 m thick.

The Lower Turonian sequence on the southeast side of the Dead Sea (figure 8a) consists of about 100 m of shale with limestone beds topped by gypsum beds, and it contains numerous ammonites, the most common of which are large *Choffaticeras*. Specimens of these ammonites collected by Professor L. Picard in Wadi Hesi are deposited at the Hebrew University. A similar sequence with abundant *Choffaticeras* occurs on the west side in the southern Negev, whereas on the southwest side of the Dead Sea the Lower Turonian consists of 20 m of limestone and marl where another ammonite, *Vascoceras pioti*, is abundant. This ammonite occurs on the east side

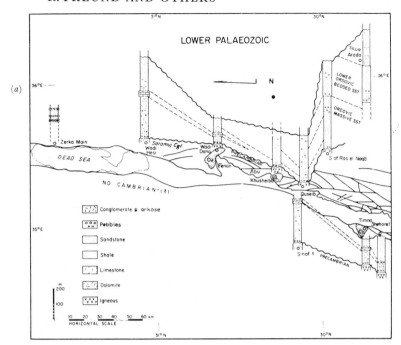

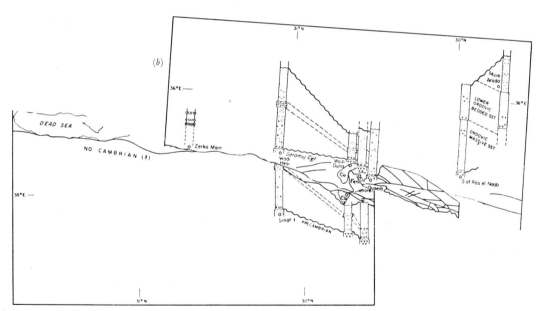

FIGURE 2. (a) Lower Palaeozoic: Marine Cambrian occurs in Transjordan from Zerka Main southwards to Wadi
Abu Khusheiba. It is most probably absent on the west side of the Dead Sea and northern Negev, and
extends from Sinaf 1 (ca. 300 m) through Timna to the vicinity of Elat. Copper and manganese sandstones
occur between Fenan and Abu-Khusheiba in Transjordan, and in Timna in Israel. The Ordovic–Silurian
sequence of southeast Transjordan is truncated towards the northwest by the Lower Cretaceous unconformity;
it disappears already near Ras el Naqb, and it does not reach the Rift. (References: Transjordan: Wetzel
& Morton 1959; Bender 1965. Israel: Weissbrod 1969b; Bartura 1966.)

(b) The reconstruction is similar to that of figure 1b. Note that the ca. 300 m Cambrian of Sinaf 1 is brought
opposite Wadi Hesi, and that the sections of Wadi Dana (Fenan), Abu Khusheiba and Timna are brought
together. It should, however, be admitted that the Cambrian of Zerka-Main, where the Permo-Carboniferous
is absent, is inconveniently brought opposite the Permo-Carboniferous (and no Cambrian) of Ramon 1 and
Hameishar 1 (cf. figure 3).

389

near Es Salt. Specimens of *V. pioti* collected in Es Salt by Dr M. Blanckenhorn are deposited at the Hebrew University. On the west side of the Dead Sea the Lower Turonian is absent, and the Lower Turonian northwest of the Dead Sea does not contain *V. pioti* but other ammonites, similar to those found in the Beqa'a on the western slopes of Mount Hermon.

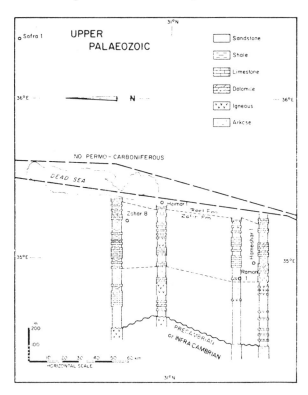

FIGURE 3. Upper Palaeozoic: the 750 m thick Permo-Carboniferous of Zohar-8 thins to the southeast to 650 m in Heimar 1, on the southwest margin of the Dead Sea, and the 750 m section of Ramon 1 thins to about 600 m (excluding the igneous material) in Hameishar 1. On the opposite east side of the Rift the Permo-Carboniferous is absent altogether. It is reported only from Safra 1 well. (References: Weissbrod 1969*a*; Bender 1968*b*.)

The *ca*. 200 m Senonian (Santonian–Campanian) sequence (figure 9*a*) on the southeast side of the Dead Sea contains flint and dolomite beds almost throughout, similar to the Senonian sequence in southern Israel, whereas on the west side of the Dead Sea the flint beds occur only in the upper half of the sequence and dolomite is absent. Section similar to the latter occurs in Transjordan farther north near Amman and Irbid, opposite which on the west side (Gilboa) the flint is very thin. The economic phosphorites of Israel occur on the southwest side of the Dead Sea; those of Transjordan beyond the northeast side.

In figures 4*b* to 9*b* the east side is shifted southwards 105 km relative to the west side with a 6° clockwise rotation. All the rock units without exception resume thereby a simple and reasonable configuration. The southward thinning out of the Triassic beds (figure 4*b*) as well as the gypsiferous belt of Mohila Formation come in line on the two sides. The southward thinning out of the Jurassic beds (figure 5*b*), the sandy facies of the Jurassic and the Kidod Shale come in line as well. The equivalent facies and thicknesses of the Lower Cretaceous (figure 6*b*) and the Cenomanian (figure 7*b*) come to juxtaposition. The various litho- and biofacies of the Lower Turonian (figure 8*b*) are placed exactly in their proper place, as do those of the Senonian along NNE trending isofacies lines (figure 9*b*).

The fact that all the dissimilarities across the rift disappear simultaneously by the 105 km shift is a strong evidence in favour of a 105 km left-hand shear along the Dead Sea rift in Post-Cretaceous times. The value of this evidence can be reduced only if an undisplaced rock unit will be encountered, or if another, simpler explanation can be proposed for the persistent changes of the rock units across the rift, or if this shear movement is shown to be impossible.

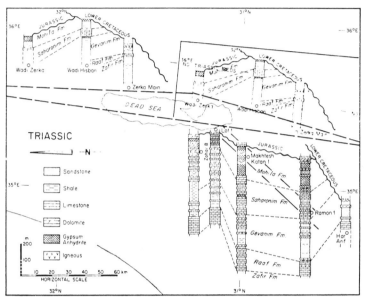

FIGURE 4. (*In black*) The Triassic section is 700 m thick in Ramon and 750 m in Makhtesh Katan 1. It thins to 550 m in Lot 1, and it is completely absent on the opposite side of the Rift. In Zerka-Main and Har Arif the Lower Triassic is truncated by the Lower Cretaceous unconformity. The gypsiferous facies of Mohila Fm extends on the west side from Ramon through Makhtesh Katan 1 to Lot 1, and disappears both to the southeast and to the northwest of this belt. It reappears in Wadi Zerka in Transjordan. (References: Transjordan: Wetzel & Morton 1959; Israel: I. Zak, I. Karcz & M. Goldberg (unpublished).)

(*In red*) The east side is displaced 105 km southwards together with the closing of the Rift. Zerka Main comes in line with Har Arif, and Wadi Zerka comes in line with Ramon to Lot-1 belt.

Another explanation (see, for example, Blake 1937 and Wetzel & Morton 1959) relates the above mentioned changes of the rock units across the rift to the original configuration of the Arabian Massive along the Dead Sea rift, reflected in the S bend of the Triassic and Jurassic coast lines. Bentor & Vroman (1960) noticed that the presumed ancient coast along the rift must have been particularly steep, and therefore suggested that the eastern fault of the rift was existing already in the Jurassic, with the Transjordan side rising relative to the west side at that time.

However, according to this explanation the *west* side must have been higher by 300 m in pre-Carboniferous times (with or without faulting) to account for the occurrence of the Cambrian beds on the eastern side only. Later, since the Upper Carboniferous and until the Cenomanian, the up and down movement was reversed and the *east* side was higher so that the 2000 excess sequence of the west side could be deposited. As the Upper Carboniferous to Ceno-

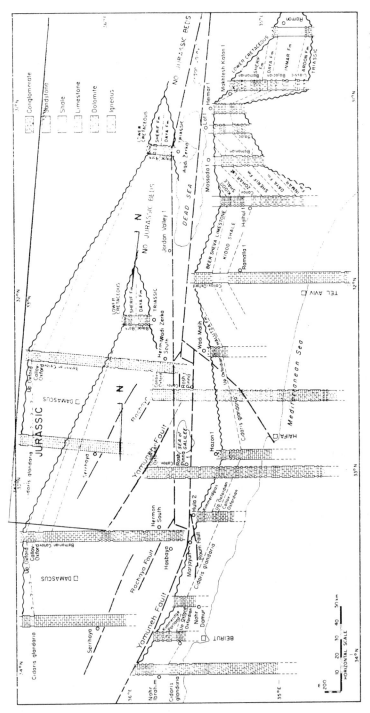

FIGURE 5. (*In black*) Jurassic: The sandy facies of the Jurassic extends in Israel from Ramon to Massada 1, opposite which in Transjordan the Jurassic is absent. It appears there in Jordan Valley 1, and in Wadi Zerka the facies is sandy. The Callovian–Oxfordian Kidod Shale occurs on the west side from the southwest side of the Dead Sea to Ramalla 1 bore hole; on the east side it occurs on the southeastern slope of Mount Hermon. Upper Oxfordian–Kimmeridgian section of 200 to 400 m occurs in the Galilee and Lebanon but is almost absent in southern Mount Hermon and Rosh Pinna 1 bore hole. It probably reappears in Anti-Lebanon from Serrhaya northwards. (References: Nahr Ibrahim: Renouard 1951; Beirut: Bishopp 1964; Nahr Damour: Heybroek 1942; Hula 2, Rosh-Pinna 1 and Hazon 1: B. Derin (unpublished); Serrhaya: Dubertret 1950; Hermon south: Dubertret 1951; M. Goldberg (unpublished); Wadi Malih: E. Aizenberg & B. Derin (unpublished); Wadi Zerka: Wetzel & Morton 1959; Northern Negev: M. Goldberg (unpublished).)

(*In red*) The east side is displaced 105 km southwards and Rosh Pinna 1 is displaced 60 km southwards. Wadi Zerka corresponds to Lot 1 and Hermon-south to Ramalla 1. A 55 km wide gap is opened along Yamuneh Fault; in this illustration the deformation of Lebanon is not restored.

manian section does not contain any conglomerate bed this upward movement of the east side must have been gradual, never forming a high relief. Afterwards, during the Turonian and the Senonian the movement was reversed again and the *west* side was rising to allow for the deposition of a thicker sequence on the east side. It is concluded that this explanation does not seem to be simpler than the shear hypothesis.

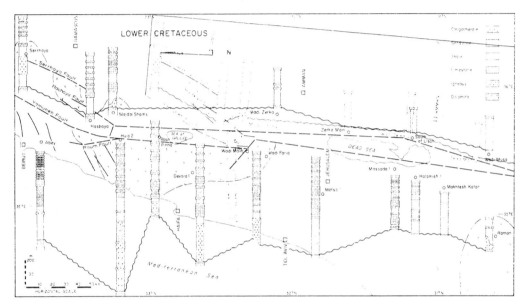

FIGURE 6. (*In black*) Lower Cretaceous: the marine Albian limestones extend in Israel as far south as Ramon, where the Lower Cretaceous starts with a base conglomerate, whereas in Transjordan the Albian limestone extends southwards only to Wadi Zerka, where the base conglomerate occurs again. Opposite Wadi Zerka, in Wadi Malih–Wadi Faria area the Lower Cretaceous section is 650 to 750 m thick, and more than two thirds of the sequence consist of marine limestones and shales. Volcanic rocks occur at the lower part of the Lower Cretaceous sequence from Wadi Faria to southern Lebanon (Marjayun), whereas on the east side they occur from Majdal Shams to Serrhaya. (References: Abey: Heybroek 1942; Serrhaya and Hasbaya: Dubertret 1950–51; Majdal Shams: U. Salzman (unpublished); Hula-2: Shiftan & Rosenthal 1963; B. Derin (unpublished); Rosh Pinna 1 and Debora 1: B. Derin (unpublished); Wadi Malih: J. Mimran & R. Freund (unpublished); Transjordan: Wetzel & Morton 1959; Bender 1968*b*; Negev and Motsa 1: Aharoni 1964; Arad 1964.)

(*In red*) The east side is displaced 105 km southwards and Rosh Pinna 1 is displaced 60 km southwards, but the displacements along Serrhaya-Rachaya faults and the Galilean and Lebanese faults are not restored, thus the gap of 55 km along Yamuneh Fault is maintained. Wadi Zerka corresponds to Ramon along northeast trending isofacies lines. Majdal Shams section corresponds to Wadi Malih section.

WADI ARAVA AND GULF OF AQABA (ELAT)

The offset of several Cambrian rock units, e.g. the copper and manganese deposits and of the southern extent of the Cambrian limestone from Fenan east of the Arava, to Timna which is about 100 km south of Fenan on the west side (figure 2*a*), was regarded by Quennell (1958,

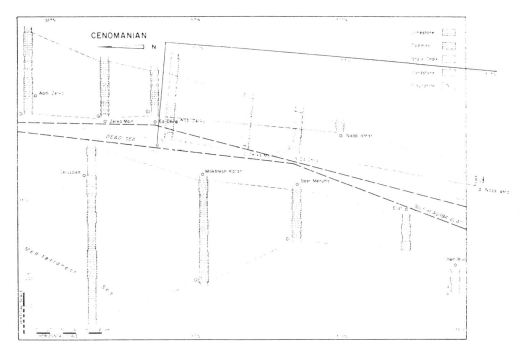

FIGURE 7. (*In black*) The Cenomanian section (including the upper part of Albian) of Jerusalem measures about 650 m, whereas that of Wadi Zerka only 280 m and those of Zerka Main and Ed Dhira about 200 m only, comparable to those of Beer Menuha (175 m). Opposite the latter the Cenomanian section of Naqb-Ishtar is less than 50 m thick. (References: Transjordan: Wetzel & Morton 1959; Jerusalem: Arkin *et al.* 1965; Makhtesh Katan: Arkin & Brown 1965; Beer Menuha: Sakal 1967; Elat and Sheih Atiya: Bartov *et al.* 1970.)

(*In red*) The east side is displaced 105 km southwards with the closing of the Rift. The thicknesses of the two sides match well along ENE trending isopach lines.

1959) as evidence for the shear movement along the rift. This was supplemented by the equal offset of an extensive body of Precambrian quartz porphyry from south of Timna on the west side to south of Fenan on the east side (figure 1*a*). Bender (1965, 1968*a*), however, argued that these data cannot be regarded as evidence for the shear movement because all these rocks occur as well in other places, such as Gharandal and Wadi Abu Khusheiba (figures 1*a* and 2*a*), forming thereby an almost continuous belt between Timna and Fenan. Bender accepted only a 25 to 35 km shear, according to the offset of a certain hornblendite-pegmatite from Elat on the west to a place opposite Timna in the east.†

Numerous parallel faults, which are anastomosing with the rift occur along this part of the rift. These faults extend in Transjordan northwards to Fenan, and perhaps to the southeast corner of the Dead Sea, and on the west side they extend from Timna southwards. These faults are nearly vertical and the upthrown and downthrown sides of these faults alternate sometimes

† Dr Bender amplifies this reasoning in his contribution to the discussion at the end of this paper (p. 127).

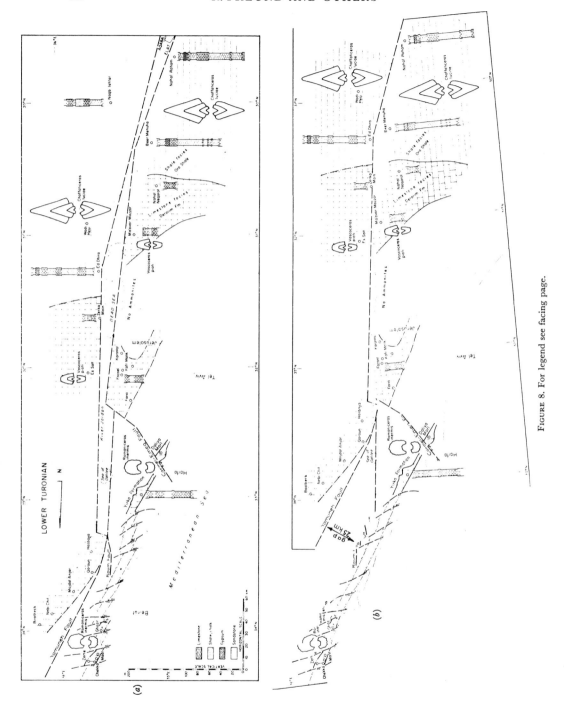

FIGURE 8. For legend see facing page.

from one side to the other along the faults. Numerous narrow fault depressions are located along these faults, where the rocks are intensively folded.

Horizontal slickensides occur on these faults in Israel and Sinai, and recent investigation has revealed that they are left-hand shears, varying in magnitude from several metres to a few kilometres. Although the faults on the Transjordanian side have not been equally studied, it seems justified to assume that they are left-hand shears as well. Moreover, an air photograph of some of these faults in Transjordan, published by Burdon (1959) shows clearly that dykes are displaced horizontally left-hand by these faults.

It seems therefore justified to propose that the shear movement is divided along this part of the rift among numerous anastomosing faults. Only a certain part of the shear is located within the rift depression, the rest occurs outside it. The reconstructions of the Precambrian (figure 1 *b*) and Lower Palaeozoic (figure 2 *b*) are carried out according to this proposal, by shifting back the two sides by 105 km, whereas the median blocks are shifted by various smaller amounts.

The Precambrian quartz-porphyry and conglomerates come thereby in line (figure 1 *b*). Moreover, the Precambrian (or Infracambrian) arkose encountered in Sinaf-1, Hameishar-1 and Ramon-1 are removed away from their present position opposite the granite and quartz porphyry of Fenan and Gharandal, towards the Saramuj series of Wadi Hesi, which they closely resemble. The Cambrian carbonates, manganese and copper of Timna, Abu-Khusheiba and Fenan are clustered in one place, and the 300 m thick section of the Cambrian of Sinaf-1 is adequately placed opposite that of Wadi Hesi.

The reconstruction requires the rotation of some of the intermediate blocks throughout their changing positions. Frequently the various blocks do not fit properly into the new places. This is most probably the reason for intensive folding and faulting which is very common along these faults.

GALILEE–LEBANON–SYRIA

The difference between the east side and west side of the rift is as large in the Galilee and Lebanon as it is in the south, but the fault pattern of the region is more complex.

The Jurassic sequence (figure 5 *a*) of the southeastern slope of Mount Hermon is truncated by Lower Cretaceous unconformity at the base of the *calcaire jaune inférieur* of Upper Oxfordian age, that of Hasbaya on the west side of Mount Hermon is truncated even lower in the section. In contrast with that, the Jurassic sections in the Galilee (except Rosh Pinna-1) and in Lebanon

---

FIGURE 8. (*a*) The Lower Turonian section on the southeast side of the Dead Sea and in the southern Negev consists of 100 m of shales with beds of concretional limestone topped by gypsum beds, with abundant ammonites, the most common of which are several species of *Choffaticeras*. On the southwest side of the Dead Sea and near Es-Salt the section consists of about 20 m of yellow-red marl and limestone, where another ammonite, *Vascoceras pioti*, is most abundant. The boundary between the shale facies (Ora Shale) and the limestone facies (Derorim Fm) trends east–west across Ramon. On the west side of the Dead Sea the Lower Turonian ammonites are missing. The Lower Turonian northeast of Jerusalem and that on the east side of Yamuneh Fault contain almost only specimens of *Thomasites* and *Choffaticeras segne*, whereas that of Mount Carmel (Daliya Marl) of the western Galilee (Yirka Fm) and north of Beirut contain several ornamented ammonites, such as *Romaniceras* and *Mammites*. (References: in Freund 1961; Freund & Raab 1969; Beer Menuha: Sakal 1967.)

(*b*) An almost complete, though schematic reconstruction of the shear along the Dead Sea–Lebanon section. The east side is displaced southwards 105 km, the Galilee is displaced 10 km southeast along the Carmel–Emeq Fault, and several small right-hand shears are restored in Lebanon. The three pairs of the Lower Turonian facies correspond to each other across the Rift and the Daliya–Yirka–Ghazir–Maat occurrences are arranged in line. The gap along Yamuneh Fault is reduced to *ca*. 25 km only.

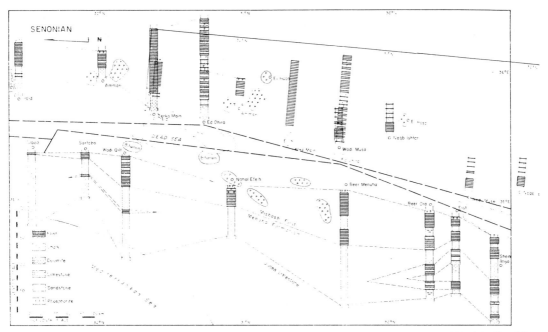

FIGURE 9. (*In black*) The *ca.* 200 m Senonian (Santonian–Campanian) sequence on the east side of the Dead Sea
contains flint and dolomite almost throughout, as does the Senonian sequence in the southern Negev. However,
the latter contains some sandstone beds which are not reported on the east side of the Dead Sea. On the west
side of the Dead Sea and in the central Negev the flint beds occur only in the upper half of the section,
similar to the section in north Transjordan. In northern Israel the flint beds are very thin or missing altogether.
The economic phosphorites occur around Amman in Transjordan and in the northern Negev in Israel, but
there is another occurrence in El Hasa in Transjordan. (References: Transjordan: Wetzel & Morton 1959;
Bender 1968*b*; Gilboa: Flexer 1959; Sartaba: A. Flexer, Y. Weiler & R. Freund (unpublished); Wadi Qilt
I. Rot (unpublished); Nahal Efeh: Kolodny 1967; Beer Menuha: Sakal 1967; Beer Ora, Elat and Sheih
Atiya: Bartov *et al.* 1970.)

   (*In red*) The 105 km southwards movement of the east side brings the corresponding facies on the two sides
in line along NNE-trending facies lines.

terminate 200 to 400 m higher in the Upper Jurassic. In Rosh Pinna-1 well which is drilled
through a block located within the rift, the Jurassic terminates within the Callovian–Oxfordian,
as it does in Hasbaya.

   The Lower Cretaceous section (figure 6*a*) of the southern slopes of Mount Hermon is about
400 m thick, whereas that of the Galilee and southern Lebanon, right across Yamuneh Fault,
is about 1000 m thick. The volcanic lower part of the Lower Cretaceous occurs on the west
side from Wadi Faria in the south, to Marjayun in the north. The volcanic sequence thins out
northwards and is almost absent in Abey. On the east side of Yamuneh Fault it extends from
Majdal Shams to Rachaya on Mount Hermon, and on the east side of Serrhaya Fault it extends
farther north. Again the Lower Cretaceous of Rosh Pinna-1 does not correspond to that of the
Galilee, but resembles those of Mount Hermon and Wadi Malih.

In figures 5*b* and 6*b*, the east side is shifted 105 km southwards, and the intermediate Rosh Pinna block is displaced southwards by about 60 km† to a location near Wadi Malih on the west side and Mount Hermon on the east side. In this way both the Jurassic and the Lower Cretaceous become reasonably similar on the two sides of the rift.

This reconstruction leads, however, to a structural problem. The southward movement of the east side leaves a 55 km wide gap along the N 30° E trending Yamuneh Fault (figure 5*b* and 6*b*). In other words, the left-hand shear movement along the diverted section of Yamuneh Fault should result in an overlap of 55 km (figure 10*a*, *b*). Such an overlap would certainly be excessively large even for the high structures of the Lebanon mountains. A closer examination of the fault pattern of the Galilee and Lebanon shows, however, that this problem can be accounted for.

It is proposed that the 55 km overlap can be reduced to about 20 to 25 km by the restoration of the deformation of the faults in this area, as demonstrated in a geometrical model (figure 10*c*):

(1) 10 km left-hand shear along the Carmel–Emeq Fault (de Sitter 1962);
(2) extensive east–west normal faults in the Galilee;
(3) small ENE trending right-hand shear faults in Lebanon (Renouard 1955);
(4) left-hand shear along Rachaya-Serrhaya fault zone.

This geometrical reconstruction is certainly schematic and, moreover, the writers of this paper are unable to examine the tectonic pattern in Lebanon. This model has therefore a qualitative value only. Still it is successful in explaining not only the dissimilar facies on the two sides of Yamuneh Fault and the fault pattern in Lebanon and the Galilee, but also the occurrence of the high Lebanon and Anti-Lebanon mountains exactly along the diverted section of Yamuneh Fault.

The western side from Haifa to Tripoli is stretched by about 30 km.through the combined deformation of the above-mentioned faults, so that the amount of shear along the rift should decrease north of Lebanon to about 75 km. This result is in line with the 70 km left-hand offset of the southern boundary of the ophiolites (roches vertes) belt of Maastrichtian age (Dubertret 1955) across the rift in northern Syria (figure 11).

The palaeogeography of the Lower Turonian is reconstructed according to the above-mentioned model (figure 8*b*). The Lower Turonian north of Beirut, in which the ammonite assemblage corresponds to that of western Galilee and Mount Carmel (e.g. *Mammites* and *Romaniceras*), is brought in this way in line with the latter locations.

STRUCTURAL EVIDENCE OF THE SHEAR MOVEMENT

Apart from the horizontal offset of the rock-bodies across the rift there are numerous structural features indicating the strike slip or shear nature of many of the rift faults.

(1) Many faults of the rift system, as those in Lebanon (e.g. Dubertret 1947, 1967) and the anastomosing faults on the two sides of Wadi Arava (Bender 1968*a*, *b*) and the Gulf of Aqaba are nearly vertical, a characteristic feature of strike slip faults.

(2) Horizontal slickensides are observed on some of these faults in southern Israel and on Mount Hermon.

† The *ca.* 60 km shear between Rosh Pinna-l and the Galilee occurs on the western fault of the Hula depression which is supposed to extend southwards on the west side of the Fuliyya and Hordos blocks beneath the Neogene sediments. This fault was apparently active only in pre-Neogene times, and the 40 to 45 km shear of post-Miocene time has apparently taken place on the fault east of Sea of Galilee and Hula depressions.

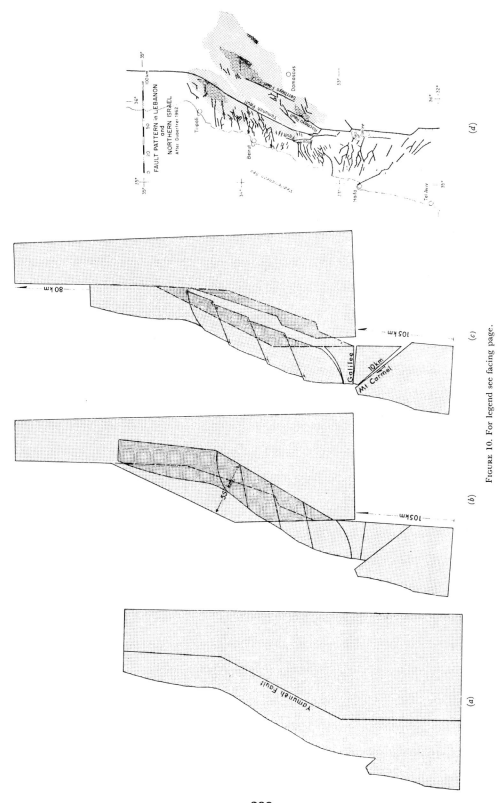

FIGURE 10. For legend see facing page.

399

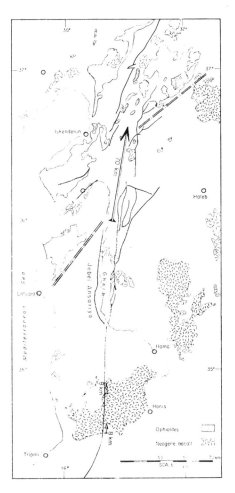

FIGURE 11. The southern boundary of the ophiolites (roches vertes) of northern Syria is offset left-hand by about 70 km across the Rift. The Pliocene (?) basalt west of Homs is offset by *ca.* 8 km. The *échelon* arrangement of the Rift faults on the two sides of the rhomb-shaped depression of Gharb, is typical of strike slip faults. (Reference: Dubertret 1962.)

FIGURE 10

FIGURE 10. A schematic geometrical model intended to show that the fault deformation in Lebanon and the Galilee can be explained by a large scale left-hand shear along the Dead Sea rift.

(*a*) Pre-shear; showing only the position of the rift faults, with the diverted section of the Yamuneh Fault.

(*b*) 105 km left-hand shear (with 6° anticlockwise rotation of the east side) results in an overlap of 55 km along the Yamuneh segment. The position of the faults in the Galilee, Lebanon and Hermon is marked.

(*c*) The following movements have taken place: 10 km left-hand shear along Carmel–Emeq fault, with normal faulting (small gaps) in the Galilee; small right-hand shears in Lebanon, and left-hand shear along Rachaya-Serrhaya fault zone. The 55 km overlap is reduced thereby to two narrow overlaps on the two sides of the Beqa'a each one being *ca.* 10 km wide. The amount of shear between the two blocks is reduced from 105 km south of Lebanon to 80 km north of Lebanon.

(*d*) Fault pattern of the Galilee and Lebanon according to Dubertret (1962). Right-hand shears in Lebanon according to Renouard (1955). Areas above topographical elevation of 1000 and 2000 m are marked. Compare the structure and topography of (*d*) with the schematic model (*c*).

(3) The up and down sides of the faults alternate frequently from one side to the other (see, for example, Dubertret 1967), a feature typical for strike slip faults.

(4) The two sides of several faults are at exactly the same elevation along certain stretches (see, for example, Dubertret 1967), yet the faults are clearly visible. This phenomenon is particularly evident along recent fault traces in the Arava and Jordan Valley. A fault without vertical throw must have a horizontal displacent.

(5) Rhomb-shaped 'grabens' such as the Hula depression in northern Israel and Gharb depression (figure 11) in Syria are conveniently explained by left-hand shear along an *échelon* offset of the fault line, as proposed by Quennell (1959) for the Dead Sea.

Other structural features, as the relation of the complex structure of Lebanon and the Galilee to the left-hand shear along the diverted Yamuneh Fault, and the anastomosing shears of Wadi Arava and Gulf of Aqaba, have already been discussed in this paper. The relation of the structure of the Dead Sea rift to the shear movement has been demonstrated by Freund (1965).

AGE AND RATE OF THE SHEAR MOVEMENT

The fact that all the rocks from the Precambrian to the Upper Cretaceous are displaced by the same amount indicates that the whole shear movement is post-Cretaceous and that no movement has taken place before it. It is, however, difficult to determine the exact age of the movements. It has recently (Freund *et al.* 1968) been reported that several rock bodies of Miocene, probably up to early Pliocene age, are offset left hand by 40 to 45 km across the rift. The 40 to 45 km movement has apparently taken place during the last 7 to 12 Ma thus the average rate of the movement is about 0.35 to 0.6 cm a^{-1}.

The younger Pliocene basalt is apparently displaced by about 10 km in the Galilee (Freund *et al.* 1968) and by about 8 km in north Lebanon (figure 11). The age of this movement has not yet been determined.

The Lissan Marl of the Jordan Valley which is older than 23 ka (Neev & Emery 1967) is displaced by about 150 m; the rate of movement is therefore 0.65 cm a^{-1} (Zak & Freund 1966).

If this rate (0.4 to 0.6 cm a^{-1}) is the average rate of the whole 105 km movement, then this movement has started some 20 to 30 Ma ago—during the Late Oligocene to Early Miocene. There are, however, indications that the last movement of 40 to 45 km was preceded by a quiet period of unknown duration.

It is most unfortunate that rocks of the age between Late Cretaceous to Miocene are rare along the Rift, except those of Paleocene and Eocene age, and even those are of little value in determining the amount of displacement: The Taqiya Shale of Paleocene age lacks distinct facies change, whereas the facies changes of the Eocene are very rapid and of local nature (figure 12).

Structural considerations (e.g. excessive large thickness of Senonian and Eocene sequences in the rift, contemporaneous compressional and tensional deformation at different places along the rift, etc.) have lead one of the present writers (Freund 1965) to advocate shear movements in several phases since Late Cretaceous. However, there is no conclusive evidence in this matter.

Several contributors have joined efforts in this paper to present as comprehensive a review of the rocks of the two sides as possible, and this paper includes results from their research projects which have not yet been published, such as that on the fault pattern on the west side of the Gulf of Elat (Aqaba) by Z. Garfunkel, on the Triassic by I. Zak, I. Karcz and M. Goldberg, on the Jurassic of southern Israel by M. Goldberg, on the Jurassic and Lower Cretaceous in northern Israel by B. Derin, on the Palaeozoic stratigraphy by T. Weissbrod and on the recent faulting of the rift by I. Zak, Z. Garfunkel and R. Freund.

Moreover, data from other unpublished works are included in this paper, and the writers wish to express their gratitude for these valuable contributions to Y. Bartov, B. Z. Begin, Y. Druckman, M. Eyal, A. Flexer, E. Gerry, Y. Mimran, I. Rot, E. Sakal, U. Salzman, G. Steinitz and Y. Weiler.

Data from deep test wells which have been recently drilled in the vicinity of the rift, the results of which have not yet been published, are presented here. The generous permission of Lapidoth Oil Co., Naphta Oil Co., Israel National Oil Co. and Israel Continental Oil Co. to publish the results of their deep test wells is gratefully acknowledged.

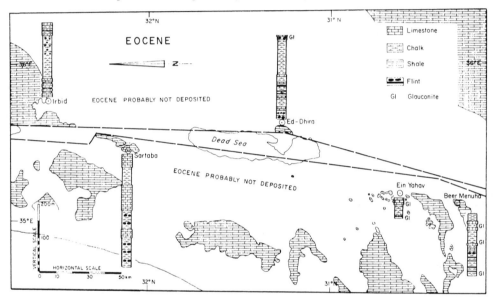

FIGURE 12. The Eocene sequences change rapidly in thickness and facies over a few kilometres, therefore they are of little value in measuring the amount and age of the shear. A reconstruction of 45 km left-hand shear would place Irbid opposite Sartaba, and Ed Dhira opposite the northeast extent of the Eocene near Ein Yahav. However, any larger offset up to a 105 km is not impossible. (References: Transjordan: Wetzel & Morton 1959; Sartaba: A. Flexer, Y. Weiler & R. Freund (unpublished); Ein Yahav: Bentor & Vroman 1957; Beer Menuha: Sakal 1967.)

REFERENCES (Freund *et al.*)

Aharoni, E. 1964 Litho-electric correlation of the 'Kurnub-Group' (Lower Cretaceous) in the northern Negev. *Israel J. Earth Sci.* **13**, 63–81.

Arad, A. 1964 The geology and hydrogeology of the Lower Cretaceous of the northern Negev and Judea mountains. Ph.D. Thesis, Hebrew University, Jerusalem (in Hebrew).

Arkin, Y. & Brown, M. 1965 Type sections of Upper Cretaceous formations in the northern Negev (southern Israel). *Geol. Surv. Israel Strat. Sect.* **2**a.

Arkin, Y., Brown, M. & Starinsky, A. 1965 Type sections of Cretaceous formations in the Jerusalem–Beit Shemesh area. *Geol. Surv. Israel Strat. Sect.* **1**.

Bartov, Y., Eyal, Y., Garfunkel, Z. & Steinitz, G. 1970 Upper Cretaceous and Tertiary of southern Israel. *Israel J. Earth Sci.* (in the Press).

Bartura, Y. 1966 Type section of Paleozoic formations in the Timna area. *Geol. Surv. Israel Strat. Sect.* **3**.

Bender, F. 1965 Zur Geologie der Kupfererzvorkommen am Ostrand des Wadi Araba, Jordanien. *Geol. Jb.* **83**, 181–208.

Bender, F. 1968*a* Über das Alter und die Entstehungsgeschichte des Jordangrabens am Beispiel seines Südabschnittes (Wadi Araba, Jordanien). *Geol. Jb.* **86**, 177–196.

Bender, F. 1968*b* *Geologie von Jordanien*. Beiträge zur Regionalen Geologie der Erde. Berlin: Gebrüder Borntraeger.

Bentor, Y. K. & Vroman, A. J. 1954 A structural map of Israel (1:250 000) with remarks on its dynamical interpretation. *Bull. Res. Coun. Israel* **4**, 125–135.

Bentor, Y. K. & Vroman, A. J. 1955–60 *The geological map of Israel*, 1:100 000. Sheet 24 *Eilat* (1955); sheet 19, *Nahal Arava* (1957); sheet 16, *Mt Sedom* (1960): *Geol. Surv. Israel*.

Bischoff, G. 1964 Die Gattung *Cytherelloidea* im Oberen Jura und in der Unterkriede, *Senckenberg leth.* **45**, 1–27.

Blake, G. S. 1937 Old shore lines of Palestine. *Geol. Mag.* **74**, 68–78.

Bogolepov, M. 1930 Die Dehnung der Lithosphäre. *Z. dt. Geol. Ges.* **82**, 206–228.

Bramkamp, R. A., Brown, G. F., Holm, D. A. & Layne, N. M. jun. 1963 *Geological Map of the Wadi as Sirhan Quadrangle, Kingdom of Saudi Arabia*, U.S.G.S. Map, 1–200–A.

Burdon, D. J. 1959 *Handbook of the Geology of Jordan to accompany and explain the three sheets of the 1:250 000 geological map east of the Rift* by A. M. Quennell, 82 pp. Colchester.

Dubertret, L. 1932 Les formes structurales de la Syrie et de la Palestine; leur origine. *C. r. hebd. Séanc. Acad. Sci., Paris* **195**, 65–67.

Dubertret, L. 1947 Problèmes de la géologie du Levant. *Bull. Soc. géol. Fr.* (sér. 5) **17**, 3–31.

Dubertret, L. 1950–51 *Carte géologique du Liban au 50.000*. Rayak, 1950, Beyrouth, 1951, Merdjayoun, 1951.

Dubertret, L. 1955 Géologie des roches vertes du Nord-Ouest de la Syrie et du Hatay (Turquie) *Notes Mém. Moyen-Orient* **6**, 5–224.

Dubertret, L. 1962 *Carte géologique du Liban, Syrie et bordure de pays voisins*, au 1:1 000 000. Mus. d'Hist. Natur., Paris.

Dubertret, L. 1967 Remarques sur le fossé de la Mer Morte et ses prolongements au nord jusqu'au Taurus. *Rev. Géog. Phys. Géol. Dyn.* **9**, 3–16.

Flexer, A. 1959 The geology of Mt Gilboa. M.Sc. Thesis, Hebrew University, Jerusalem (in Hebrew).

Freund, R. 1961 Distribution of Lower Turonian ammonites in Israel and the neighbouring countries. *Bull. Res. Coun. Israel* **10** G, 79–100.

Freund, R. 1965 A model of the structural development of Israel and adjacent areas since Upper Cretaceous times. *Geol. Mag.* **102**, 189–205.

Freund, R., Zak, I. & Garfunkel, Z. 1968 Age and rate of the sinistral movement along the Dead Sea Rift. *Nature, Lond.* **220**, 253–255.

Freund, R. & Raab, M. 1969 Lower Turonian ammonites from Israel. *Sp. Papers in Palaeontology* **4**.

Garfunkel, Z. 1966 Problems of wrench faults. *Tectonophysics* **2**, 457–474.

Geological map of Egypt, 1:1 000 000 (1928). Cairo.

Henson, F. R. S. 1956 Tectonic problems of the Middle East. *20th Int. Geol. Congr. Mexico*.

Heybroek, F. 1942 La géologie d'une partie du Liban Sud. *Leid. geol. Meded.* **12**, 251–470.

Kolodny, Y. 1967 Lithostratigraphy of the Mishash formation, northern Negev. *Israel. J. Earth Sci.* **16**, 57–73.

Lartet, L. 1869 *La géologie de la Palestine*. Paris.

Neev, D. & Emery, K. O. 1967 The Dead Sea, depositional processes and environments of evaporites. *Bull. Geol. Surv. Israel* **41**, 147 pp.

Picard, L. 1943 *Structure and evolution of Palestine*. Jerusalem.

Picard, L. 1966 Thoughts on the Graben System in the Levant. *Geol. Surv. Pap. Can.* 66–14, 22–32.

Picard, L. 1968 On the structure of the Rhinegraben with comparative notes on Levantgräben features. *Israel Acad. Sci. Hum.* **9**, 34 pp.

Quennell, A. M. 1958 The structural and geomorphic evolution of the Dead Sea Rift. *Q. J. geol. Soc., Lond.* **114**, 1–24.

Quennell, A. M. 1959 Tectonics of the Dead Sea Rift. *20th Int. Geol. Congr., Mexico* (1959), pp. 385–405.

Renouard, G. 1951 Sur la découverte du Jurassique inférieur (?) et du Jurassique moyen au Liban. *C. r. hebd. Séanc. Acad. Sci., Paris* **232**, 992–994.

Renouard, G. 1955 Oil prospects of Lebanon. *Bull. Am. Ass. Petrol. Geol.* **29**, 2125–2169.

Sakal, E. 1967 The geology of Rekhes Menuha. M.Sc. Thesis, Hebrew University, Jersualem (in Hebrew).

Seidlitz, W. von 1931 *Diskordanz und Orogenese der Gebirge am Mittelmeer*. Berlin:

Shalem, N. 1954 The Red Sea and the Erythrean disturbances. *19th Int. Geol. Congr. Alger, C.R.* 15, pt. 17, 223–231.

Shiftan, Z. L. & Rosenthal, E. 1963 *Emeq Hula water sources development*. Geol. Surv. Israel Hydro Report 63/4 (in Hebrew).

Sitter, L. U. de 1962 Structural development of the Arabian Shield in Palestine. *Geologie Mijnb.* **41**, 116–124.

Swartz, D. H. & Arden, D. D. 1960 Geologic history of the Red Sea area. *Bull. Am. Ass. Pet. Geol.* **44**, 1621–1637.

Vroman, A. J. 1957 Strike-slip movements, their associated features and their occurrence in Israel. *20th Int. Geol. Congr., Mexico*, §5, **3**, 399–408.

Vroman, A. J. 1961 On the Red Sea rift problem. *Bull. Res. Coun. Israel* **10** G, 321–338.

Vroman, A. J. 1963 The Ramon fold in its Levantine Setting. *Israel J. Earth Sci.* **12**, 86.

Vroman, A. J. 1967 On the fold pattern of Israel and the Levant. *Bull. Israel Geol. Surv.* **43**, 23–32.

Weissbrod, T. 1969*a* The subsurface Palaeozoic stratigraphy of southern Israel. *Bull. Geol. Surv. Israel* **47**.

Weissbrod, T. 1969*b* Palaeozoic outcrops of southwestern Sinai and their correlation with those of southern Israel. *Bull. Geol. Surv. Israel* **48**.

Wetzel, R. & Morton, D. M. 1959 Contribution à la géologie de la Transjordanie. *Notes Mém. Moyen-Orient* **7**, 95–191.

Willis, B. 1938 Wellings' observations of the Dead Sea structure (with discussion). *Bull. geol. Soc. Am.* **42**, 659–668.

Zak, I. & Freund, R. 1966 Recent strike slip movements along the Dead Sea Rift. *Israel J. Earth Sci.* **15**, 33–37.

DISCUSSION

F. Bender (*Bundesanstalt für Bodenforschung, Hannover, Germany*). Referring to Dr R. Freund's discussion on 'The shear along the Dead Sea rift' I should like to list below a number of observations made in Jordan. These observations, mainly mapping results, allow a simpler explanation of the anomalies in geological features across the rift than does the assumption of a post-Cretaceous 105 km left-hand shear. They also throw more light on the age of the Dead Sea geosuture. They have been described and interpreted in some more detail in a paper titled: 'Über das Alter und die Entstehungsgeschichte des Jordangrabens am Beispiel seines Südabschnittes (Wadi Araba, Jordanien)' by F. Bender (*Geol. Jb.* **86**, 177–196, Hannover 1968).

(1) At the west side of the southern Wadi Araba, the Precambrian 'Eilat-Gabbro' and the 'Timna-Gabbro' are exposed together with two small occurrences of a pegmatitic hornblendite (lateral facies of the Eilat–Gabbro complex; Bentor 1961). Meanwhile, also at the *east* side of the southern Wadi Araba, gabbro has been mapped with small occurrences of the very typical hornblendites. According to the locations of these rocks, a lateral displacement of about 25 to 35 km (sum of all lateral movements since the Precambrian) may be assumed. It is, however, just as reasonable to explain these observations with an approximate NNE–SSW-tending complex of gabbroide rocks within the basement, cut by a 'normal' 'graben' (here *ca.* 15 km wide).

(2) Quartzporphyries have been mapped along the central east side of the Wadi Araba, where they occur for about 70 km in a regional NNE–SSW striking direction (Geol. Map of Jordan, 1:250 000, sheet Aqaba-Ma'an, published by the Geol. Surv. of the Federal Republic of Germany, Hannover, 1968). They do not show evidence of a piece-by-piece cutting and movement to the north by shear forces. Their SSW-continuation points exactly to the quartz-porphyries occurring at the west side of the southern Wadi Araba, and therefore does not imply a 105 km lateral displacement.

(3) The facies borders of Cambrian sediments have been determined quite accurately and continuously along the east side of the Wadi Araba, due to very favourable outcrop conditions. They run in a NNE–SSW direction. ESE of the Wadi Araba, only continental Cambrian sediments have been found. The SSW-continuation of the facies border between continental and marine Cambrian sediments points to the central portion of the southern Wadi Araba; therefore, the existence of near-shore marine Cambrian at the west side of the southern Wadi Araba can be explained without any lateral displacement.

(4) The copper ore mineralization of parts of the Cambrian sediments has also been followed continuously for about 70 km from Feinan to the south along the Wadi Araba east side. No

copper mineralization is observed in the continental Cambrian; the copper mineralization is concentrated in the shallow marine facies and consequently occurs along a NNE–SSW tending belt, clearly and without shift pointing to the copper ore mineralized Cambrian in Timna at the west side of the southern Wadi Araba.

(5) Shorelines, isopachs and facies borders of the Triassic and Jurassic run WNW to ENE to the west side of the central Wadi Araba; they leave the rift in the Dead Sea area in a NE to ENE direction. They may have been shifted; they may as well have been influenced by the palaeogeographic west border of an elevated 'Transjordan Block' (as it happened with the Cambrian facies borders). Because of the thick cover of younger sediments in the rift valley, this question remains yet unanswered. The sudden increase in thickness and the rapid facies changes of Triassic and Jurassic sediments as observed across the rift are in other parts of the world quite a common feature over tectonical hinge lines or zones of weakness.

(6) Marine sediments of Albian age have been determined near Jerusalem; they also exist opposite the rift at the east side of the Jordan valley.

(7) No detailed investigation on Lower Turonian ammonites has yet been made in south Jordan east of the rift. It can therefore not be decided whether or not there is a NE continuation of a certain ammonite zone which Freund (1961) found west of the Wadi Araba. Recent studies have shown that facies and thicknesses of the Upper Cretaceous are very much changing in south Jordan and are to a great extent controlled by the SE–NW striking El Jafr sedimentary basin.

(8) Bituminous limestones and marls are observed in the Maestrichtian through Eocene sequence, as far south as El Hasa in south Jordan. Those occurrences are definitely not restricted to Nebi Musa at the west side of the southern Jordan valley, and to the Yarmuk area at the east side of the northern Jordan valley (which would ask for a 107 km shift according to Quennell 1958).

(9) The Lisan peninsula in the Dead Sea is not formed by fluviatile deltaic sediments shifted for 45 km (Quennell 1956, 1959) but is formed by Upper Pleistocene lacustrine marls overlying thick evaporites.

(10) Along the entire east side of the rift there is an overwhelming evidence of dip-slip movements along hundreds of faults and fault zones with vertical throws up to 1000 m. Direct evidence of lateral displacements, however, (horizontal slickensides, etc.) is very rare (observed at three places) and in the order of centimetres up to a few metres. These minor lateral movements and some minor folding due to tangential compression are explained as secondary structural phenomena: taphrogenesis due to vertical movement demands a lengthening of the crust in a right angle to the direction of the graben (H. Cloos 1939).

As to the age of the Wadi Araba–Dead Sea rift, the existence of a meridional zone of weakness (or hinge linge, or geosuture) preceding the formation of the rift may be inferred from:

(i) The disposition of the Precambrian structural pattern, and of parts of the Precambrian joint and dyke system in Jordan.

(ii) The pronounced relief of the surface of the Precambrian basement complex covered by thick conglomerates of the Lower Cambrian along both sides of the Wadi Araba; the peneplanation of the Precambrian rock complex and absence of conglomerates east of the present rift in south Jordan.

(iii) The quartzporphyry-volcanism during uppermost Proterozoic to Lower Cambrian, occurring along the zone of weakness.

(iv) The Cambrian facies borders, running for more than 70 km almost parallel to the present rift, the rapid change of thicknesses and facies characteristics across the hinge line of Triassic, Jurassic and Cretaceous sediments.

(v) The taphrogenetic structural movements which initiated the formation of the graben followed this meriodional zone of weakness and occurred in the (?) Upper Eocene–Oligocene. In the Upper Oligocene, the block east of the southern part of the 'graben' had been already elevated structurally and was subject to continental erosion, and, locally, to continental deposition. In the 'graben' itself, marine sediments still occur in the Oligocene part of the section. During the Neogene and Pleistocene, several phases of major structural (taphrogenetic) movements are recorded in the stratigraphical column.

F. E. Wellings (*Iraq Petroleum Company*). Drs Dubertret and Freund have mentioned my modest contribution to the Dead Sea problem 35 years ago in the form of a southerly displacement of Cambrian and Jurassic shorelines on the west (Palestine–Sinai) side of the Jordan–Dead Sea–Akaba trough.

The originator, Lartet in 1869, did not base his 160 km slip along the eastern fault on geological evidence from the Dead Sea but from trying to make the two granite sides of the Red Sea fit together, like a jig-saw puzzle. He would be as gratified to see the new structural and stratigraphic data produced by Dr Abel-Gawad as I was to hear Dr Freund's exposition using new well logs and giving more detail than Quennell in 1956.

Geologists investigating Dead Sea tectonics have to decide between three theories of origin, namely: (1) the rift or graben faulting of Suess and Blanckenhorn, (2) Lartet's earlier longitudinal slip, (3) Bailey Willis's 1938 ramp or reverse faults observed along Mount Gilboa and Mount Carmel.

The then Government Geologist, G. S. Blake, said I was the first geologist he knew of to make a trek from Beersheba down to the south end of the Dead Sea and on down the Wadi Araba to Akaba in 1933. I was impressed with the huge NE–SW asymmetric and faulted anticlines exposing Jurassic in their cores as well as by the lack of rift faulting along the west side of the Wadi Araba until nearing Akaba.

I gave Dubertret my shoreline data because the year before (1932) he had revived Lartet's slip which he renounced today.

In 1934 I sent Bailey Willis a letter and map giving my field evidence of compressive folding oblique to the trough, the way boundary faults swing outwards at both ends of the Dead Sea as crescentic·and reverse faults and die out along the flanks of plunging folds and the fact that for long distances along the Palestine side of the Jordan and Wadi Araba there are no border faults, only dip slopes.

Willis published this in 1938.

The present trough is probably due to rift faulting of Pleistocene age but the main folding in post-Eocene times that formed the trough in which the Neogene (Usdum) conglomerates, clastics and evaporites were deposited could well have had elements of ramp and thrust faulting.

We need more field investigation of these alternative rift, ramp and wrench faulting at fault contacts.

R. Freud (written reply). Dr Bender's comments are somewhat misleading. For example, the marine Albian (comment 6) does indeed occur in Wadi Zerka, but it is very different from that of the Jerusalem area (see figure 6). On the other hand it resembles that of the Negev, about 100 km south of Wadi Zerka on the west side. Comment 9 concerning the Lisan Peninsula is most probably correct, but Dr Bender should direct it to Quennell, not to us. We made our opinion clear that the post-Lisan movement is 150 m, not 45 km. Dr Bender is selective about the occurrence of the pegmatitic hornblendite in Transjordan (comment 1); his 1968 geological map of Transjordan shows that the same rock occurs not only opposite Timna, but in several other places along a NNE direction. These are separated one from the other by the anastomosing faults east of the Arava, having been sheared into several pieces along a 100 km stretch of the shear zone. Dr Bender's other comments are fully dealt with in the text.

The major weakness in not accepting the shear hypothesis is the necessity to provide a separate *ad hoc* explanation for every single dissimilarity across the rift. Whereas the restoration of the 105 km shear (*a*) gives one single explanation to all the dissimilarities, (*b*) places all the equivalent features (more than twenty) in adjacent positions, and (*c*) does not create a single new dissimilarity across the rift. The fact that Dr Bender in his lengthy discussion could not point out a single failure to these quantitative tests speaks for itself.

R. W. Girdler (written comments). Independent evidence for *recent* strike-slip motion along the Dead Sea rift comes from earthquake magnitude–frequency studies. Miyamura (1962, *Proc. Jap. Acad.* **38**, 27–30) suggested that the values of the constants in the relation lg $N = a - bM_b$ (N is the number of events of body wave magnitude M_b or greater) may be dependent on the tectonics of the region where the earthquakes occur. Francis (1968, *Earth & Plan. Sci. Lett.* **4**, 39–46) in a detailed study of the seismicity of mid-ocean rifts finds the value of b is significantly different for rift zones (regions of normal fault motion) from that of fracture zones (regions of strike-slip motion). Using 154 events from 1963 to May 1967 associated with the mid-Atlantic part of the world rift system Francis found $b = 1.72$ for 79 events in regions of rifting and $b = 0.99$ for 75 events associated with fracture zones.

In a review of the seismicity of Israel and adjacent areas Arieh (1967, *Bull. geol. Soc. Israel* **43**, 1–14) finds $b = 0.8$ for 25 events of magnitude $M_b \geqslant 4.5$ for the period 1919 to 1963. As nearly all these events are likely to be associated with the Jordan rift, Arieh concludes that the seismicity of the Jordan Valley–Baka'a region may be described to a first approximation by lg $N = 4.9 - 0.8M_b$. The low value of b would suggest that the events are associated with strike-slip motion and is consistent with the geological evidence given above.·

AUTHOR CITATION INDEX

SUBJECT INDEX

About the Editor

A. M. QUENNELL is currently engaged in writing and editing, having retired in 1974 from an editing post with the Institute of Geological Sciences, London. He was graduated with a B.Sc. in physics and an M.Sc. in geology in the 1930's from Otago and Victoria universities, New Zealand. He is an associate of the Otago School of Mines. He served before and after World War II with the New Zealand Geological Survey and the Shell Company. He was assistant director of the Trans-Jordan Lands and Surveys Department. He made the first definitive geological survey of Trans-Jordan, which included the eastern flank of the Dead Sea Rift. He was with the Tanganyika (now Tanzania) Geological Survey until 1960, and for the last years he was director. He then practiced as consultant geologist in New Zealand and, in the late 1960s, was manager of United Nations projects in Nigeria and the Sudan.

In the course of his service in the Middle East and Africa, Mr. Quennell had unusual opportunities for the study of the Afro-Arabian Rift System along most of its length. On the basis of comparative geology, he further advanced the strike-slip hypothesis of earlier writers for the formation of the Dead Sea Rift and the opening of the Red Sea. In the course of this, he became the first to recognize in 1956 and publish (1959) the location of what is now known as a pole of rotation—that of the Arabian plate in relation to Sinai-Palestine.

Mr. Quennell is the author and editor of publications on the stratigraphy of New Zealand; on the geology, especially of the Precambrian, of eastern Africa, Jordan, and the CENTO countries.

He is a fellow of the Geological Society, London, and of the Institute of Mining and Metallurgy, and is a member of the Institution of Civil Engineers.

Benchmark Papers
in Geology

Series Editor: Rhodes W. Fairbridge
Columbia University